Papers presented in May, 1980, at a two-day meeting on 'Interaction Between Corrosion and Mechanical Stress at High Temperatures' in Petten, The Netherlands, organized by the Commission of the European Communities, Joint Research Centre, Petten Establishment, in co-operation with the European Federation of Corrosion's Working Group on Corrosion by Hot Gases and Combustion Products, and Nederlands Corrosie Centrum.

CORROSION AND MECHANICAL STRESS AT HIGH TEMPERATURES

Edited by

V. GUTTMANN and M. MERZ

*Materials Division, CEC, JRC Petten Establishment,
Petten (N.H.), The Netherlands*

APPLIED SCIENCE PUBLISHERS LTD
LONDON

APPLIED SCIENCE PUBLISHERS LTD
RIPPLE ROAD, BARKING, ESSEX, ENGLAND

British Library Cataloguing in Publication Data

Corrosion and mechanical stress at high
temperatures.
1. Materials at high temperature—Congresses
I. Guttmann, V. II. Merz, M.
620.1'1217   TA418.24

ISBN 0-85334-956-8

WITH 33 TABLES AND 254 ILLUSTRATIONS

© ECSC, EEC, EAEC, Brussels and Luxembourg 1981

Publication arrangements by: Commission of the European Communities, Directorate-
General for the Information Market and Innovation, Luxembourg

EUR 6984

LEGAL NOTICE
Neither the Commission of the European Communities nor any person acting on behalf of
the Commission is responsible for the use which might be made of the following information.

Printed in Great Britain by Galliard (Printers) Ltd, Great Yarmouth

PREFACE

The European Symposium on the Interaction of Corrosion and Mechanical Stress at High Temperatures was organised by the Commission of the European Communities, Joint Research Centre, Petten Establishment, in cooperation with the European Federation of Corrosion working party: Corrosion by Hot Gases and Combustion Products and the Nederlands Corrosie Centrum.
It was the ninth conference of the working party since its formation in 1966.

The Symposium dealt with the aspects:
- influence of corrosion on the mechanical behaviour of high temperature metallic materials under static and dynamic load,
- influence of mechanical stress on corrosion and in particular on the cracking of scales,
- testing methods to determine the interactions between corrosion and stress.

A panel discussion concluded this Symposium.
For convenience papers are arranged in the same order of presentation as at the Symposium, each followed by the discussion.

The editors wish to express their gratitude to the authors for the preparation and the presentation of their papers and to the Symposium participants for providing their discussion contributions. The excellent direction of the Symposium discussion periods exercised by the session chairmen is highly appreciated as well as the effort of all others which assured a successful Symposium from which these proceedings evolved.

The Editors

TABLE OF CONTENTS

OPENING OF THE SYMPOSIUM

P.J. van Westen

CEC, Joint Research Centre

Petten Establishment, The Netherlands

Ladies and gentlemen, It is with pleasure that I welcome you to this symposium at the Petten Establishment of the Joint Research Centre. By now, it is probably known to many of you that our programme on High Temperature Materials was started about five years ago in support of the European Commission's Energy policy which is aimed at the efficient exploitation of existing sources of energy and the development of new ones.
The object of our programme is in the first place to survey technological and industrial requirements in order to pinpoint future needs, secondly to establish research projects in selected key areas, finally to build and maintain multilateral communications with manufacturers, users of these materials and with the research and development sector. By means of this three-pronged attack we collect and disseminate information and encourage research by focussing attention on particular problems.
This symposium clearly makes a valuable contribution to a very important aspect of the application of materials at elevated temperatures.
The use of high temperature materials is fundamental to energy production and utilization, and the search for improved efficiency and for novel methods of energy generation or conversion necessitates the development of improved materials to withstand higher temperatures, more severe stresses or more aggressively corrosive environments. The particular topic of this symposium – the interaction of corrosion and mechanical stresses – has become of paramount importance during recent years because of the increasingly severe conditions which the materials are called upon to withstand. For the major installations of generating and conversion plant, typified by electrical power stations and petro-chemical processing plant, long troublefree life is required, extending to as much as

30 years. A full understanding of the degrading mechanisms and
their interactions is therefore necessary.
Of important significance at present, in view of the world-wide
threats to the supply of oil and of strategic metallic minerals,
are the processes currently being explored for the conversion of
coal to a more convenient energy form. These include fluidized-bed
combustion, gasification and liquefaction processes, all of which
may involve corrosive attack of the plant materials by carburization, sulphidation or ash-corrosion with specific influences on
the mechanical strength of the parts. Similarly in the atomic
energy field, the effects of cooling media such as steam, carbon
dioxide, helium or sodium on the mechanical reliability of the
plant must be assessed, with special attention being required
because of the danger of leakage of radioactive species. Even the
so-called renewable energy sources often involve high temperature
materials, and problems of mechanical strength in relation to the
environment are presented. Solar energy, if used in a focussed
mode to generate high temperatures presents such difficulties
and the other thermal processes such as urban refuse combustion,
biomass conversion or geothermal power, all present particular
problems associated with the specific medium involved. Only wind
energy and water power from tidal wave or terrestrial reservoir
sources can be excluded from the high temperature field. The more
hypothetical sources such as nuclear fusion and magneto-hydrody-
namics will involve extremely high temperatures and the materials
problems are not yet clearly defined but will undoubtely pose
severe difficulties.
This symposium has been organized by the Petten Establishment of
the Joint Research Centre in co-operation with the European
Federation of Corrosion and the "Nederlands Corrosie Centrum", and
it is to be hoped that it will provide an ideal medium for the
exchange and clarification of views on the whole problem of high
temperature materials in corrosive environments.
I wish you a very successful meeting and hope that the discussions
will be fruitful to each of you.

INTRODUCTION TO SYMPOSIUM

M. Van de Voorde

CEC, Joint Research Centre

Petten Establishment, The Netherlands

Mr. van Westen has indicated the extent to which high-temperature materials technology is relevant to the development of existing and potential sources of energy and to other industrially important processes. In power generation by thermal processes a Carnot cycle operates, so that the laws of thermodynamics dictate that higher efficiencies must be sought by using higher maximum temperatures. Similarly chemical processes are controlled by thermodynamic stability of the compounds involved and this often dictates that higher temperatures must be used. Hence containment materials capable of operating at higher, and higher temperatures are required. It is this requirement which has led the current broad topic - the relationship between the serviceability of materials and the environment in which they operate - to be the subject of a number of recent conferences or symposia, including the one held here at Petten last October. In the last few years much testing and development work has been carried out with the aim of selecting and assessing the merit of different materials under particular operating conditions. While such work is necessary to provide answers to pressing engineering problems it is desirable to support these largely empirical studies with well-planned and directed basic research. It is the purpose of the present symposium to discuss all sides of the broad problem but particularly to concentrate on the more fundamental aspects in the hope of developing a better understanding of the mechanisms operating and providing guidance for further progress. I must emphasize that this meeting is to be a true symposium. The papers to be presented are to be regarded as providing a foundation for free and open discussion of the subject and speculation as to future trends. The discussion must not be restricted to the topics of the papers.

The types of plant involved in high-temperature processes are
becoming more complex and expensive, and may involve lives in the
order of 25 years or more, so that the degradation mechanisms of
the materials employed need to be studied and categorized to enable
reliable predictions of serviceable life to be made. High temperature materials technology has largely followed the needs of power
generation and until steam temperatures exceeded about $400^{o}C$
low-alloy steels were adequate for most of the highly-stressed
components of steam engines. The corrosion rates of these steels
were low enough to pose no serious problems. With progressively
increasing steam temperatures in the search for higher thermal
efficiencies creep deformation was recognised as a factor and during
the 1930s considerable progress was made in developing creep-
resistant steels. These still maintained reasonable corrosion resistance at the temperatures experienced and studies of mechanical
strength were invariably carried out in an atmosphere of air, while
studies of corrosion rates in appropriate environments were carried
out by metallographic and gravimetric methods on unstressed samples.
The assumption was made that there was no interaction between the
two processes. It was generally considered that it was sufficient to
ensure that the stress-bearing cross section of a component, calculated from unstressed corrosion-test results, was not reduced
below the safe limit dictated by the results of mechanical tests
carried out in air. Within the gasturbine industry it was only
about 20 years ago that the significance of the surrounding medium
on these properties was recognised and since then work involving
creep and fatigue testing of alloys in different atmospheres has
been carried out. It must be recognised that although thermodynamic
stability of the metals and their reaction products is of importance
in determining what corrosive reactions are possible under certain
circumstances, almost all high-temperature corrosion phenomena
of practical significance are controlled by kinetics. The longterm
resistance of an alloy to high temperatures corrosion is provided
by elements which readily react with the environment to form a
stable protective scale. A number of helpful theories relating to
the properties of corrosion scales have been developed, mainly
based on studies of oxidation of pure metals and simple alloys,
but these theories only apply to adherent continuous films, while,
cracking and spalling of the scale may occur, disturbing the direct
application of rate laws. The incidence of cracking depends on
many factors including volume change from metal to oxide, the
strength and plasticity of the scale, differential thermal expansion
and the shape of the specimen. In practice it is found that many
metals experience what is termed 'breakaway' corrosion, in which
an initial parabolic rate law reverts at a certain point to a
less protective law - even a linear one. The cause is usually some
form of disruption of the oxide scale involving visible or sub-
microscopic cracking, porosity or recrystallisation. In some
cases metal or oxide volatilization may play an important role.

The cracking and spalling of scales is caused by the stresses
developed between the scale and the substrate and these stresses
may lead to creep distortion of the substrate with thin-walled
components.
The mechanical strength of a high-temperature component is normally
assessed by consideration of the properties of small samples
tested in air, but it is now clear that reactions between the
alloy and its environment may have a marked influence - usually
detrimental, but sometimes favourable.
Conversely the imposition of steady or fluctuating stresses may
affect the rate of corrosive attack by the medium. Under high-
temperature conditions the effects are usually less dramatic
than those of stress-corrosion under wet conditions, but nevertheless corrosive attack may be accelerated by periodic cracking of
the normally protective scale under the influence of the applied
stresses.
The papers to be presented to the symposium are divided into
three groups. Those dealing with the influence of corrosion on
mechanical properties show a preponderence of interest in carbu-
rising/sulphidizing environments, relevant to various energycon-
version systems and petrochemical processes, but particularly to
coal-conversion systems, in view of the current world-wide interest in these possibilities. Other contributions deal with the
effects of combustion gases including ash and sea-salt contamination on the properties of nickel-base gas-turbine blade alloys.
The effects of steam corrosion are dealt with in two papers. It
might be noted here that steam is a relatively innocuous medium
to most ferritic or austenitic steels at temperatures up to about
600°C, but some troubles have been experienced due to scale spalling leading to blockage of superheater or reheater tubes in power
plant.
The papers dealing with the influence of mechanical stress on
corrosion mainly refer directly to the effects of creep on the
structure and morphology of oxide scales produced on steels and
nickel-base alloys. Other papers deal with the effects of imposed
stresses on sulphidation, carburisation and other corrosion
processes on specific high-temperature alloys. No contributions
are offered dealing with the effects of fatigue stresses, and I
consider this unfortunate as stress reversals would appear to be
particularly effective in promoting scale spalling and consequent
acceleration of corrosion processes.
The three papers dealing with testing methods refer to specific
experimental observations but may give useful concepts of more general applicability. It is to be hoped that informative discussion
will develop in this area since many problems are posed by test
procedures.
I hope that these introductory comments will serve to set the
scene for a most interesting and stimulating symposium and I look
forward to participating fully in the proceedings.

CREEP OF 9 Cr-1 Mo STEEL IN A CARBURISING ENVIRONMENT

C.J. BOLTON and I.R. McLAUCHLIN

CENTRAL ELECTRICITY GENERATING BOARD,
BERKELEY NUCLEAR LABORATORIES,
BERKELEY, GLOUCESTERSHIRE, U.K.

SYNOPSIS

Creep behaviour at 833K of 9 Cr-1 Mo steel carburised by pre-treatment or during testing in CO_2 is described and compared with that of uncarburised material. Carburisation is found to reduce both the creep rate and the ductility, but not the failure time. A theoretical model has been developed which considers the carburising specimen as a composite made up of thin layers each of different carbon content and creep properties. The model predicts satisfactorily the strain-time behaviour of specimens tested in CO_2, and failure times in CO_2 tests are predicted to be greater than those in vacuum tests.

1. INTRODUCTION

Oxidation of 9 Cr-1 Mo steel in CO_2 results in carbon being released at the oxide-metal interface. The ensuing carburisation of the underlying metal is known to improve tensile proof stress and UTS, whilst ductility is reduced[1]. The purpose of the present work was to investigate the effect of the carburising environment on creep properties of 9 Cr-1 Mo.

Diffusion of carbon into the specimen would be expected to strengthen the material progressively, resulting in a continuously decreasing creep rate and a consequent increase in rupture life. However, failure of the embrittled carburised surface layers at a low strain will increase the net stress and hence the strain rate. The specific objective of this work was to examine whether effects on ductility could outweigh those on strain rate and produce reduced failure times.

The most severe effects of carburisation may occur at long times.
Tests in CO_2 were therefore supplemented by tests on specimens
given accelerated pre-carburising treatments, and a theoretical
model has been developed to help to predict the effect of long
exposure times.

2.   PROPERTIES OF PRE-CARBURISED SPECIMENS

All specimens tested throughout this work were from a single batch
whose composition is given in Table 1. The specimens, of cross
section $\sim$3 mm x 0.8 mm, were punched from strip and given a
standard normalising treatment of $\frac{1}{2}$ h at 1253K (air-cool),
followed by a 2h temper at 1023K (air-cool). Constant load creep
tests were performed in rigs with specimen chambers which could be
evacuated or supplied with flowing CO_2 at 1 atmosphere as desired.

Pre-carburising treatments were performed using a proprietary
pack-carburising compound. The effect on microstructure of two
different exposures at 923K is illustrated in Figure 1, and the
profiles of total carbon content as determined by nuclear
microprobe analysis[2] are shown in Figure 2. The Figures reveal
an increasing depth of penetration and an increase in peak carbon
level with increasing exposure time.

Results of creep tests at 200 $N.mm^{-2}$ and 833K in vacuum on these
specimens are shown in Table 2. Increasing degrees of near
surface carburisation cause a 25-fold decrease in creep rate
after a 5h pre-carburisation, while rupture strain falls from
40.6% to 29.4%. Strengthening outweighs the embrittlement and
rupture time increases from 3.3×10^5 s to 4.2×10^6 s.

The failed specimen given a 5h pre-carburising treatment was
examined metallographically after etching in 2% nital to show the
carburised layer. The microstructure is compared in Figure 3
with that of an uncarburised control, electrolytically etched in
oxalic acid. Cracking of the carburised layer clearly illustrates
the reduced ductility of carburised material. However, the core
material remains uncarburised and hence intrinsically ductile, and
therefore cannot contribute to the reduced total failure strains
of carburised specimens. The latter effect is thought to arise
from localised necking brought about by cracking of surface
carburised layers.

3.   TESTING IN A CARBURISING ENVIRONMENT

Figure 4 compares the creep curves at 200 Nmm^{-2} and 833K of
specimens tested in CO_2 and vacuum. Strain rate and ductility
are reduced while rupture time is increased by carburisation.

Also plotted in Figure 4 are curves illustrating the effect of
increasing stress during constant load tests. These were

calculated by assuming that the secondary creep rate law (with
n=12.2 obtained from vacuum tests) describes all the strain after
a chosen point late in the linear portion of a creep curve and
that straining is uniform.

The strain-time curve of the vacuum test is predicted well on this
basis except for a small deviation in the final stages ascribed
to specimen necking. Necking must also contribute to the CO_2
curve, but the first deviation, at a relatively low strain, is
thought to be due to the onset of surface cracking which raises
the specimen stress and creep rate. Further cracking and eventual
necking ensue.

Metallography has confirmed that necking has occurred at the
fracture of both samples and that the carburised case of the
CO_2-tested specimen shows surface cracking (Figures 3a and 5a).
No cavitation is found, which accounts for the high ductility of
uncarburised 9 Cr-1 Mo.

Creep results in CO_2 and vacuum for a range of stresses at 833K,
producing endurances up to 4 years, are given in Table 3 and
plotted in Figure 6. Strain rate and ductility are generally
lower, and failure time higher in CO_2, especially at lower
stresses where more carburisation would be expected. The
strengthening in CO_2 appears despite the loss of section due to
oxidation in these tests. Relative to air tests, the strengthening would be expected to be even greater, and this is suggested
by the ISO rupture data included in Figure 6. No air data is
available on the particular material tested here.

Figure 5(b)-(d) shows metallography of the remaining CO_2-tested
specimens. The oxide thickness gradually increases with exposure
time as expected, but the variability in the degree of carburisation is puzzling, especially as all the specimens were polished
and etched together. The relative weakness of the 120 N.mm^{-2}
specimen compared with extrapolations from higher stress may be
due to the apparent lack of carburisation remote from the fracture
area. Nevertheless the failure time of this specimen was
considerably greater than that obtained from a similar test in
vacuum.

The cause of the variability in degree of carburisation is not
understood. Although it is possibly rig-dependent, it seems more
likely to arise from differences in initial surface condition,
particularly since there appears to be a correlation with oxide
structure, the most heavily carburised specimens having the most
markedly duplex oxide. From autoclave experiments on CO_2 oxidation of 9 Cr-1 Mo it is known that a porous oxide is necessary
for carburisation to occur, and that the degree of carburisation
is influenced by local oxidation rates, local silicon content and

variations in surface work.

All carburised specimens were found to contain surface cracks (Figure 5). Oxidation of crack faces suggests early crack formation and slow crack growth. As with the pre-carburised specimens, the presence of cracks is thought to cause localised necking and reduce the total elongation at failure.

4. A MODEL FOR THE CREEP STRAIN AND FAILURE BEHAVIOUR OF CARBURISING SPECIMENS

An earlier model of 9 Cr–1 Mo creeping in a carburising environment treated the specimen as a strong carburised case and a weaker core[3]. Advance of the interface resulted in progressive redistribution of the load, such that the stresses in both case and core fell during the test. The predicted decrease in creep rate agreed well with experiment. Here this treatment is developed to consider the carbon concentration gradient in the case, and to include effects of carbon on ductility. The specimen is treated as a large number of thin shells, each of uniform carbon content and with creep strength and ductility related to that carbon level. The specimen strain rate at a given time is governed by the combined response to loading of the shells. Failure of each shell is considered to occur when its creep ductility is exhausted. Since the carbon content of each element, and therefore its ductility, changes with time, a strain fraction rule is invoked, based on the strain to failure at each carbon content. Failure of successive shells causes a gradual increase in stress, culminating in specimen failure. Details of the model are as follows.

Consider a specimen of cross section w x b, comprising 2N shells of equal thickness dx. The area A_i of the ith shell is

$$A_i = 2dx(b + 2dx(N + 1 - 2i)) \qquad \qquad \ldots \ldots (1)$$

The carbon content C_i of the ith shell is assumed to be that at its centre, and calculated from the following equation[4] for the carbon content C_x at time t and distance x from the centre of a plate of thickness w:

$$C_x = C_B + (C_S - C_B)\left\{ \sum_{n=0}^{\infty} (-1)^n \, \mathrm{erfc}\left(\frac{(2n+1)^w/2 - x}{2\sqrt{Dt}}\right) + \right.$$

$$\left. \sum_{n=0}^{\infty} (-1)^n \, \mathrm{erfc}\left(\frac{(2n+1)^w/2 + x}{2\sqrt{Dt}}\right)\right\} \qquad \ldots \ldots (2)$$

where D is the diffusion coefficient. C_S and C_B are the surface and bulk carbon concentrations, assumed constant throughout; particular applications may require different boundary conditions, which will give rise to different expressions instead of equation (2).

The strain rate $\dot{\varepsilon}_i$ of each shell is considered to be dominated by secondary creep processes. Primary creep in carburising samples is found experimentally to be of relatively short duration, and in uncarburised 9 Cr-1 Mo the 'tertiary' stage is due predominantly to the effect on secondary rates of the increase in true stress accompanying specimen strain (e.g. Figure 4). $\dot{\varepsilon}_i$ is therefore assumed to obey the familiar secondary creep law

$$\dot{\varepsilon}_i = K_i \sigma^n \qquad \ldots\ldots (3)$$

where $K_i = f(C_i, T)$ and $\sigma = \sigma_o (1 + \varepsilon_t)$.

Here σ_o is the initial stress, T the temperature, n the creep stress exponent, and ε_t the current specimen strain. To maintain specimen integrity the creep rate of each shell must equal $\dot{\varepsilon}_t$, and it follows that

$$\dot{\varepsilon}_t = \left(\frac{\sigma wb}{\sum\limits_{i=J}^{N} A_i K_i} -1 \middle/ n \right)^n \qquad \ldots\ldots (4)$$

where J is the current number of shells fractured.

Account is also taken of metal loss due to oxidation in high temperature CO_2, using kinetics of metal loss established by measurements on the specimens illustrated in Figure 5. Specimen stress is modified accordingly at each time step in the computation.

Failure of successive layers is assumed to occur when

$$\int \frac{\dot{\varepsilon}_i dt}{\varepsilon_{F_i}(C_i)} = 1 \qquad \ldots\ldots (5)$$

where $\varepsilon_{F_i}(C_i)$ is an assumed relation between failure strain ε_F and carbon content.

Clearly, if the exact nature of the dependence of strain rate and ductility on carbon content is known then using equations (2), (4) and (5) it is possible to calculate the creep rate as a function of exposure time, the onset and rate of crack growth, and the failure time and strain of a specimen under a constant load.

Vacuum tests on samples with different uniform carbon content constitute an obvious direct method of establishing the exact relationships. Attempts have therefore been made to produce suitable samples by pack carburising specimens of normalised and tempered steel, or by exposure to high carbon activity liquid sodium. However, the method is complicated by several factors.

To achieve significant carbon uptake in a reasonable time the required carburising temperatures are somewhat higher than those used for testing. Exposure at such temperatures eventually leads to overtempering of the structure and the full effect of carburisation is not attained. For example, although short carburisation treatments cause reduced creep rates (as shown in Table 2), carburising times above 5h cause minimum creep rates to rise again, towards that of uncarburised material. To illustrate this, Figure 7 shows the data of Table 2 supplemented by results on specimens with longer carburising treatments. The effect is due to thermal ageing, as confirmed by testing samples aged, but not carburised, for long times at 923K.

Sufficiently short carburisations avoid significant thermal ageing, but the specimens retain a carbon gradient, and thus cannot be used directly to provide the data required. Attempts to homogenise the specimens by various heat treatments failed to reproduce the original structure.

So far it has not been possible to produce specimens of uniform carbon content. There remains the possibility of testing specially prepared alloys of appropriate composition, and this is being pursued. Meanwhile, it has been necessary to adopt an indirect approach using data from samples with a carbon gradient.

The creep rate of a specimen with a carbon gradient, and a surface carbon level C_s, can be taken as an upper limit for the creep rate of a specimen of uniform carbon content C_s. The greater the concentration gradient in the sample and the greater the degree of thermal ageing, the greater will be the extent by which that upper limit will exceed the real value. Upper strain rate limits at 833K and 200 N.mm^{-2} obtained from samples pre-carburised for short times at 923K are shown in Figure 8 together with rates measured on two samples of low uniform carbon level. The upper limit curve indicates a large effect on creep strength at low carbon levels, and agrees qualitatively with the effects of carbon on the tensile properties of 9 Cr-1 Mo[1]. The shape of the curve indicates a relationship of the general form:

$$\dot{\varepsilon}_i = \frac{K'}{C_i + K''} \sigma^n \qquad\qquad \dots\dots(6)$$

where K' and K'' are constants at constant temperature.

A closer approximation to the real curve can be obtained, again indirectly, by using the model to predict the creep rate of a number of samples, each with a carbon concentration gradient, by optimising the parameters K' and K''. When this is done for specimens creep tested in CO_2, the second curve in Figure 8 results. With n=12.2 measured from tests performed on uncarburised material in vacuum,

$$K' = 5.10^{-38}$$

and $K'' = -0.0873$

when $\dot{\varepsilon}$, σ and C_i have units of s^{-1}, $N.mm^{-2}$ and wt.% respectively.

The relationship between ductility and carbon content can also be estimated indirectly from tests on pre-carburised samples. The more brittle surface layers of such a specimen will fail at a lower strain than the specimen as a whole. Indeed oxide seen on the faces of cracks in failed carburised specimens examined metallographically indicates that the cracks formed some time before total specimen failure. When the effects of true stress are taken into account for pre-carburised samples, the measured curve departs from the calculated curve at a much lower strain than does that for an uncarburised sample. It is postulated that the strain at this point is the failure strain of the surface layer of known carbon content.

Values obtained in this way from tests on samples pre-carburised at 923K for short times, together with that for uncarburised material, are plotted in Figure 9. Here, primary strain is ignored to maintain compatibility with the treatment of strain rate. Again a large effect is apparent at low carbon levels, in qualitative agreement with that found for tensile ductility[1,5] and the following relationship can be fitted through the points (curve 1):

$$\varepsilon_{F_i} = \frac{Z'}{C_i + Z''} \qquad\qquad \dots\dots(7)$$

When C_i is in wt.% and ε_{F_i} in %, $Z' = 5.44$ and $Z'' = 0.044$.

An effective diffusion coefficient, related to the combined effect of diffusion and carbide precipitation was estimated from metallographic measurements of the depth of the carburised case at

failure. Penetration measured in this way agrees well with that
measured by nuclear microprobe and microhardness methods. The
approximation used was

$$D_{eff} = Y^2/4t \quad (mm^2 s^{-1}) \qquad \qquad \ldots\ldots(8)$$

where Y is the case depth produced in time t. D_{eff} was found to
lie between 10^{-9} and 10^{-10} $mm^2 s^{-1}$ at 833K. CO_2 carburisation at
833K has been found to produce a surface carbon level, C_s, of
~ 0.6 wt.%[6], while analysis of the material employed in this study
shows $C_B = 0.09$ wt.%

Predictions of the model have been compared with measured creep
curves from the CO_2 tests on 9 Cr-1 Mo steel at nominal stresses
in the range 155-200 $N.mm^{-2}$ and at 833K. When a single (mean)
value is used for D_{eff} at all stresses, relatively poor agreement
is found. This is not surprising because the creep behaviour
appears to be quite sensitive to carbon level. When values of
diffusion coefficient appropriate to the observed carbon profiles
in each test are employed (Table 4), agreement is very much better
(Figures 10 to 13). The variation in D_{eff} between apparently
similar tests is not understood, but differences in conditions at
the oxide/metal interface may be responsible.

Figures 10 to 13 show that the model predicts the creep rate and
failure times of most of the specimens reasonably well, though the
strain at failure is always overpredicted. The reduced failure
strain measured experimentally was attributed in Sections 2 and 3
to the effects of localised necking, which the model ignores.
The failure time is perhaps more meaningful under these circumstances, and indeed is the parameter of most interest here.

Other characteristics of the model can be outlined briefly. Creep
rates at a given stress and temperature are governed solely by
carbon level. Thus high diffusion coefficients produce a rapid
fall in creep rate. The strain will then remain small, and
failure of the first layer may be greatly delayed. For small
diffusion coefficients, little carbon penetration occurs, creep
rates are much larger and the failure criterion is satisfied much
earlier for each shell. As expected, behaviour then approaches
that observed in a vacuum test.

5. DISCUSSION

Carburisation causes an increase in the creep rupture life of
9 Cr-1 Mo steels but a decrease in the failure ductility. The
effects are greater when exposure to the carburising environment
is prolonged by testing at lower stresses, and such tests often

cannot be completed on a realistic time scale. Attempts have
therefore been made to model the effects of carburisation on creep
properties. Using data deduced from tests on partially carburised
samples, model predictions support the experimental result at
higher stresses that the creep life of carburised samples will
always be greater than that of uncarburised material, and indicate
that the same is true at lower stress. However, it is appropriate
to examine the confidence which can be attached to these predictions and to the data on which they are based.

When no carburisation occurs, the model predicts vacuum behaviour
and, neglecting the primary stage, creep rates increase steadily
due to the increase in true stress with strain, until failure is
indicated by the strain fraction criterion. The predicted strain
vs. time curve matches experiment very closely. The effect of
carbon on strain rate has been deduced from experimental tests in
CO_2, and so when carburisation is included in the model, predicted
strain rates naturally agree closely with experiment. By contrast,
the onset of cracking is calculated using ductility data derived
independently from pre-carburised specimens. Using these data,
the model predicts the failure times of samples tested in CO_2
reasonably well. This constitutes independent confirmation of the
validity of the failure ductility vs. carbon relationship used
(Figure 9).

In principle earlier failures in CO_2 could be produced either by
large creep rates combined with reduced ductilities, or by reduced
creep rates accompanied by very severe embrittlement. Experiment
has indicated that the former combination is most unlikely, but
ductility effects are not so closely defined. To examine the
sensitivity of the predictions to the assumed relationship between
ductility and carbon content, calculations using more severe,
hypothetical, ductility relationships (Figure 9, curves 2 to 5)
were performed. Earlier failure in CO_2 is not predicted for curves
2 and 3, while curves 4 and 5 do cause predicted failure to be
earlier than in vacuum. This confirms that with suitable assumptions the model is indeed capable of predicting reduced lives in
CO_2. However, curves 4 and 5 are regarded as being unrealistically
severe and it is concluded that failure of a carburising specimen
is always likely to occur later than that of a vacuum tested
sample.

The prediction of reasonably accurate rupture lives for CO_2-tested
specimens has only been possible because carburisation kinetics
could be measured by examining the tested specimen. A single, mean,
diffusion coefficient could not describe accurately the rate of
carburisation in all specimens, and hence could not produce
accurate strain-time predictions. Thus, if a mean coefficient is
used, of necessity, in a calculation to predict the creep behaviour

of an unseen sample, there will be considerable uncertainty in the
result arising from uncertainties in carburisation kinetics.
Moreover, laboratory tests of components in environmental conditions may similarly not always reproduce the effects of service
exposure.

Thermal ageing treatments of up to 5000h at 923K produce increased
creep rate and ductility compared to normalised and tempered
specimens. Also, pack carburisations longer than 5h have less
effect on creep properties than shorter carburisations would
indicate (Figure 7). This suggests that prolonged exposures in
CO_2 at 833K (or shorter times at higher temperatures) may diminish
the effects of carburisation assumed here in the model. Any
reduction due to ageing of the effect of carbon on ductility will
be beneficial provided the associated effect on creep rate does
not overcompensate. Pre-carburisation in high carbon activity
sodium for 5000h at 923K (giving a uniform 2 wt.% of carbon)
resulted in both ductilities and creep rates much greater than
expected for unaged material of this carbon level, but endurance
remained much greater than that of a control tested in vacuum.
While this supports the earlier conclusion that carburisation
always leads to greater endurance, it is apparent that, to allow
more precisely for thermal ageing effects, alterations to the model
are required, which can be qualitatively described as follows.

Tests indicate that thermal ageing displaces the $\dot{\varepsilon}_i(C_i)$ and $\varepsilon_{F_i}(C_i)$
curves to higher values of strain rate and ductility. It is
anticipated that a family of iso-structural $\dot{\varepsilon}_i(C_i)$ and $\varepsilon_F(C_i)$
curves exists, each curve characterised by a
time-temperature parameter. Computations would generally follow
the procedures outlined earlier, but in addition would allow for
concurrent thermal ageing by appropriately stepping from curve to
curve with increasing exposure time. This aspect of the work will
be pursued.

Finally, although the model has been applied here only to the
effects of carburisation the approach can clearly be adapted to
applications where the creep properties of a material are non-
uniform through the section. Examples are the effects of nitriding,
decarburisation, or even surface cold work.

6. <u>CONCLUSIONS</u>

6.1 Specimens given a short accelerated pre-carburising treatment
     at 923K showed reduced creep rates and failure strains
     relative to uncarburised specimens, but always had greater
     failure times.

6.2 Specimens tested in CO_2 to produce continuous carburisation also generally exhibited reduced creep rates and failure strains, and always showed increased failure times, relative to vacuum tests.

6.3 A theoretical model has been developed in which a carburising specimen is considered as a composite made up of thin layers each of different carbon content. The model predicts satisfactorily the strain-time behaviour of specimens tested in CO_2.

6.4 Failure time in CO_2 tests is predicted to be greater than that in vacuum tests. The only way to reverse this prediction is to assume a relationship between specimen ductility and carbon content which is considered unrealistically severe.

6.5 The effects of thermal ageing would probably combine to lessen still further the likelihood of early failures in CO_2, but further work is required in this area.

6.6 Carburisation in CO_2 has been found to be subject to an unexplained variability which should be the subject of further work.

7. ACKNOWLEDGEMENTS

The nuclear microprobe analysis was performed by Dr. J.W. McMillan of AERE Harwell. This paper is published by permission of the Central Electricity Generating Board.

REFERENCES

1. P.J. Jeffcoat and A.W. Thorley, To be Published.

2. J.A. Cookson, J.W. McMillan and T.B. Pierce, J. of Radioanalytical Chemistry, 1979, <u>48</u>, 337-357.

3. I.R. McLauchlin, Unpublished work.

4. J. Crank, The Mathematics of Diffusion, p. 48, 2nd Ed$^{n.}$, 1975, Oxford University Press, England.

5. A. Thorley and C. Tyzack, Proceedings of International Conference on Corrosion, London, 143-154, 1971, British Nuclear Engergy Society, London.

6. J. Soo and P.C. Rowlands, Proceedings of Conference on Environmental Degradation of High Temperature Materials, Isle of Man, 1980, Institution of Metallurgists, London.

TABLE 1.

Chemical composition (wt.%) of the 9 Cr-1 Mo Steel

C	S	P	Mn	Si	Cr	Mo
0.09	0.008	0.011	0.51	0.69	8.3	1.05

TABLE 2

Effect of prior carburising treatments on creep properties.
All tests at 200 Nmm^{-2} and 833K in vacuum

Pre-carburising time (h)	t_f (s)	$\dot{\varepsilon}_{min}$ (s^{-1})	ε_f (%)
0	3.3×10^5	4.8×10^{-7}	40.6
1	1.8×10^6	3.9×10^{-8}	19.0
5	4.2×10^6	1.9×10^{-8}	29.4

TABLE 3

Creep Properties in Vacuum and CO_2 at 833K

Environment	Stress $N.mm^{-2}$	Minimum creep rate s^{-1}	Failure time s.	Failure strain %
Vacuum	120	$5.0.10^{-10}$	$4.6.10^{7}$	38.0
"	140	$4.2.10^{-9}$	$1.6.10^{7}$	23.5
"	155	$1.5.10^{-8}$	$6.5.10^{6}$	33.1
"	165	$3.1.10^{-8}$	$3.3.10^{6}$	36.9
"	180	$1.0.10^{-7}$	$1.4.10^{6}$	40.0
"	186	$8.9.10^{-8}$	$1.2.10^{6}$	37.5
"	200	$4.8.10^{-7}$	$3.3.10^{5}$	40.6
"	220	$8.3.10^{-7}$	$1.5.10^{5}$	40.0
"	220	$8.5.10^{-7}$	$1.6.10^{5}$	48.0
CO_2	120	$6.0.10^{-10}$	$1.2.10^{8}$	13.8
"	140*	$\leq 8.4.10^{-10}$	$>6.7.10^{7}$	>13.6
"	155	$3.1.10^{-9}$	$2.8.10^{7}$	19.6
"	165	$7.9.10^{-9}$	$1.6.10^{-9}$	20.8
"	180	$6.4.10^{-8}$	$1.8.10^{6}$	19.1
"	200	$1.8.10^{-7}$	$6.9.10^{5}$	23.2

* Test continuing

C. J. BOLTON and I. R. McLAUCHLIN

<u>TABLE 4</u>

<u>Effective Diffusion Coefficient Measured for Tests in CO_2</u>
at 833K

Stress $N.mm^{-2}$	D_{eff} $mm^2.s^{-1}$
155	$1.5.10^{-10}$
165	$4.0.10^{-10}$
180	$3.3.10^{-10}$
200	$9.0.10^{-10}$

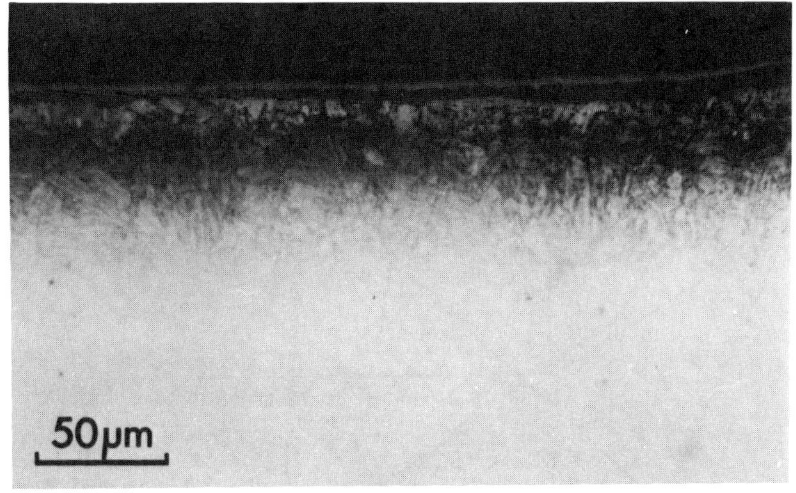

(a)

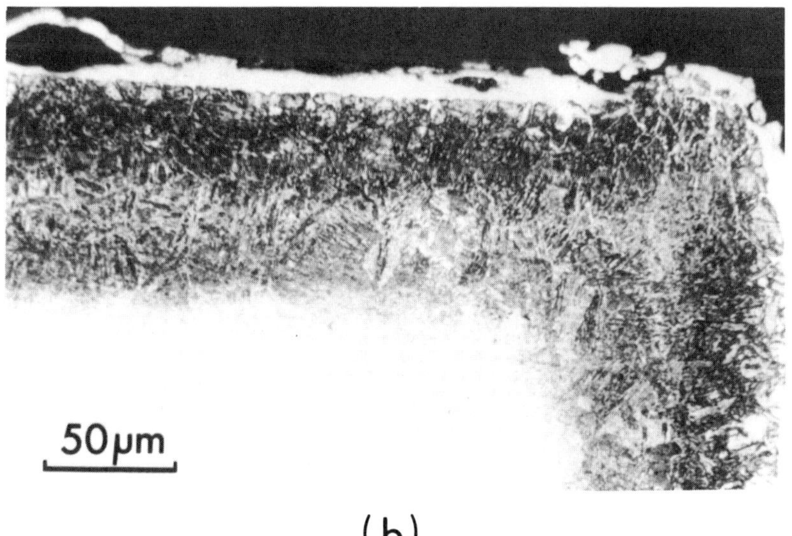

(b)

Figure 1.   Effect of two different pre-carburising treatments on microstructure. (a) 1h at 923K, (b) 5h at 923K. 2% nital etch.

               C.J. BOLTON and I.R. McLAUCHLIN

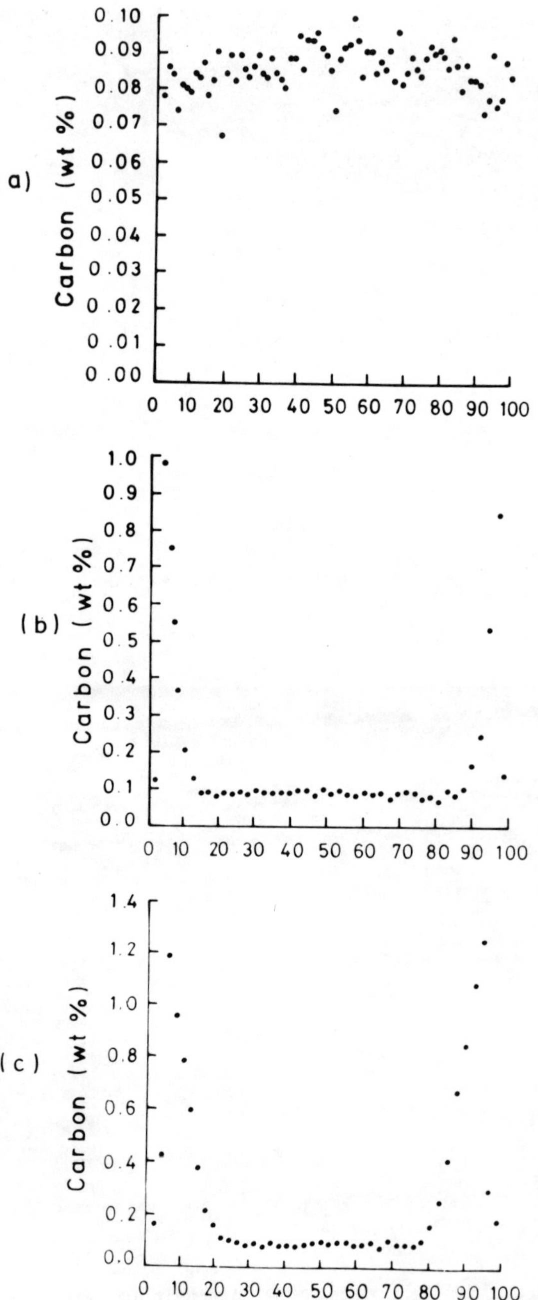

Figure 2.   Carbon profiles measured by NPMA. (a) normalised and tempered control, (b) 1h at 923K, (c) 5h at 923K.

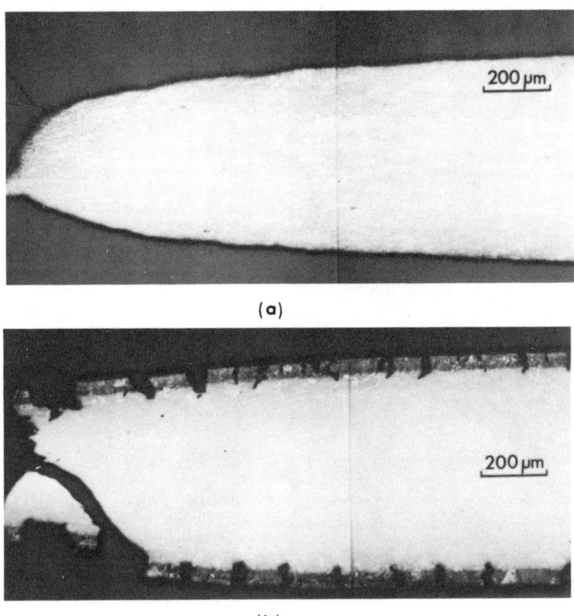

(a)

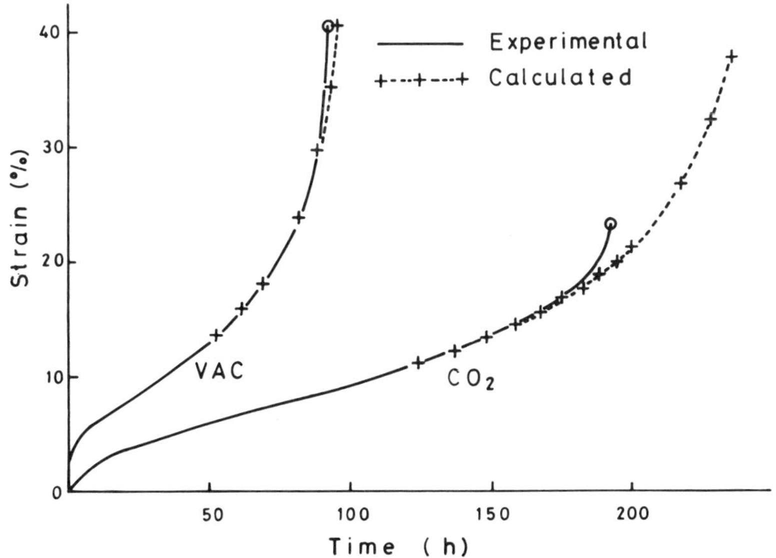

(b)

Figure 3.   Micrographs of creep specimens tested at 200 Nmm^{-2} and 833K in vacuum. (a) uncarburised, (b) 5h at 923K.

Figure 4.   Comparison between creep curves for tests in CO_2 and vacuum at 200 Nmm^{-2} and 833K.

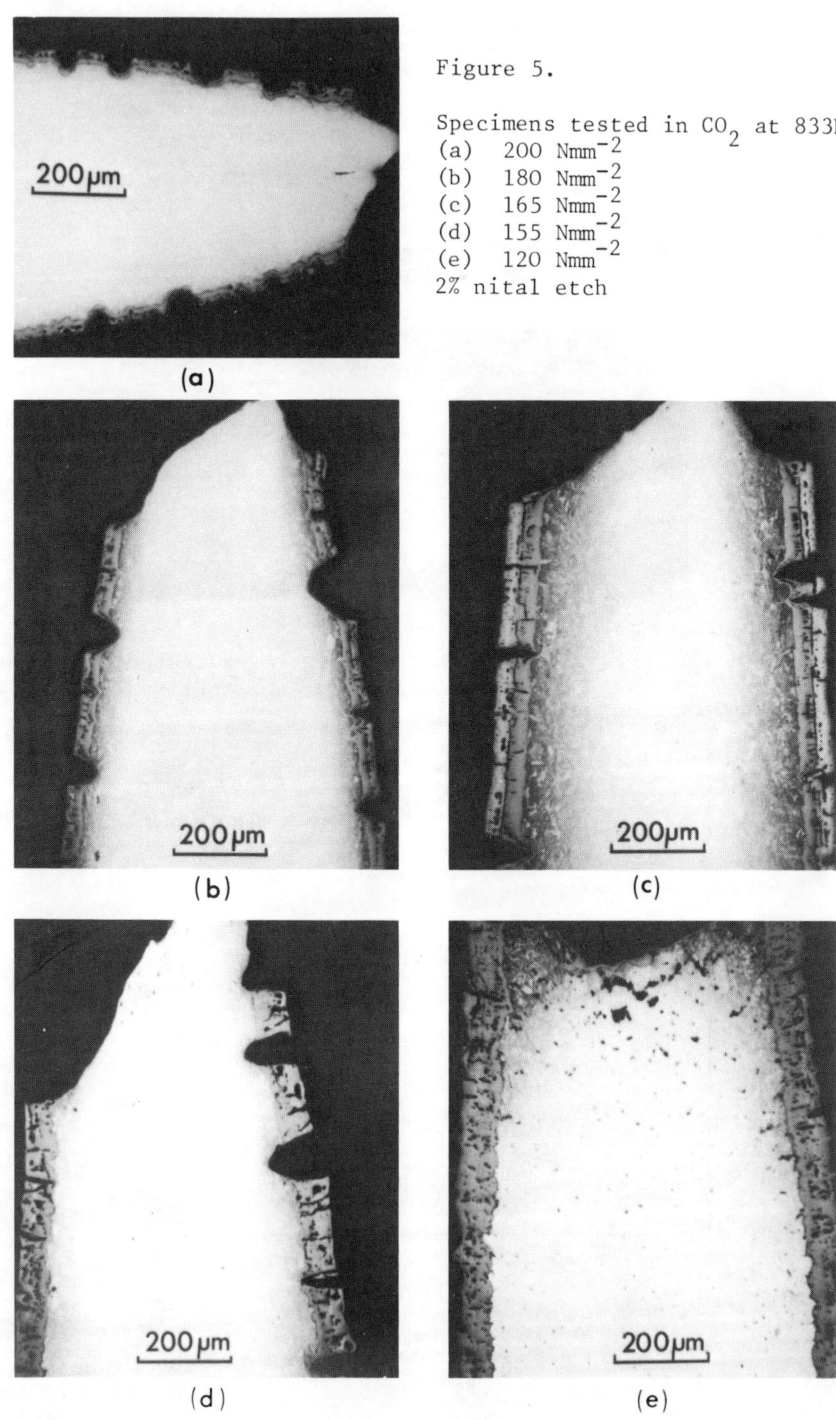

Figure 5.

Specimens tested in CO_2 at 833K
(a)   200 Nmm^{-2}
(b)   180 Nmm^{-2}
(c)   165 Nmm^{-2}
(d)   155 Nmm^{-2}
(e)   120 Nmm^{-2}
2% nital etch

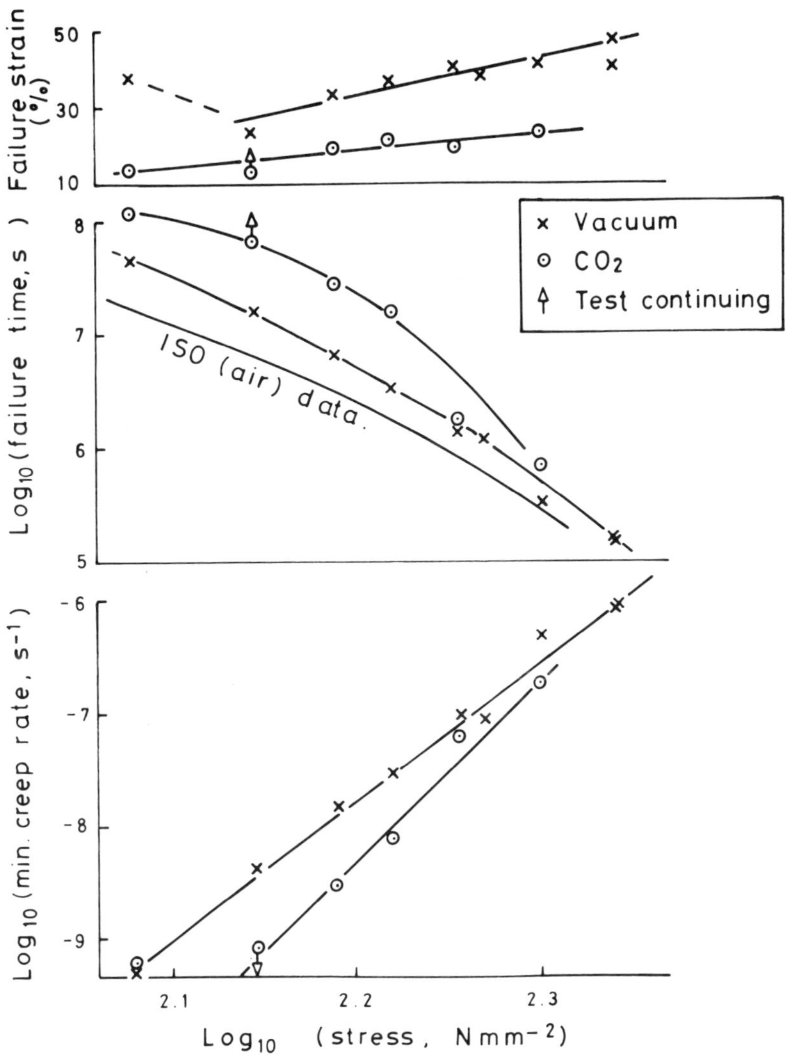

Figure 6.   Comparison between creep properties in CO_2 and vacuum.

C.J. BOLTON and I.R. McLAUCHLIN

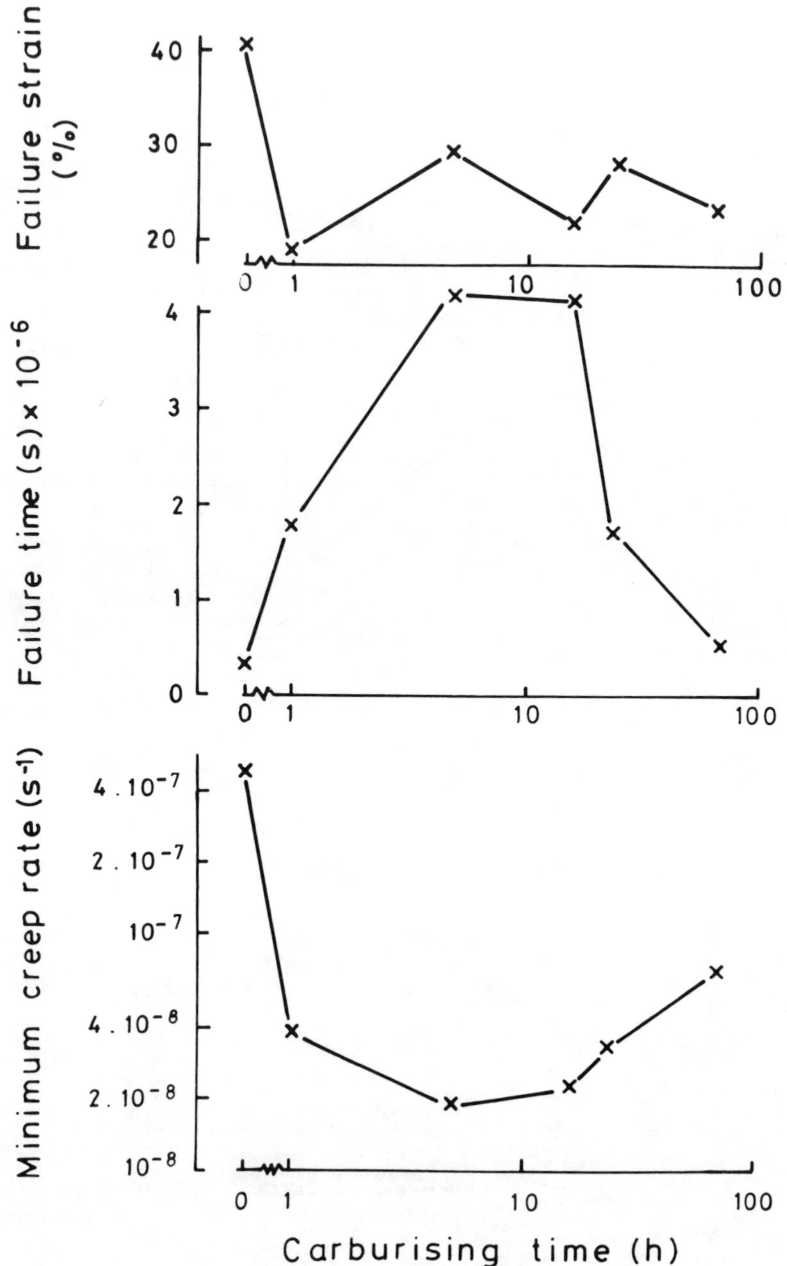

Figure 7   Creep properties as a function of pre-carburising time. All tests at 200 Nmm^{-2} and 833K in vacuum.

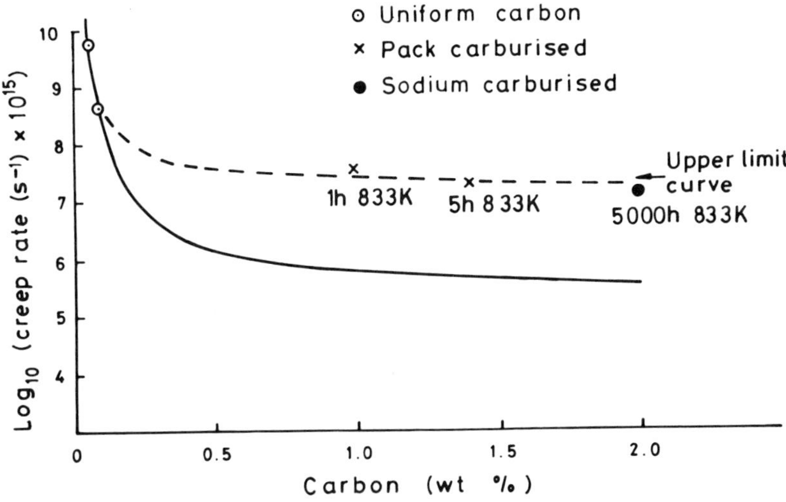

Figure 8.   Relationship between strain rate and carbon content.

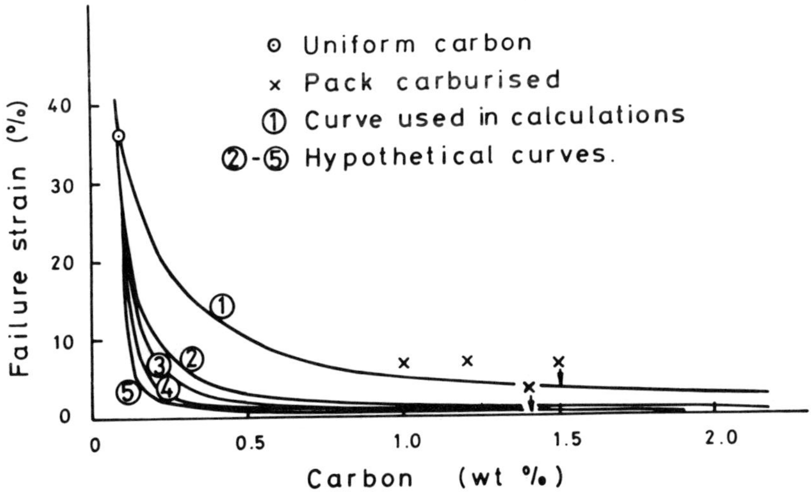

Figure 9   Relationship between failure strain and carbon content.

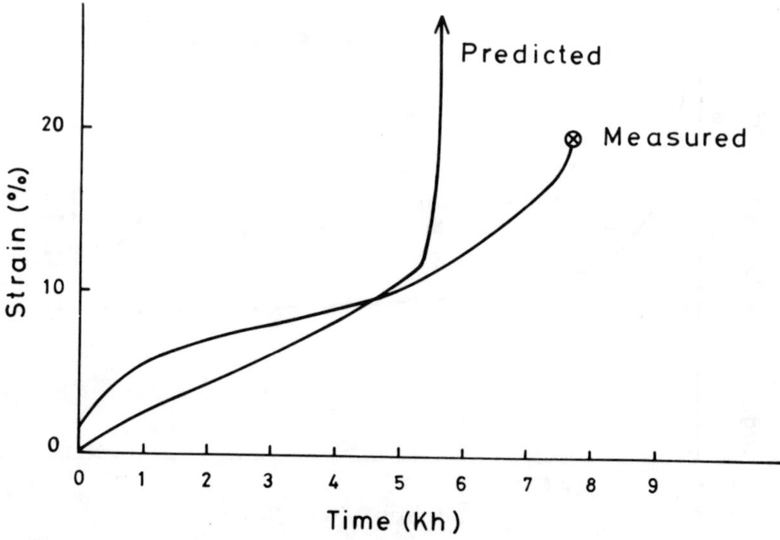

Figure 10.   Measured and predicted creep curves at 155 Nmm^{-2} and 833K in CO_2.

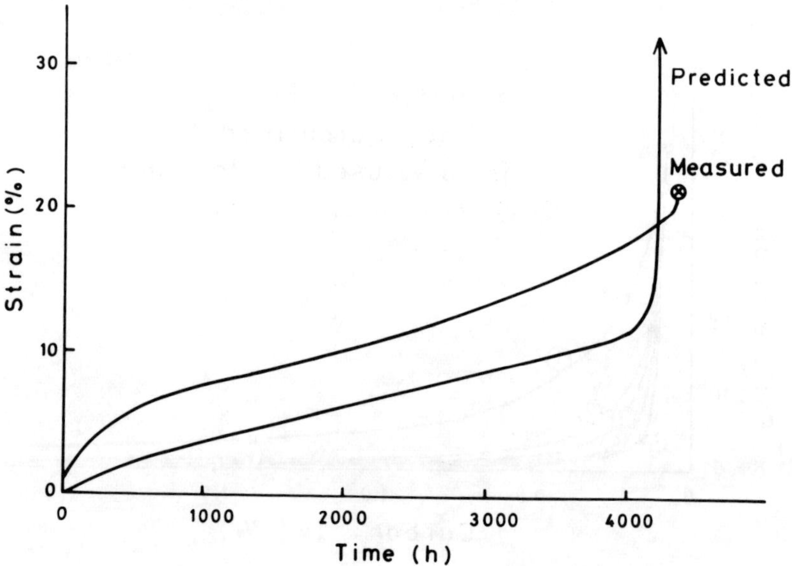

Figure 11.   Measured and predicted creep curves at 165 Nmm^{-2} and 833K in CO_2

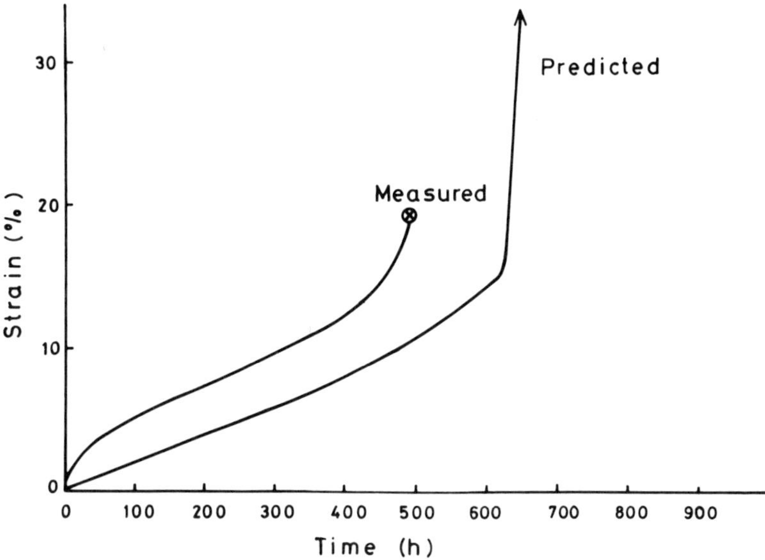

Figure 12.   Measured and predicted creep curves at 180 Nmm^{-2} and 833K in CO_2.

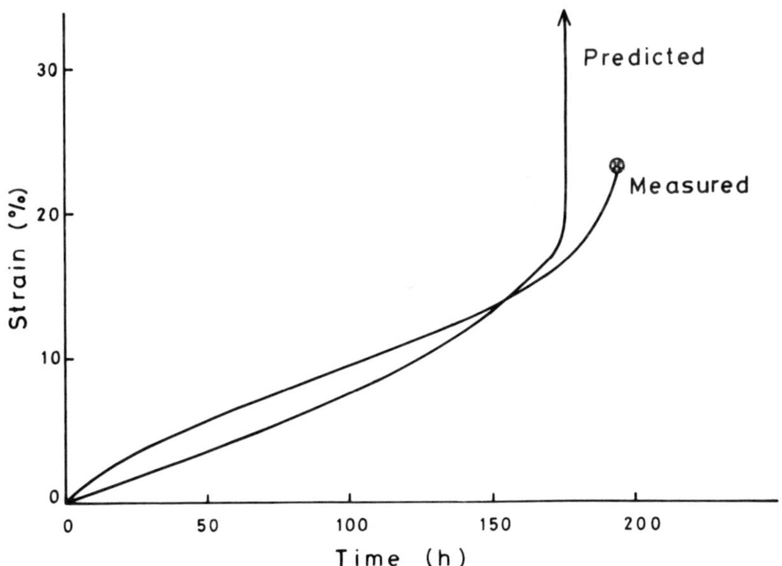

Figure 13.   Measured and predicted creep curves at 200 Nmm^{-2} and 833K in CO_2.

DISCUSSION

T. Boniszewski: You said that the carburised layer had failed
intergranularly. You also said that the reduction in creep rate
outweighed the reduction of ductility. Did you look at the effect
of grain size to find out whether with the increasing grain size
the reduction of ductility will be much greater? During welding
large grains, eg. 200-400 μm, can be formed in the heat affected
zone.

I. R. McLauchlin: An effect of grain size on ductility is indeed
possible. However, we have confined our tests to boiler tube
material in the normalised and tempered condition with grain
diameter $\sim$ 50 μm. We therefore have no information on the precise
influence of grain size on failure.

B. Ilschner: Could any of the authors of this session comment on
the effect of strain rate on carbon uptake? Is, for example,
"D" in McLauchlins paper a diffusion coefficient for the creep
rate? Is accelerated carbon penetration conceivable for $\dot{\varepsilon} > 0$?
May I add that my question is not related to the indirect effect
mentioned by Hemptenmacher and Grabke, viz., acceleration of
carbon uptake by oxide cracking.

I. R. McLauchlin: The values of D_{eff} which we used were estimated
from the microstructure of tested specimens, and thus include any
contribution from deformation rate. However, we have found
considerable variability in effective diffusion coefficient for
tests in CO_2. This may be due to effects of local silicon content,
surface cold work or oxidation rate. Also, our experiments, whether
in CO_2 or on pre-carburised specimens, are complicated by ill-defined
boundary conditions and precipitation. We have therefore been
unable to draw useful conclusions about any effect of stress or
strain rate on D or D_{eff}.

V. Guttmann: The effect of stress (without a correlation to
the strain rate) on the carbon uptake will be reported in a later
paper (see session 2 :"Interaction between Creep Deformation and
Carburisation", by H. Penkalla, V. Guttmann and J. Timm). I would
briefly like to mention here that preliminary results have shown
the interesting feature that for HK 40 and Alloy 800 H the carbon
uptake in materials under low stresses is smaller compared to the
unstresses condition.

INFLUENCE OF CARBURISATION ON THE

PROPERTIES OF FURNACE TUBE ALLOYS

by D.M. Ward

Inco Europe Limited

European Research and Development Centre, Birmingham, England.

Heat resisting steels and alloys are liable to be carburised in service when the carbon potential of the process gas is high enough to prevent the formation of protective oxides. Besides impairing oxidation resistance, severe carburisation is reported to have a detrimental effect on rupture strength and to produce changes in volume, expansion coefficient and conductivity which superimpose additional tensile stresses on tubes. This can cause cracking during cooling for shut-down or prior to decoking and reduce the life of tubes in ethylene pyrolysis furnaces. Reductions in low temperature ductility can occasionally prevent successful nipping of pigtails to isolate individual reformer tubes.

In the present work the properties and structures of carburised samples of HK 40 and Alloy 800, and cast samples of 25Cr-20Ni steel with high carbon contents have been studied. Ring specimens from internally pack carburised tubes of HK 40 and Alloy 800 have been tested under hoop stress conditions to determine the effects of various depths of carburisation on stress-rupture properties at 1000°C, tensile properties at RT, and to study fracture under simulated service conditions. The results show that after equivalent carburisation, HK 40 and Alloy 800 behave the same, but there are differences in properties and structures between carburised samples and alloys cast with the same carbon contents.
Furthermore, carburised HK 40 specimens failed in rupture by cracks initiated in the mid wall, unaffected zone behind the carburised layer, whereas carburised Alloy 800 failed in rupture by cracking of the carburised zone on the inside

of the tube.   It is proposed that this difference in failure
mechanism is attributable to differences in high temperature
ductility between carburised and unaffected zones of the two
alloys.

INTRODUCTION

The petrochemical industry is the major consumer of alloys to resist carburisation.   This highly competitive industry uses centricast tubes for thermal cracking hydrocarbons.   Production demands high throughput and high cracking temperatures.   Moreover, the feedstock creates a highly carburising environment and sometimes produces severe carburisation on the inside of the tubes. Also carburisation may occur in expansion loops which connect catalyst tubes to the outlet headers in steam hydrocarbon reformers.   However, operating temperatures in reformer furnaces are lower than in cracking furnaces and carburising potential is lower also.   The major problems caused by carburisation are thus in centricast tubes used for thermal cracking and far less of a problem exists with wrought tubes used for ('pigtails') expansion loops.   However, as temperatures in cracking furnaces rise and more advanced steam-reformers start-up, resistance to carburisation will become more important in the choice of construction materials for both processes.

Many laboratory tests, both pack and gas carburising types have been used previously to confirm carburising effects observed in practice and to rank proprietary alloys.   Little insight had been given into the mechanisms of carburisation or the effects of carburisation on alloy properties, and alloy ranking lists vary according to individual test procedures.   On mechanisms progress has been made by development of a new set of laboratory test conditions for evaluating carburisation under equilibrium conditions[1].   Also there have been attempts to measure and understand the effects of carburisation on mechanical and physical properties of furnace tube alloys[2,3,4,5,6,7].

In the current work a study has been made of the effects of carburisation on some mechanical and physical properties of centricast HK 40 (25Cr/20Ni/0.4C steel) tubes and annealed Alloy 800 (20Cr/32Ni/0.05C/Fe) tube.   For both materials a procedure has been developed which produces internal carburisation of these tubes.

Centricast HK 40 specimens were carburised to the level which has been observed in highly carburised areas in actual pyrolysis furnace tubes.   Wrought 20Cr/32Ni/Fe specimen were carburised to lower levels of carburisation as expected in the lower temperature and lower severity conditions of steam reformer expansion loops.   Pack carburisation was used to produce carburisation because it simulates effects of solid carbon deposition on internal surfaces of tubes without being overly aggressive like hydrogen-methane gas mixtures.   Although carburising potential varied throughout the exposure an attempt

was made to minimise this variability by replacing the activator
at predetermined intervals.

        Specimens in the form of rings were machined from tubes and
tested initially under hoop stress conditions.   The purpose was
to simulate practical conditions of stress and carburisation, and
to reproduce the type of cracking that often occurs in practice
in uncarburised metal adjacent to the carburised layer at the
internal diameter of furnace tubes.   Additionally test
specimens were machined from centricast tubes and carburised, and
alloys with carbon levels up to 2.75% were cast and tested so that
comparisons could be made with properties from ring specimens and
with those published previously.

EXPERIMENTAL PROCEDURE

        Centricast tubes of HK 40 (25Cr/20Ni/0.4C steel) 87mm OD
60mm ID of compositions listed in Table 1 were used.   Cylindrical
creep and tensile specimens, gauge diameter 6.5mm, were machined
with their axes parallel to the cast tube axes.   Some of these
bars were stress-rupture tested at 1000°C and tensile tested at
RT and some were similarly tested after carburising for 100, 300
and 500 hours at 1050°C in sealed boxes containing 'Pearlite[*]'
carburising compound.   Also, complete tube sections were bored
out to 67mm ID to remove porosity, and then carburised internally
by packing with Pearlite carburising compound and plugging both
ends of the tube to produce a sealed box.   Ring test pieces
12.5mm wide x 67mm ID x 80mm OD were machined from these tubes
after carburising for 100, 300 and 500 hours at 1050°C.

        Some additional heats of HK 40 were melted and cast into
tapered bar moulds to produce bars with carbon levels of 0.95,
1.95 and 2.75% carbon.   Some of these bars were used for
comparison with carburised test pieces.

        An expansion loop (pigtail) of Alloy 800 28mm OD 3.7mm wall
thickness was machined to remove slight oxidation.   By using
plugs at both ends of the tube to contain the carburising
compound, sections of this tube were carburised internally for
100 hours at temperatures from 900° to 1050°C.   After
carburising the tube was sectioned into rings 7.5mm wide x 22mm ID
x 28mm OD (3mm wall) for testing.

        All ring samples from the cast and wrought tubes were
tested by using two inserts to form a yoke (Figure 1) which
initially applied a hoop stress to the ring.   These specimens
were tested in air in conventional creep machines at 1000°C for

[*]Trade mark

HK 40, and at 780° and 1000°C for Alloy 800.   Both materials
were tensile tested at RT after carburising.   For HK 40 tests
were done on bars whereas for Alloy 800 tests were done on the
tube sections.

The depth of carburisation was determined on bars and tube
specimens by layer analysis using an 'Autocarb'[x] instrument in
which consecutive layers of turnings were burnt to carbon dioxide.
Each sample weighed 1 gram.

Metallographic examination was made of adjacent rings for
carburised tube specimens and on the one half of carburised bars.
Volume fraction of carbides was estimated by point counting.   A
5mm square grid was placed upon the screen of a Vickers projector
microscope.   An accuracy of at least $\pm5\%$ of the true value was
obtained by making 400 point counts on each sample according to
the formula:

$$\text{No. of point counts, } P = \frac{1}{SD^2} \times V(1-V) \qquad\qquad (1)$$

where V = estimate of max. volume fraction

SD = standard deviation allowed

Magnetic moment measurements were made on a Sucksmith
balance using a small amount of the turnings from the carbon layer
sample.   The deflection in the coil θ is proportional to the
force on the sample as a result of its mass and permeability

$$\theta = \frac{\sigma \cdot M}{K} \qquad\qquad\qquad (2)$$

where K is a constant determined using an iron standard.

Some of the samples were checked for magnetic response with
an Elcometer[x] which is normally used to measure non-magnetic
material thickness on a ferromagnetic substrate but can be used
also to measure percent ferrite in austenitic stainless steels.

Density measurements were made by immersion and displacement
in ethanol of solid samples of cast 25Cr/20Ni steels with different
carbon contents and on cylindrical carburised samples after
machining to remove consecutive layers for analysis.   For the
latter specimens weights were obtained by difference after removing
the next layer.   Static and cyclic oxidation tests were made on
as-cast and carburised samples of 25Cr/20Ni steel at 1000°C for up
to 300 hours.   The weight lost after each cycle was recorded and

[x]Trade mark

after 300 hours all the samples were sectioned to determine
maximum penetration of oxides and the depth of decarburised zone.

<u>RESULTS</u>

a) <u>Structure</u>

Pack carburisation produced uniform carburisation as
determined by layer analysis and metallographic examination.   In
HK 40 etching with Murakami's reagent produced an easily
recognisable band of carburisation (Figure 2) which increased in
depth with increase of carburising time or temperature.   In
front of the etching band was a slight increase in carbon content
and volume fraction carbide, but behind the band the carbon content
had risen to such an extent that another form of carbide
developed.   This carbide was not etched by Murakami's reagent but
was etched electrolytically in KOH.   The carbide was identified
by X-ray diffraction as of the type Cr_7C_3 whereas the other
carbide was of the type $Cr_{23}C_6$.   The change in carbide type
occurred at carbon levels of about 1.75% for HK 40 when the volume
fraction had risen to about 30%.   This etching band was seen also
at about 0.6% carbon in Alloy 800 when etched electrolytically in
KOH.   Full structures were shown only after etching potentio-
statically in KOH at 0.4 volts for 10 seconds.   This etching
effect correlated with changes in matrix composition rather than
change of carbide composition.   For the carbon levels studied
the carbide was always type $Cr_{23}C_6$ in Alloy 800.

As shown by others[4] the carbides formed in HK 40 and Alloy
800 by carburisation at high temperatures are globular and in the
size range 5-10μm (Figure 3).   In contrast those produced in 25Cr-
20Ni steel cast with 2.75% carbon content are massive hexagonal
carbides about 200-300 μm in size (Figure 3b).

Volume fraction of carbides obtained by point counting gave
a useful correlation with increase of carbon content although
inevitably as the carbide changed in composition there was a dip or
plateau in the curve of volume fraction carbide against depth
below surface as shown in Figure 4 for HK 40.   Attempts to
quantify volume fraction of carbides more easily were unsuccessful.
Quantimet techniques coupled to standard etching, Pepperhoff
etching, or contrast enhanced photography were subjective and thus
can underestimate the carbide volume fraction.

b) <u>Carbon Layer Analysis</u>

Carbon profiles for HK 40 and Alloy 800 are shown in
Figures 5 and 6 respectively.   Carbon layer analysis was a
reliable reproducible quantitative measurement of carburisation but
only a few determinations could be made because of the limiting

wall thickness of centricast tubes (about 15mm) and the weight of sample required for analysis. Microprobe technique of analysis was tried without success because of the similar size of the carbides compared with the 10um square scanning raster used.

c) <u>Magnetic Moment</u>

The results show a positive correlation between magnetic moment and carbon content (Figure 7). There was a threshold carbon level for each alloy above which the relation exists and below which the permeability is very low. This limits the method to high carbon levels of $>$ 2% for HK 40 (25Cr/20Ni steel) and to $>$ 1% for HP 40 (25Cr/35Ni steel). Elcometer readings (per cent ferrite) increased also with carbon content but the values obtained varied non-uniformly with carbon content (Table 2) and depended on mass of magnetic material available: thus tubes gave higher readings than rings machined from these tubes. As with magnetic moment values, the Elcometer readings were higher for the higher nickel alloy than for HK 40.

Samples of carbides extracted from HK 40 after carburisation were checked for magnetic properties. Carbides obtained from the edge and middle of the sample carburised 500 hours at 1050°C gave very low magnetic moments of 2.4 and 3.9 emu /g respectively. However, microprobe measurements of matrix composition from these zones gave Cr depletion from 25 to 11% and nickel enrichment from 20 to 28%.

High magnetic readings may result also from surface oxidation rather than internal carburisation of high nickel alloys. In one sample of an Alloy 800 pigtail, giving high magnetic readings with an Elcometer, it was possible to show indirectly and then directly that oxides were the cause of the high readings. Firstly the oxides were machined from the surface causing the Elcometer readings to disappear also; secondly permeability readings of the oxides gave magnetic moments of 11.6 compared with 1 emu/g for metal swarf from just below the surface. Microprobe examination confirmed the presence of two oxides in the scale: Cr_2O_3 and a mixed nickel rich iron oxide. In this case there was virtually no change in matrix composition beneath the oxide scale. Furthermore when carburisation produced a magnetic effect with Alloy 800 the matrix Ni increased from 32 to 40% and Cr was depleted from 20 to 8%.

d) <u>Density</u>

In agreement with others[2,3] density was found to decrease with increas of carbon content of as-cast specimens (Figure 8),

but measurements made on carburised specimens of HK 40 at
intervals during layer analysis showed little or no variation of
density with carbon content over the range 0.4 to 2.3%.   The
results on cast specimens of different carbon levels are in
agreement with Roach[2]   and followed the same trend as reported
by Ohtomo[3] , but the carburised specimen results are in
disagreement with all the as-cast bar results.

e) <u>Oxidation Resistance</u>

        Static and 24 hour cyclic oxidation tests at 1050°C showed
that carburised specimens of HK 40 had slightly poorer oxidation
resistance than as-cast specimens on the basis of weight loss
and oxide penetration measurements (Table 3).   In both as-cast
and carburised samples there was a decarburised layer and surface
oxides at previous carbide sites but the differences between the
samples were quite small and would therefore have little direct
influence on high temperature creep properties of similar
duration to the oxidation test.   At worst the surface oxide
would produce a slight notch effect.   Longer time tests[6] have
shown that oxidat-on resistance of carburised samples depends on
carbide type and that samples with Cr_7C_3 type carbides oxidise
more slowly than those with $Cr_{23}C_6$ type carbides in the surface
layer

f) <u>Stress-Rupture Properties</u>

        Rings from centricast tubes of HK 40 gave shorter stress
rupture lives than bar specimens machined from the tube wall.   In
terms of stress the rings gave about $6N/mm^2$ lower strength than
bar specimens at 1000°C (Figure 9).   However both specimens
gave good stress/log time plots from results at different
stresses for different levels of carburisation (Figure 10).
Carburisation treatments at 1050°C first raised the rupture
strength of HK 40 and then reduced strength slightly while
markedly increasing rupture ductility.   For ring specimens, the
highest strength was produced by carburising for 300 hours,
whereas for bars this occurred after only 100 hours.   This may
be explained by comparing the different carbon profiles for the
two specimens.   As rings were carburised only on the inside tube
wall they had a lower average carbon content than bars which were
carburised all round.   When compared in terms of average carbon
content both specimens showed a similar effect of carburisation
on rupture properties (Figure 11).

        For the greatest degree of carburisation (from Figure 4) of
2.75% carbon average (3.2% close to the surface) the fall off in
strength was only $3N/mm^2$ for 100 hour life, whereas ductility
increased from 5 to 50%.   On bars cast with this level of

carbon (2.75%) the fall-off in strength measured was greater ($8N/mm^2$ lower for 100 hour life) and the rise of ductility was lower (20%) (Figure 11).

Ring specimens of Alloy 800 gave higher rupture lives at $50N/mm^2$ 780°C and $20N/mm^2$ 1000°C as carburisation was increased by raising the tube carburising temperature (Table 4).   In this case rupture ductility was high to begin with and was gradually reduced with increase in average carbon content (from Figure 6). After carburising for 100h 1050°C rupture elongation in the 780°C was much lower than on as-received tube.   On the basis of average carbon content both HK 40 and Alloy 800 behaved at 1000°C in a similar manner (Figure 12).   The level of strength was higher for the cast alloy than the wrought alloy but the ductility was lower.

g) Mode of Stress-Rupture Failure

After stress-rupture testing, rings of both HK 40 and Alloy 800 were examined at the point of failure and on the opposite side of the ring where cracking was well advanced. Failure in cast samples of HK 40 was initiated from the outside of the rings.   After carburisation this cracking still occurred but there was some additional internal cracking in the un-carburised region just below the carburised zone (Figure 13). There was no cracking in the carburised zone apart from minor surface cracking.   In contrast for Alloy 800 failure was initiated from the inner bore of rings in both uncarburised and carburised tubes.   Cracks were initiated in the carburised zone and propagated to failure (Figure 14) but some less intensive minor cracking occurred on the outside uncarburised surface also.

h) Tensile Properties at RT

Tensile tests at RT on bar specimens of HK 40 show that carburisation reduces strength and ductility.   Similar trends were found on Alloy 800 ring specimens after caburisation (Table 5).

DISCUSSION

There are many examples published[2,3,8] which show that cracking occurs in the uncarburised metal adjacent to the carburised zone on the ID of HK 40 furnace tubes.   The present work shows that this cracking behaviour can be reproduced in HK 40 by testing ring samples machined from internally carburised tubes.

In previous work[2,3] results from cast samples of 25Cr/20Ni

steels with high carbon contents have been used to suggest that
this type of mid-wall cracking occurs because of additional
stresses caused by significant changes in conductivity, thermal
expansion and specific volume, and because of large reductions in
creep strength in the carburised zone.   The present work
confirms marked effects of carbon on density and creep strength
only when using cast 25Cr/20Ni steel specimens cast with high
carbon contents.   Tests on carburised samples of HK 40 suggest
that results on cast samples over estimate these effects in
practice and examination of the structures showed markedly
different carbide structures between carburised samples and cast
samples of the same carbon content.   Stress rupture tests on
carburised bar and tube specimens show that the strength of HK 40
is first raised and then decreased to slightly below as-cast
values.   High temperature rupture ductility is markedly
increased whereas RT tensile ductility is just as markedly
reduced by carburisation.   Thus, it is the ductility rather than
creep strength of carburised HK 40 which is completely different
to that of uncarburised material.   A simplistic view is that for
HK 40 failure occurs in the uncarburised zone of a tube wall
because this region has very much lower ductility, and only
slightly higher (or lower) rupture strength than carburised
material.   Thus large creep strains can be accommodated easily
in the carburised zone but not in the adjacent uncarburised layer.
In practice temperature cycling during shut-down or decoking
would make these matters worse and could cause premature failure
at lower temperatures.   Under these conditions differences in
thermal conductivity and expansion between the carburised inner
layer and the adjacent carburised zone may make a major
contribution to the failure.   Previous results[5] on Alloy 800
showed that carburisation increased tensile strength at RT, 871
and 927°C but reduced ductility with the greatest effects at RT,
and with longer carburising times.   The present results show that
as well as this effect on tensile properties, the situation with
creep rupture properties is opposite to that with HK 40 for the
carburisation treatments examined.   Carburisation of Alloy 800
increases rupture strength progressively and produces only a small
adverse effect on rupture ductility.   Thus, in contrast with
HK 40, the rupture ductility of uncarburised Alloy 800 is higher
than for carburised material.   In the case of carburised rings
of Alloy 800, therefore, uncarburised material though slightly
weaker is able to withstand high creep strains, and cracks less
easily than carburised material.   In Alloy 800 cracking
occurs mainly in the carburised zone on the inside of the
carburised tube specimens in contrast to mid-wall cracking in
HK 40 (Figure 14) as might be expected from the properties.

     Low RT ductility produced by carburising is a problem with
some designs of expansion loops (pigtails) because it will
prevent successful pigtail nipping to isolate failed reformer

tubes without furnace shut-down.   It is common practice
therefore to take frequent Elcometer readings on tubes where
carburisation is suspected so that new tubes can be fitted as
required.   Magnetic techniques are very useful for detecting
and measuring degree of carburisation, but are crude and the
magnetic balance method requires machined samples or swarf.   By
far the biggest drawback is the   limit of 1%C for 25Cr/35Ni
and 2%C for 25Cr/20Ni steels below which the magnetic moment
method is ineffective.

Comparing methods of measuring carburisation showed that
layer analysis is the most reliable way and works best when the
samples are large to minimise the layer depth and maximise the
number of levels obtained.   Estimation of volume fraction
though as informative as layer analysis is non-destructive but is
tedius and not adaptable to automated processing with present
equipment.   Microprobe techniques may be automated and are
non-destructive also but need to use a much larger roster area
than the carbide particles being scanned.

REFERENCES

1.   R.H. Kane, Steel Foundry Facts, 1979.

2.   D.B. Roach, Corrosion 76, NACE, Paper No. 7.

3.   Akira Ohtomo et al. Ishikawayuma-Harima Engineering Review
     1976, 16, No. 3 pp. 239-245.

4.   N.G. Persson, Sandvik Lecture No. 56-6E FSI, Oct. 1976.

5.   D.E. Wenschhof and J.A. Harris.   Corrosion 77 NACE
     Paper No. 9.

6.   A. Schnass and H.J. Grabke.   Werkst. Konos.   1978, Oct. 29
     (10)   pp. 635-644.

7.   R. Petkovie - Luton, Can. Metall. Q. 1979 Apr-June 18 (Z)
     pp. 165-170.

8.   M.W. Clark.   Annual Conference NACE Philadelphia, March 1970.

ACKNOWLEDGEMENT

    The author is indebted to Inco Europe Limited for permission
to publish this paper.

TABLE 1   Analyses of Materials Used

	C	Cr	Ni	Si	Mn	Fe	
ZHMT	0.40	24.8	20.8	1.61	0.71	Bal.	HK 40 Centricast tube
ZJMT	0.45	24.9	20.6	1.70	0.79	Bal.	" " " "
ZHUO	0.42	25.5	20.3	1.46	0.69	Bal.	HK 40 cast bars
ZHUP	0.94	25.5	20.1	1.49	0.70	Bal.	HK 100 cast bars
ZHUS	1.95	26.2	20.5	1.68	0.77	Bal.	HK 200 cast bars
ZHUT	2.75	25.3	20.4	1.55	1.03	Bal.	HK 300 cast bars
ZJCH	0.05	21.8	32.5	0.95	1.08	Bal.	Alloy 800 wrought tube also 0.43Ti 0.23Al

TABLE 2   Elcometer Reading and Magnetic Moment

Alloy	Elcometer (% ferrite)	Sucksmith Balance Magnetic Moment (emu/g)	Carbon Content (%)
HK 40	0	1.8	0.54
25Cr/20Ni	0	2.4	0.57
	0	3.0	0.65
	0-1	3.6	0.73
	0-1	6.1	0.88
	0-1	7.3	2.1
	0-1	8.9	2.35
	0-1	11.3	2.55
	3-4	30.2	2.75
HP 40	0	5.3	0.81
25Cr/35Ni	1-2	8.5	1.33
	6.5	19.3	1.75
	9	41.4	1.90
	15	43.2	2.25

TABLE 3　　　　Oxidation Resistance at 1050°C

Material Condition	316 hours (continuous)		316 hours (cyclic)	
	Wt. change (mg/cm^2)	penetration[*] (μm)	Wt. change (mg/cm^2)	penetration[*] (μm)
HK 40 as cast	− 2	40	− 2	20
HK 40 carburised	− 5	40	− 8	50
300h 1050°C	+ 2	45	− 9	57

[*]Determined metallographically at worst place.

TABLE 4   Effect of Carburisation on Stress-Rupture
                    Properties of Alloy 800

Rupture Test Results

Condition before Testing	$50N/mm^2$ 780°C Life(hrs)	RA(%)	$20N/mm^2$ 1000°C Life (hrs)	RA(%)
As machined	336	36	40	37
Carburised 100h 900°C	478 564	30 30	32	30
Carburised 100h 950°C	1204	22	67	19
Carburised 100h 1000°C	1534	26	89	30
Carburised 100h 1050°C	1569	9	187 212	24 20

Ring specimens   7.5mm wide x 3mm wall thickness
                 x 28mm O.D. tube.

D.M. WARD

TABLE 5 | Tensile Properties of HK 40 and Alloy 800 after Carburisation

Alloy	Treatment (hours/°C)	Average Carbon Content (%)	RT Tensile Properties		
			0.2%PS N/mm^2	UTS N/mm^2	E1/RA %
HK 40	as cast	0.4	265 264	606 598	27 28
	100/1050	0.95	338 317	466 450	1.7 1.4
	300/1050	1.95	– 376	396 396	1.3 0.6
Alloy 800	as annealed	0.05	–	534	27
	100/900	0.08	–	483	38
	100/950	0.16	–	434	44
	100/1000	0.23	– –	332 342	20 20
	100/1050	0.42	– –	273 335	8 11

HK 40 bars   6.4mm gauge diameter x 33mm long.

Alloy 800 rings 28mm diameter x 3mm wall x 7.5mm wide.

BG 1253/9

FIGURE 1.

Stress-rupture testing of ring specimens using two inserts which form a yoke of the same diameter as the inside tube diameter.

D.M. WARD

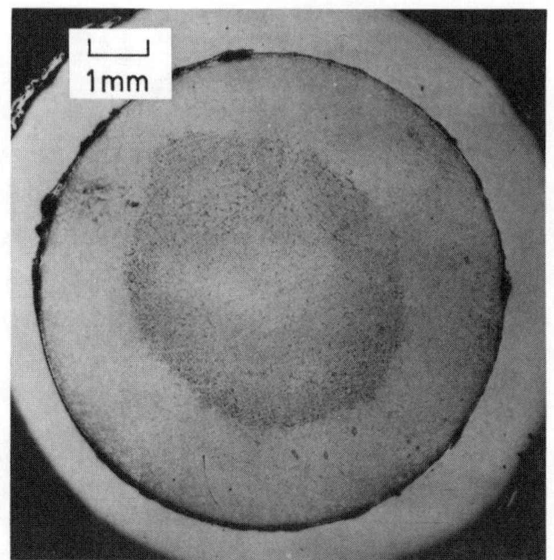

x 8 B.24517

<u>FIGURE 2a.</u> Band of carburisation in HK 40 after 300 hours carburisation at 1050°C. Etched in Murakami's reagent.

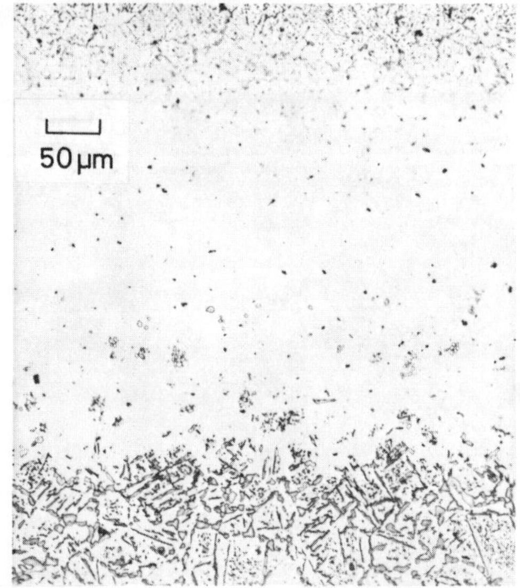

x 150 B.26566

<u>FIGURE 2b.</u> Band of carburisation in Alloy 800 after 100 hours carburisation at 1000°C. Etched electrolytically in KOH.

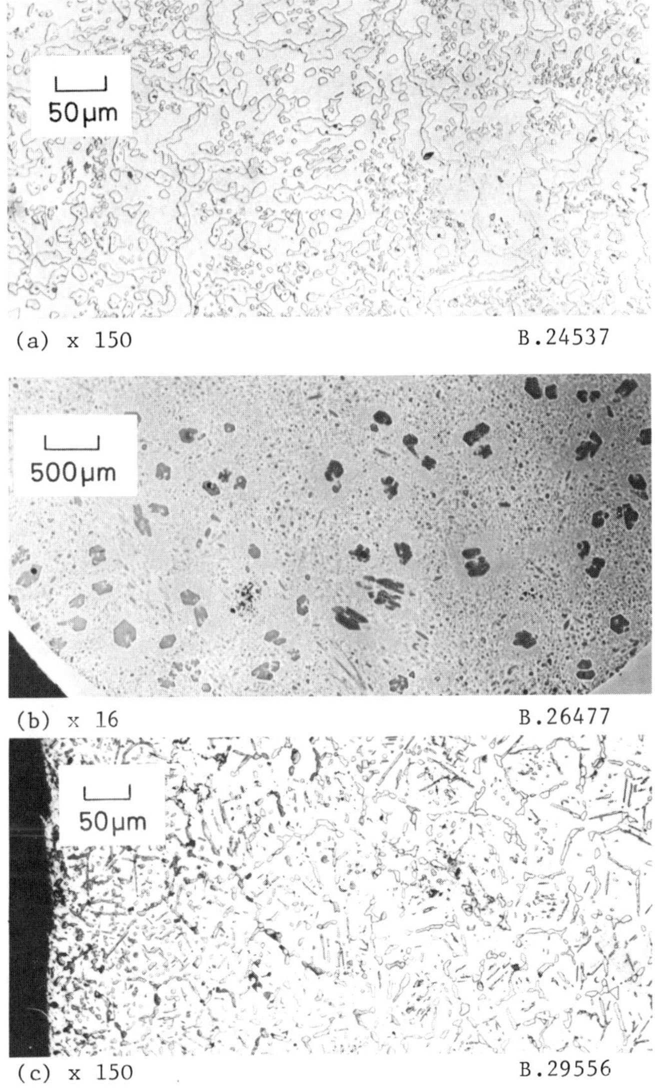

(a) x 150   B.24537

(b) x 16   B.26477

(c) x 150   B.29556

FIGURE 3. Microstructures of (a) HK 40 after carburising 500 hours 1050°C to 3% carbon, (b) HK 275 as cast sample with 2.75% carbon, (c) Alloy 800 sample carburised 100 hours at 1050°C.

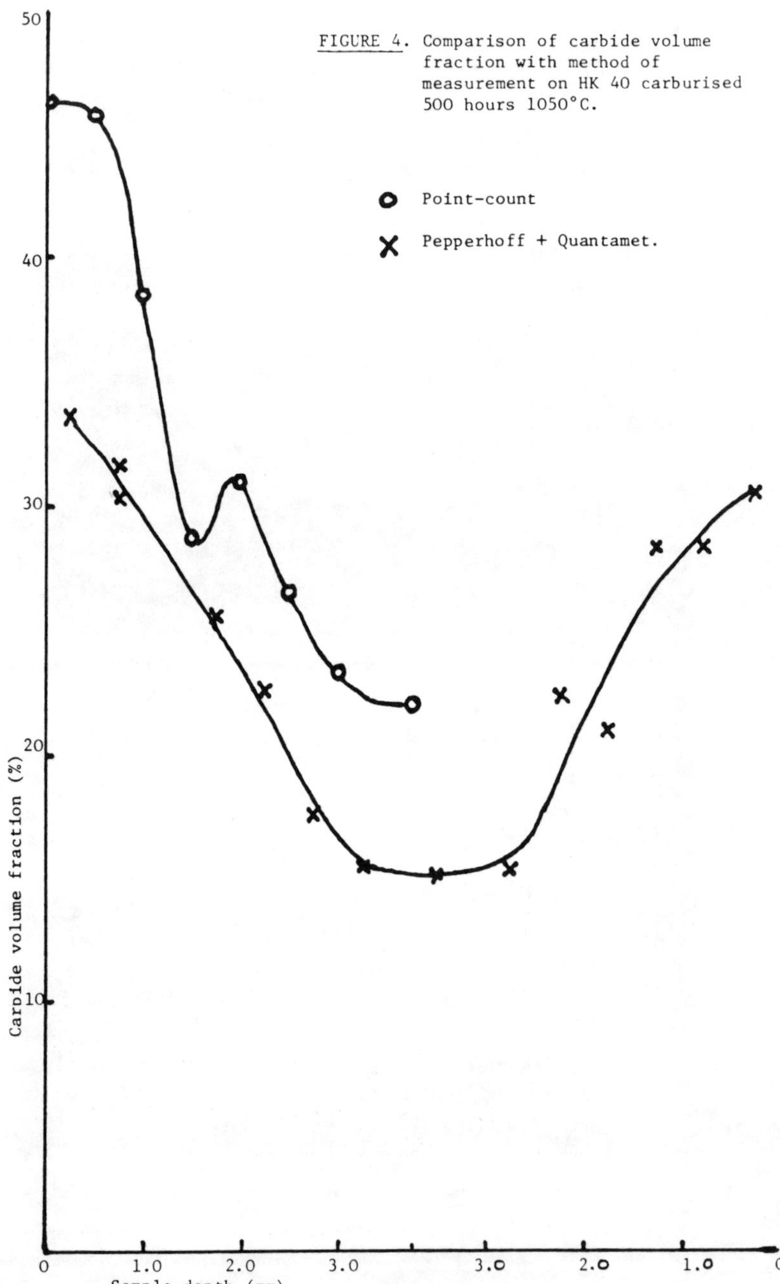

FIGURE 4. Comparison of carbide volume fraction with method of measurement on HK 40 carburised 500 hours 1050°C.

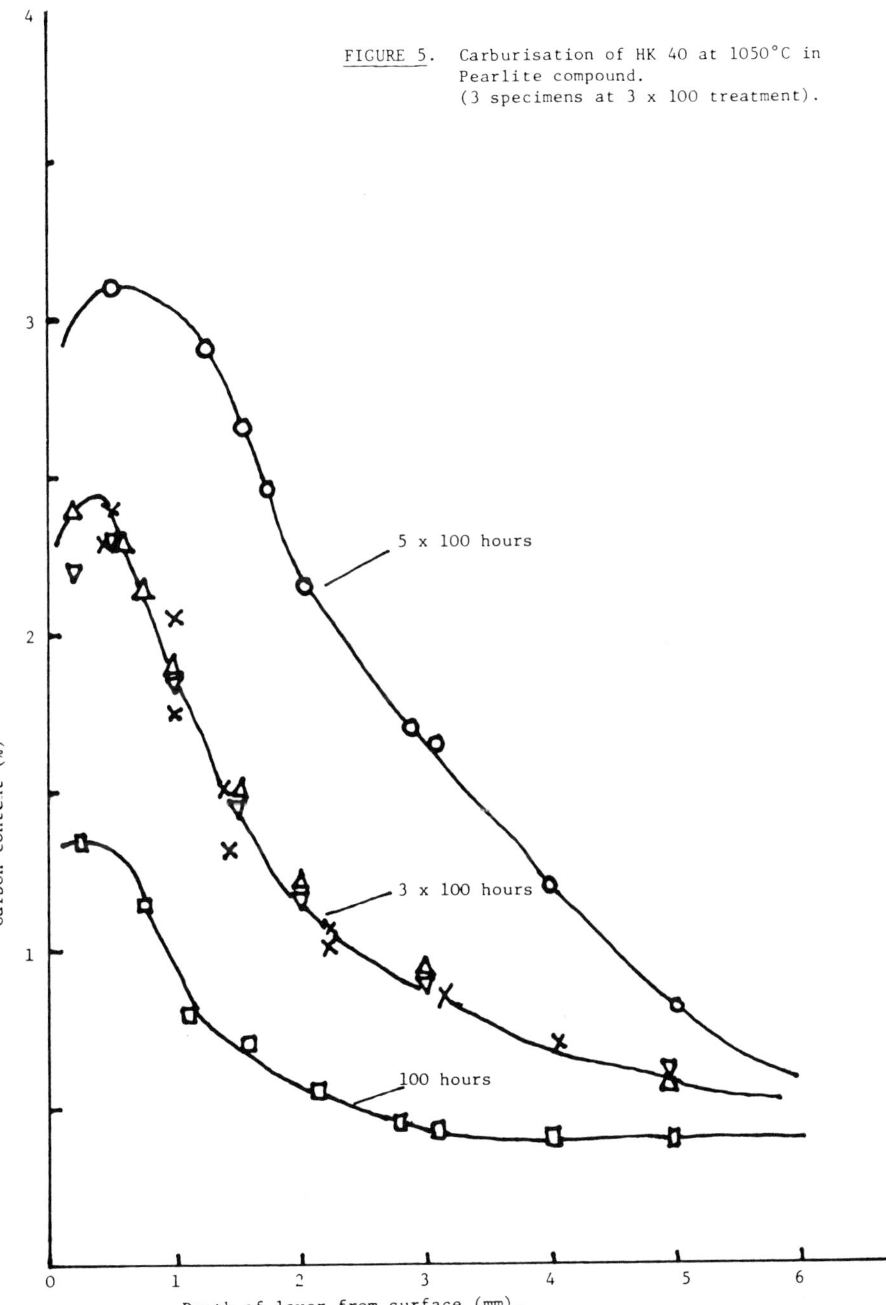

FIGURE 5. Carburisation of HK 40 at 1050°C in Pearlite compound. (3 specimens at 3 x 100 treatment).

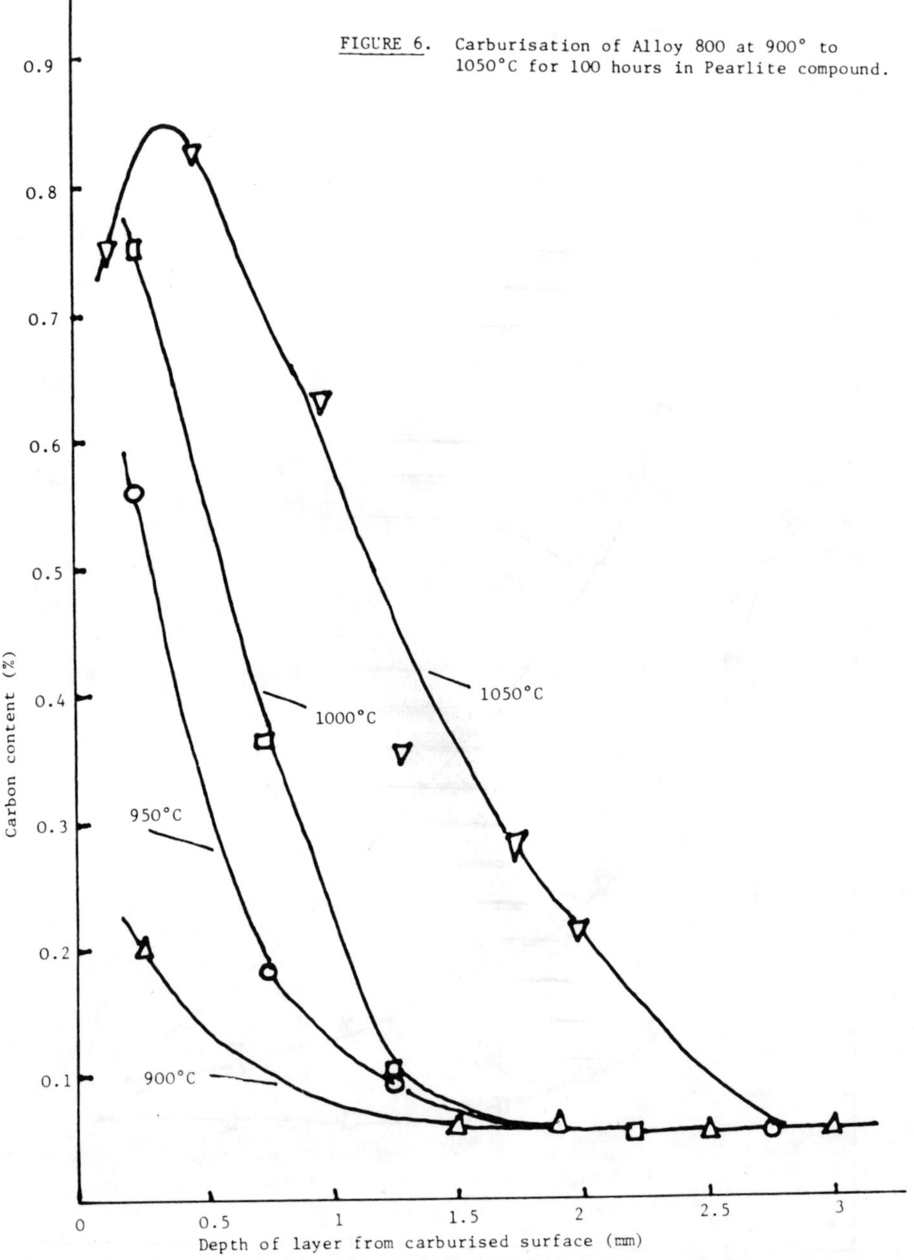

FIGURE 6. Carburisation of Alloy 800 at 900° to 1050°C for 100 hours in Pearlite compound.

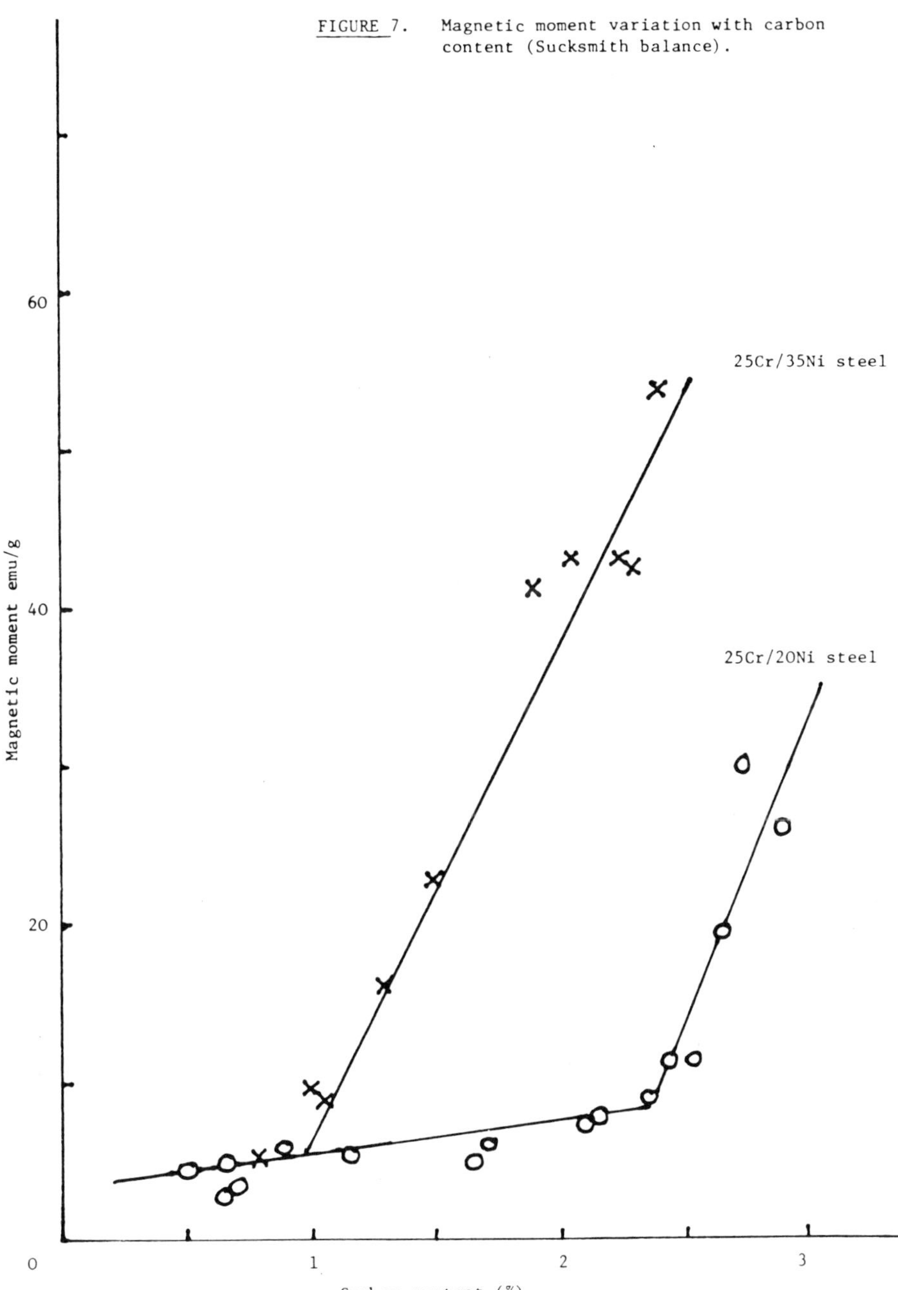

FIGURE 7. Magnetic moment variation with carbon content (Sucksmith balance).

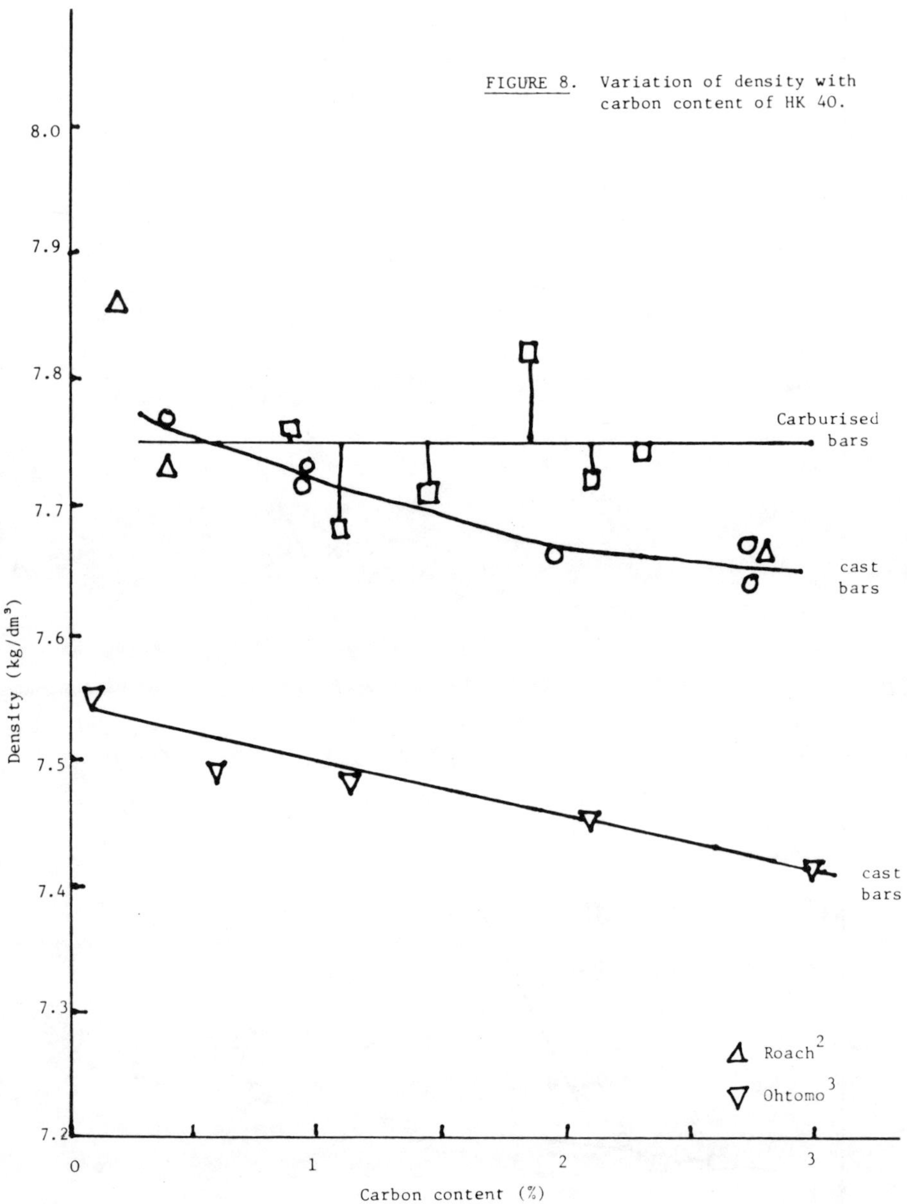
FIGURE 8. Variation of density with carbon content of HK 40.
Density (kg/dm³)
Carburised bars
cast bars
cast bars
Carbon content (%)
Roach
Ohtomo

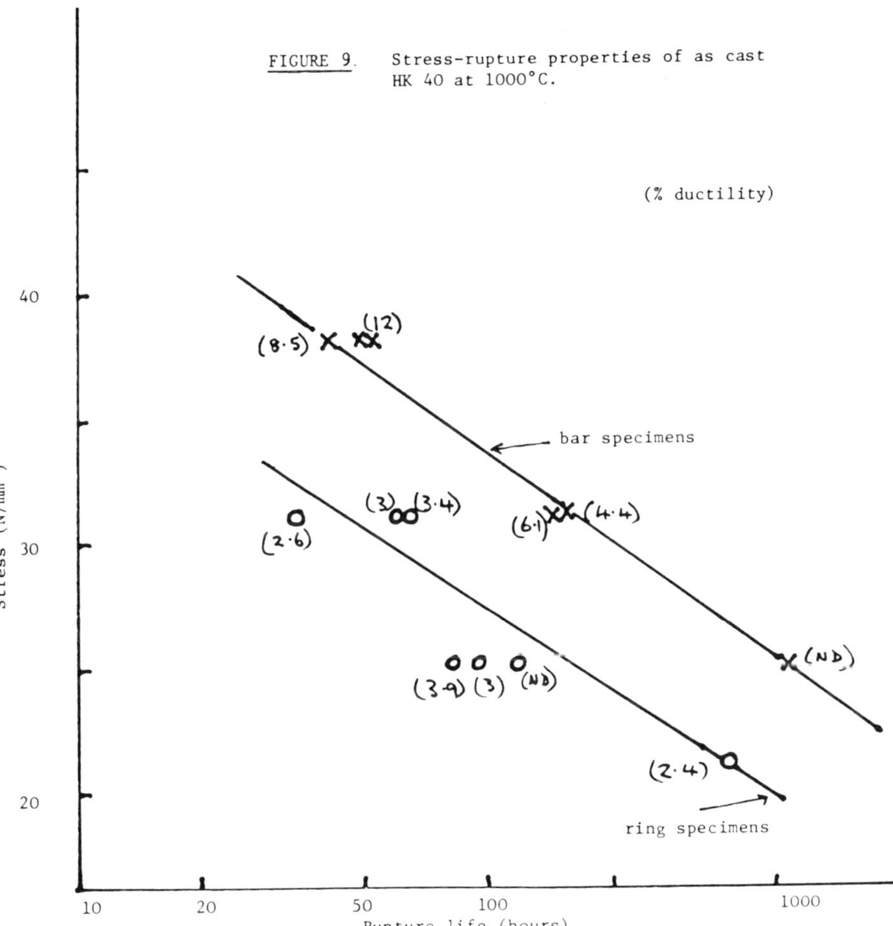

FIGURE 9. Stress-rupture properties of as cast HK 40 at 1000°C.

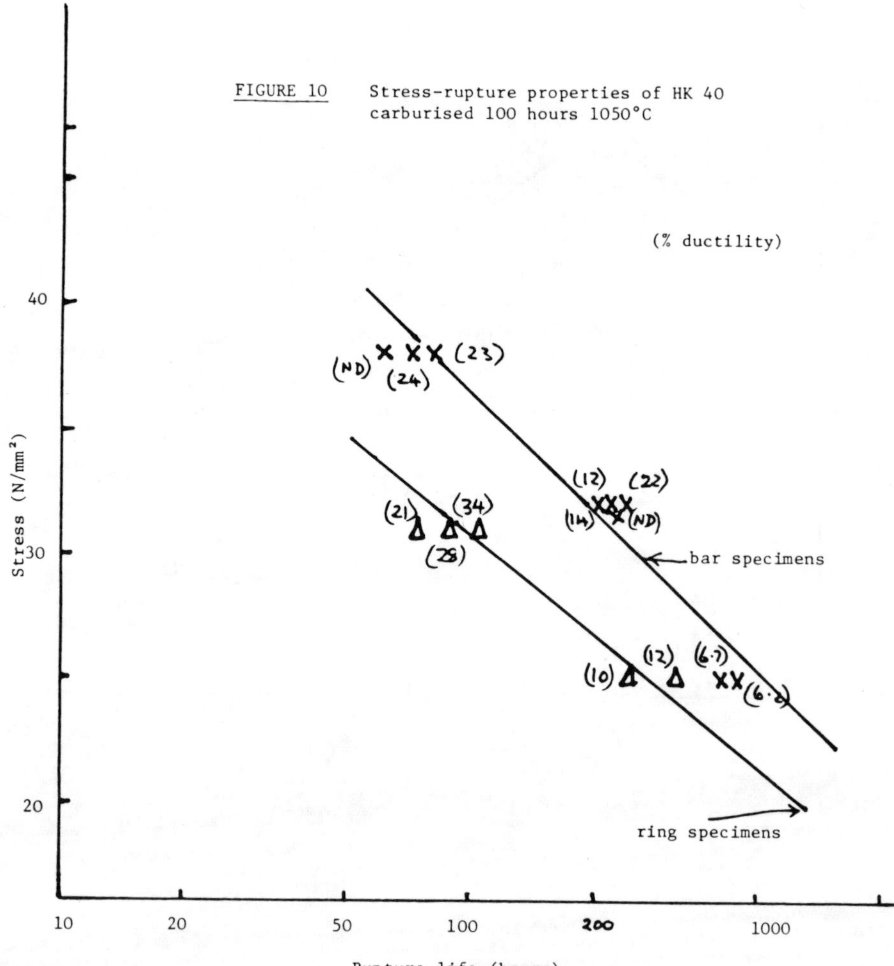

FIGURE 10   Stress-rupture properties of HK 40
carburised 100 hours 1050°C

(% ductility)

40

Stress (N/mm²)

30

20

10   20   50   100   200   1000

Rupture life (hours)

(ND) X X X (23)
      (24)

(21)   (34)
(28)

(12)   (22)
(14)   (ND)
bar specimens

(10)   (12) (6.7)
            X X (6.8)

ring specimens

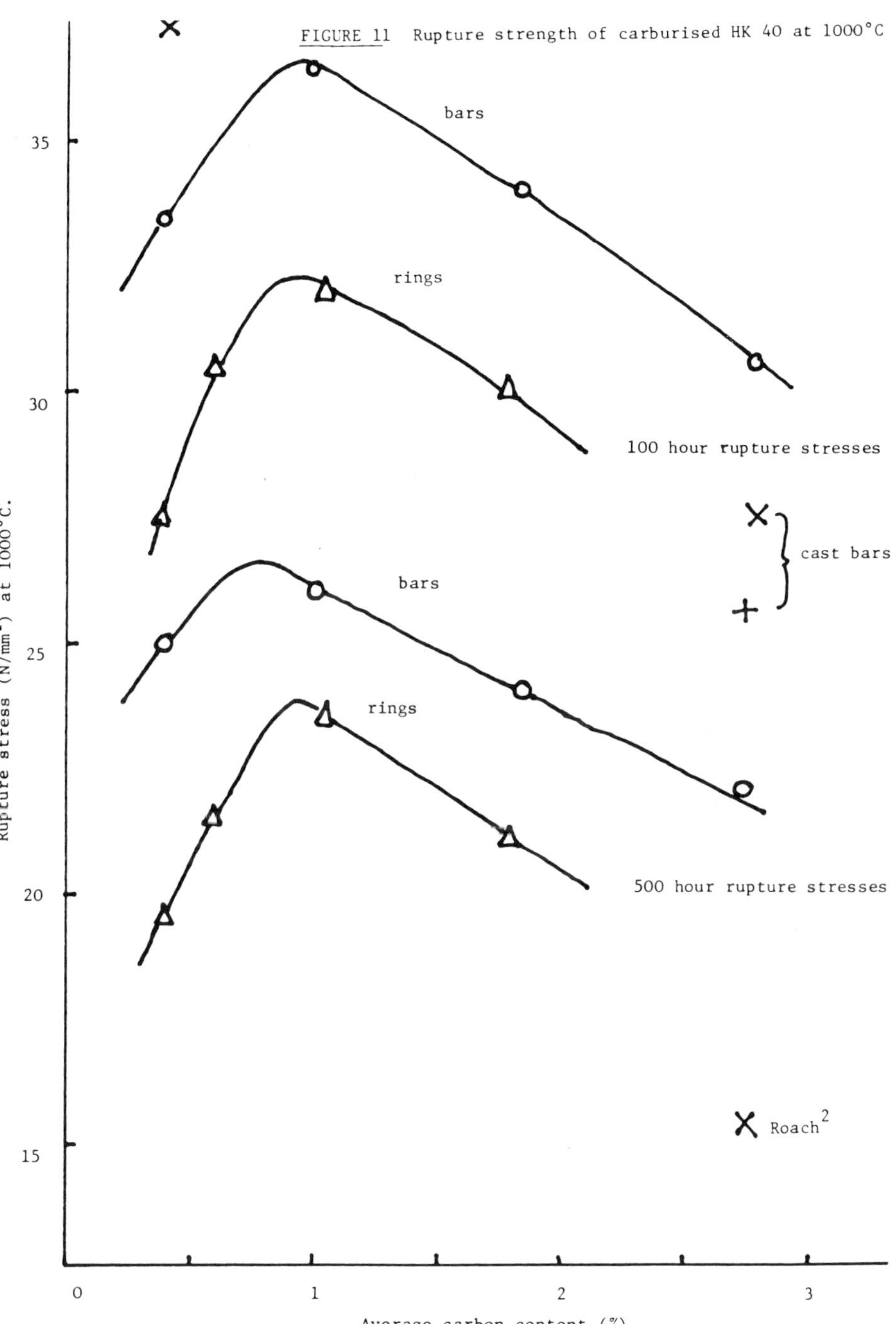

FIGURE 11   Rupture strength of carburised HK 40 at 1000°C
bars
rings
100 hour rupture stresses
cast bars
bars
rings
500 hour rupture stresses
Roach
Rupture stress (N/mm²) at 1000°C.
35
30
25
20
15
0
1
2
3
Average carbon content (%)

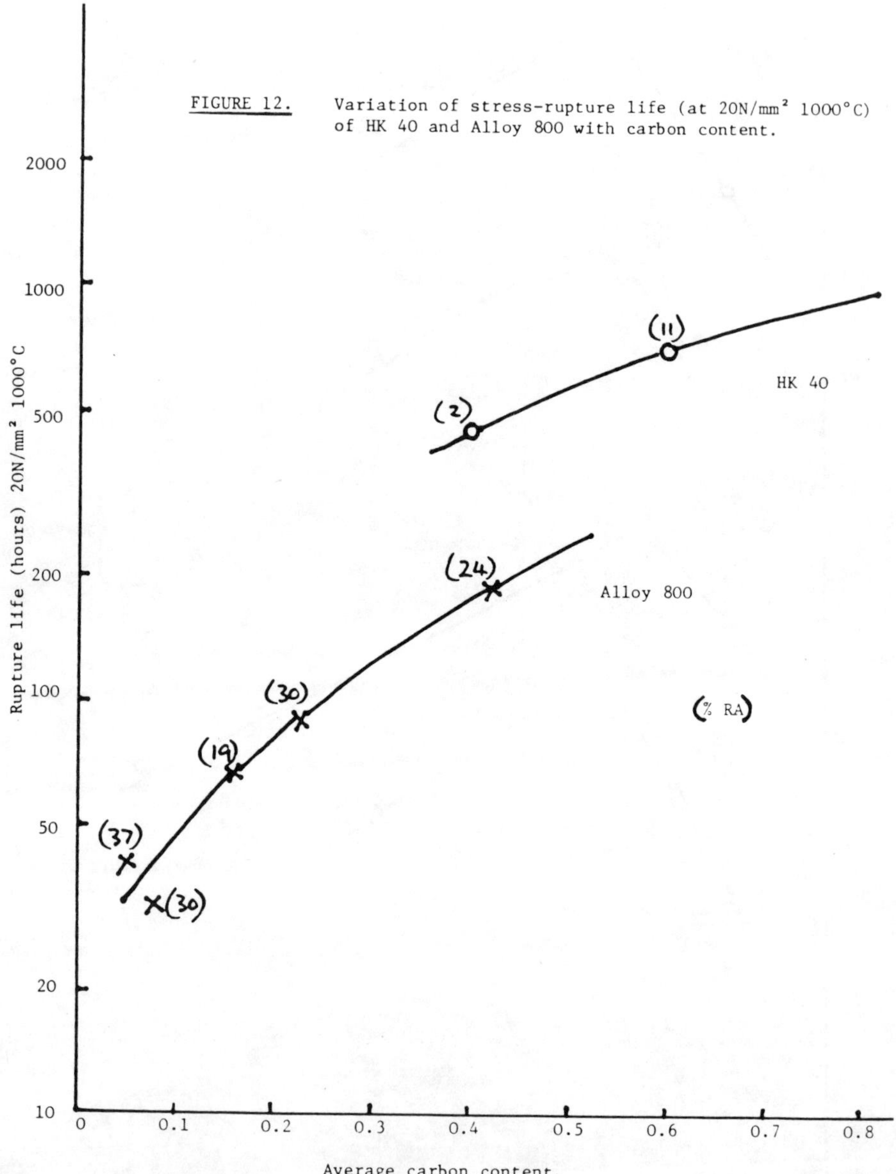

FIGURE 12.   Variation of stress-rupture life (at 20N/mm² 1000°C) of HK 40 and Alloy 800 with carbon content.

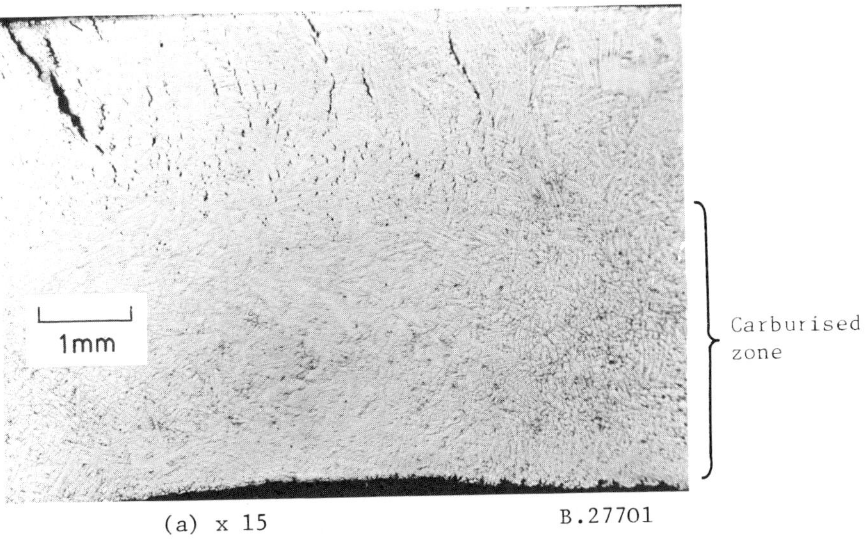

(a) x 15   B.27701

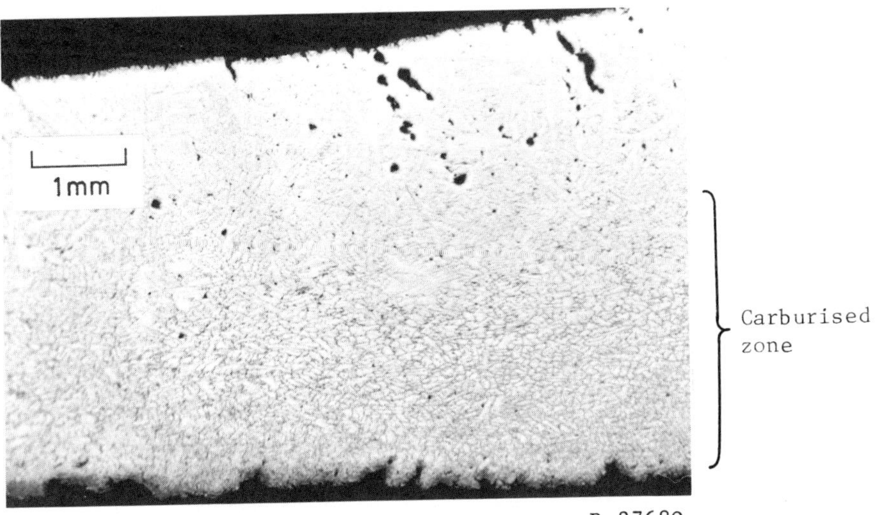

(b) x 15   B.27680

<u>FIGURE 13</u>   Sections through rings of HK 40 stress
rupture tested after carburisation.

(a) Carburised 300h 1050°C gave 62 hour life
at $31N/mm^2$ 1000°C.

(b) Carburised 500h 1050°C gave 354 hour life
at $25N/mm^2$ 1000°C.

                    D.M. WARD

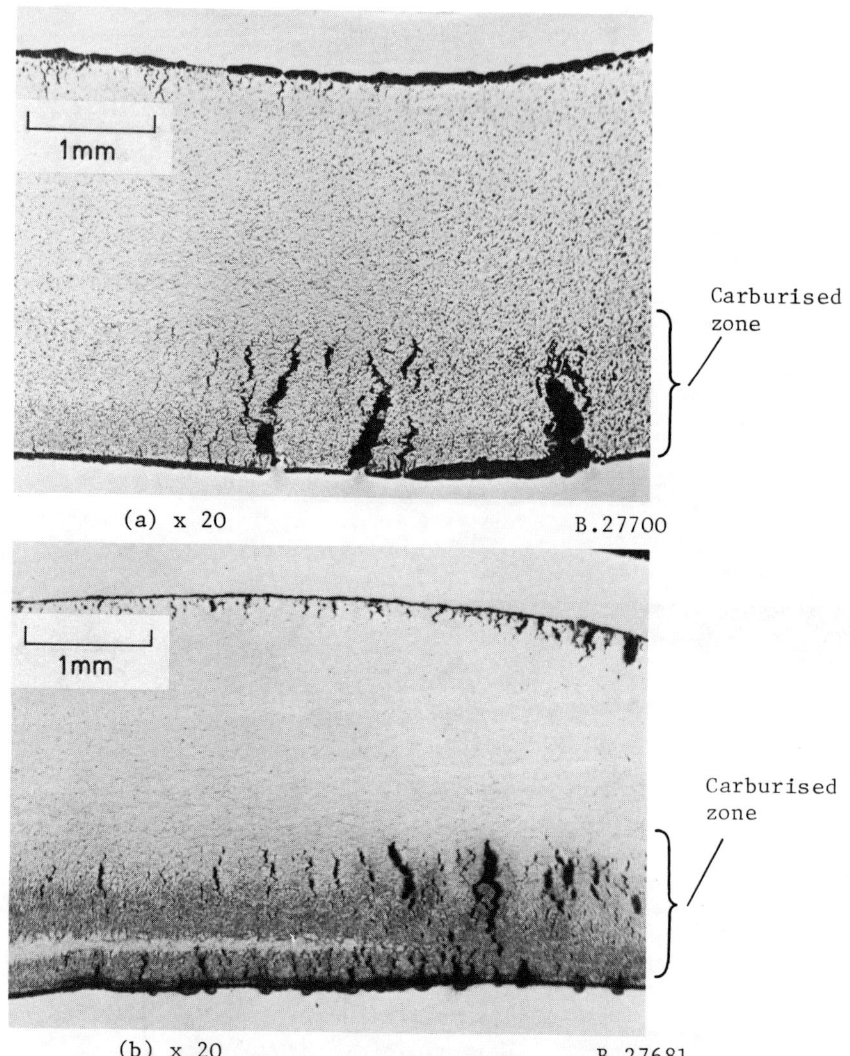

FIGURE 14.   Sections through rings of Alloy 800 stress-
rupture tested after carburisation.

(a) Carburised 100h 1000°C gave 89h life 30% RA at
20N/mm² 1000°C.

(b) Carburised 100h 1050°C gave 187h life 24% RA at
20N/mm² 1000°C.

Corrosion and Creep of Alloy 800 with Nb-additions in $CO-H_2O-H_2$ Atmospheres

J. Hemptenmacher, H.J. Grabke, K. Önel

Max-Planck-Institut für Eisenforschung GmbH

Max-Planck-Str. 1, 4 Düsseldorf, F.R.Germany

On the basis of Alloy 800 ingots were prepared with additions of 0.35, 0.7 and 1% Nb. The ingots were solution annealed at $1150^{\circ}C$ and water quenched. Creep experiments in oxidizing and carburizing $CO-H_2O-H_2$ atmospheres were performed at $1000^{\circ}C$ and constant load: $\sigma = 8.2$ N/mm^2. The creep rates increase with increasing Nb-content. In region of high strain the oxide scales have cracked and heavy internal carbide formation has occurred. The character of the carburization changes with increasing Nb-content, from prevailing chromium carbides at grain boundaries to prevailing NbC in the matrix.

Introduction

The structural requirements for high creep resistance have been well documented (1-2). A fine distribution of a second phase increases the creep strength provided this is combined with other structural features.

In heat resistant austenitic alloys the addition of Nb facilitates to obtain a fine distribution of MC type particles on dislocation and stacking faults (3) and improves the creep properties substantially (4). The precipitates of this kind are very effective strengtheners at intermediate service temperatures but do not seem likely to be effective at high temperatures due to particle coarsening. Nevertheless even at high temperatures a certain distribution of a second phase can be effective in controlling the creep strength by reducing the rate of grain boundary sliding (5,6,7) and probably by decreasing the efficiency of boundaries as vacancy sources (8).

One important requirement for high temperature creep resistance is the maintenance of homogeneous deformation throughout the material and structural integrity through the service life. In industrial applications with the effect of the stress, temperature and a certain atmosphere the structural integrity cannot be maintained. High chromium containing alloys such as Alloy 800, in certain industrial practices are exposed to oxidizing and caburizing atmospheres at high temperatures ($\simeq$1000°C). With the effect of creep strain the protective scale cracks, carbon diffuses into the material and forms Cr-rich large carbides mainly at the grain boundaries (9). During creep the cavities are initiated at these large grain boundary particles, ductility and rupture life is decreased. By adding a strong carbide former into this alloy the formation of carbides due to the carburization during the service may be altered, ductility may be increased. By obtaining a certain particle distribution in the starting material overall creep properties may be improved.

In the present work it was intended to study the effects of Nb additions on the creep properties and the carburization characteristics of Alloy 800. Preliminary experiments (10) had indicated that Nb improves the creep and corrosion resistance of Alloy 800. Nb additions are used in several high temperature alloys for application in cracking furnaces, methane-reforming etc. (11,12). However, the role of Nb in the corrosion and creep processes was not yet studied in detail.

Experimental

As a basis material for the preparation of ingots Alloy 800 was used with the following composition:

Fe	Ni	Cr	C	Si	Mn	Ti	Al
44.3	32,8	21.3	0.05	0.3	0.6	0.3	0.35 wt%

By melting in a vacuum-induction furnace cast alloys were obtained with additions of 0.35, 0.7 and 1.0 % Nb. The ingots were solution annealed at 1150°C and water quenched.

Creep experiments were performed in oxidizing and carburizing $CO-H_2O-H_2$ atmospheres (9,10). A carbon activity of $a_c = 0,6$ was established. The oxygen activity was high enough to form chromium oxide (Cr_2O_3).

A constant load was applied to give a stress of $\sigma = 8.2$ N/mm^2 at a temperature of 1000°C. This value corresponds to the rupture stress for 1000 h life time in air, as reported for Alloy 800.

Creep curves

The creep curves measured for the specimens with different Nb-additions are shown in Fig. 1. All creep experiments were stopped after 1300 h. The tests were repeated and similar creep curves could be reproduced. The creep rate increases with increasing Nb-content.

Specimens with the highest Nb-content (1 %), showed the highest elongation after 1300 h and the highest creep rate:

$$\dot{\varepsilon} = 3.8 \cdot 10^{-5} \text{ s}^{-1} \text{ and } \dot{\varepsilon} = 2.0 \cdot 10^{-5} \text{ s}^{-1}$$

Specimens with the lowest Nb-content (0.35 %) showed the lowest elongation and creep rate:

$$\dot{\varepsilon} = 0.3 \cdot 10^{-5} \text{ s}^{-1} \text{ and } \dot{\varepsilon} = 0.2 \cdot 10^{-5} \text{ s}^{-1}$$

Intermediate values of elongation and creep rate were observed with the alloy with 0.7 % Nb: $\dot{\varepsilon} = 0.6 \cdot 10^{-5} \text{ s}^{-1}$ Alloy 800 without Nb showed a creep-strain similar to the strain of the alloy with the lowest Nb-content (see dashed line in figure 1): $\varepsilon = 13$ %. The creep rate, however, after 1300 h was higher:

$$\dot{\varepsilon} = 0.7 \cdot 10^{-5} \text{ s}^{-1}$$

Microstructure of the starting materials

The three alloys with different concentrations of Nb show cast structures and contain niobium carbide precipitates.

The X-ray diffraction results show that the precipitates are cubic and the lattice constant is slightly smaller than that of Nb C. In addition to Nb some Ti was detected in these particles by the microprobe. These carbides are homogeneously distributed at the grain boundaries and in the grains. Their number and size increase with increasing Nb content.

After solution heat treatment some carbides were dissolved and some were coarsened (Fig. 2). Their shape is spherical and some of those at the grain boundaries are elongated. In addition to these coarse particles there are very fine ones in clusters along the grain boundaries (Fig. 2). These fine precipitates have been extracted by replicas and were identified by TEM as carbides of the type MC (NbC). At the grain boundaries of the alloy with the lowest Nb-content, some precipitates of the type $M_{23}C_6$ were identified.

The distribution of fine precipitates near and at the grain boundaries is much more dense in the alloy with 0.35 % than in those with a higher amount of Nb. Some grain boundaries of the 1 % Nb alloyed material are without any precipitates.

The creep behaviour is influenced by the size and distribution of the carbides. The results of the creep tests are related to the carbide-distribution in the following way: Fine precipitates at the grain boundaries improve creep resistance, coarse carbides induce lower creep resistance.

<u>Oxidation</u>

In the $CO-H_2O-H_2$ atmosphere during creep tests an oxide scale is formed on the surface. It consists of spinels of the type $FeCr_2O_4$ and a small amount of Cr-oxides (Cr_2O_3). Nb plays no role in this process, microprobe studies showed no Nb enrichment in or below the oxide layer. At the phase boundary between oxide and alloy there is an enrichment of Si, this suggests an interfacial layer of SiO_2. Internal oxidation occurred in the surface zone. Fine oxide particles are situated under the scale.

The oxide scale is porous; primarily at grain boundaries the scale is fractured, this is obviously caused by the creep of the sample. Along the grain boundaries, vertical to the stress direction, many cracks were formed, which penetrate into the bulk of the material. The cracks are diffusion paths for molecular transportation of oxygen and carbon by CO and H_2O. This leads to internal carbide formation and subsequent oxidation. Most of these cracks are filled with oxide.

Carburization and Internal Carbide Formation

The carburization is closely related to the creep rate and to the deformation of the sample. The time dependence of the carburization must not be considered since the experiments were stopped after the same time.

The C-content was analysed as the average content of the whole cross-section. In the diagram (Fig. 3) the carbon intake is plotted against the total elongation of the sample. From the diameter of the section, where C was analysed, the local elongation was calculated. These values are also given in the figure. The C-intake increases with increasing total and local elongation.

Heavy carburization obviously took place in the range of tertiary creep. Specimens, tested to fracture, show the highest C-intake, even when they had been tested for a shorter time (for example "A", in Fig. 3).

After the tests the distribution of the carbides depends on the carbon intake and on the concentration of Nb in the material.

In the centre of the specimens the number and the size of carbides have obviously increased. Especially in the grains of the 1 % Nb containing material the number of coarse carbides increased. Due to the high affinity of Ti and Nb to C during carburization only Nb-Ti carbides will be formed up to a certain level of C intake in the matrix.

When heavy carburization occurred (Δ C $>$ 0.1 %), for example in necked areas, a second type of carbide is formed (Fig. 2). These carbides are very coarse, they look like prisms or plates, which are distributed at the grain boundaries as a bulky layer. Investigations by microprobe show that these particles are rich in Cr. Polarized light microscopy showsan optically isotropic, i.e. cubic structure, suggesting that these carbides are $M_{23}C_6$ (M = Cr,Fe). The heavy carburization forms coarse $M_{23}C_6$ carbides primarily at the grain boundaries (See Fig. 2). Following this internal carbide formation, internal oxidation takes place at the carbide-matrix interface. By the applied stress voids are created at the internal carbides and oxides and grow into cracks. This process initiates rupture of the specimens.

Conclusions

Alloying the basis material Alloy 800 with small additions of Nb (0.35 %) causes increased creep resistance, higher concentrations (0.7 and 1 % Nb) cause relatively high ductility. The creep properties can be correlated to the microstructure after the solution heat treatment. In the case of low Nb-content, there are many very fine NbC particles in a cloud near the grain boundaries - in the case of high Nb content the NbC is distributed as rather large cubic or rounded particles at grain boundaries as well as throughout the matrix.

The favorable influence of Nb on the behaviour of creep specimens in oxidizing and carburizing atmospheres ($CO-H_2O-H_2$) seems mainly be caused by the improved creep resistance of alloys with appropriate NbC-particle distribution. The carburization of the creep specimens could clearly be correlated to the total and local strain of the specimens. In regions of highly reduced cross section area, i.e. high local strain, the carbon content was relatively high. The mechanism of degradation can be deduced from the metallographic cross sections (Fig. 4): Grain boundary sliding leads to repeated cracking of the oxide layer, mainly at intersections of grain boundaries with the surface. Carbon uptake occurs through the crack in the oxide layer before it heals by oxidation. The carbon uptake causes chromium carbide formation at the grain boundary and chromium depletion at the grain boundary. After repeated cracking the oxide layer above such chromium depleted grain boundary cannot heal and heavy internal carburization and oxidation start. By oxide formation the cohesion is lowered and cracks will open at the grain boundaries near the surface (Fig. 4). This process of material degradation has been described before (9,10) and is influenced by the presence of Nb mainly according to its influence on creep rate and ductility. The presence of Nb does not play a role in the formation and stability of the oxide layer. However, since in case of high Nb-alloys mainly NbC is formed in the first stages of carburization, the chromium depletion and internal oxidation should be somewhat retarded. The character of the carburization changes with increasing Nb-content, from prevailing chromium carbides at grain boundaries to prevailing NbC in the matrix.

Acknowledgement

The authors are grateful for the support provided by the Forschungs- und Entwicklungsprogramm 'Korrosion und Korrosionsschutz', Bundesministerium für Forschung und Technologie, Fed. Rep. Germany. We are indebted to Mannesmann-Forschungsinstitut for supplying the base alloy.

References

1. J. Nutting and J.M. Arrowsmith
    Symposium on structural processes in creep,
    ISI 1969, p. 147

2. E. Hornbogen
    high temperature materials in gas turbines, p. 187
    Ed. P.R. Sahm and M.O. Speidel, Elsevier 1974

3. J.M. Silcock
    J. ISI 1963, 201, 409

4. K.J. Irvine, J.D. Murray and F.B. Pickering
    J. ISI, 1960, 196, 166

5. R. Raj and M.F. Ashby
    Met. Trans. 1971, 2, 1113

6. R. Raj and M.F. Ashby
    Met. Trans. 1972, 3, 1937

7. C.A.P. Horton
    Acta Met. 1972, 20, 477

8. M.F. Ashby
    Scripta Met. 1969, 3, 837

9. A. Schnaas and H.J. Grabke
    Oxidation of Metals 1978, 12, 387

10. A. Schnaas
    Veränderung der Eigenschaften hochwarmfester
    Stähle durch Aufkohlung und Karbidausscheidung,
    Dissertation, Universität Dortmund, 1977

11. U. Gravenhorst und W. Steinkusch
    Arch. Eisenhüttenwes. 1975, 46, 397

12. H.J. Grabke, U. Gravenhorst und W. Steinkusch
    Werkstoffe u. Korrosion, 1976, 27, 291

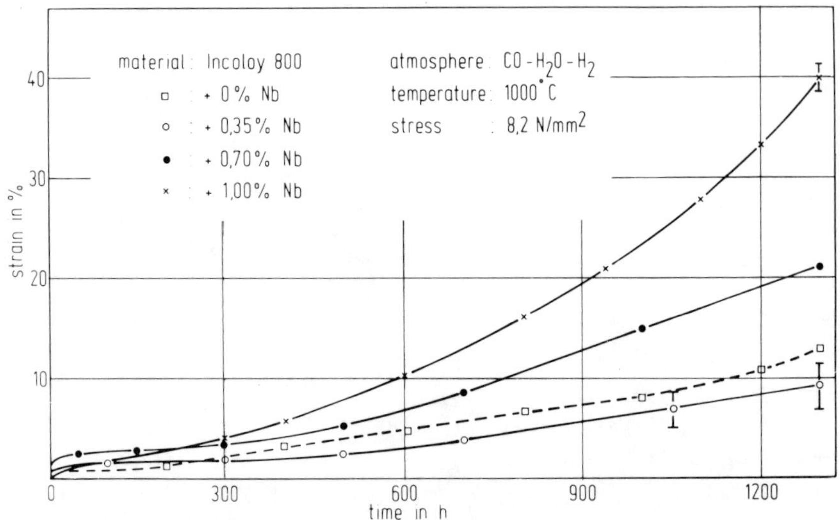

Fig. 1 Creep curves for Alloy 800 with different contents of niobium, tested in oxidizing and carburizing atmospheres (a_C = 0.6)

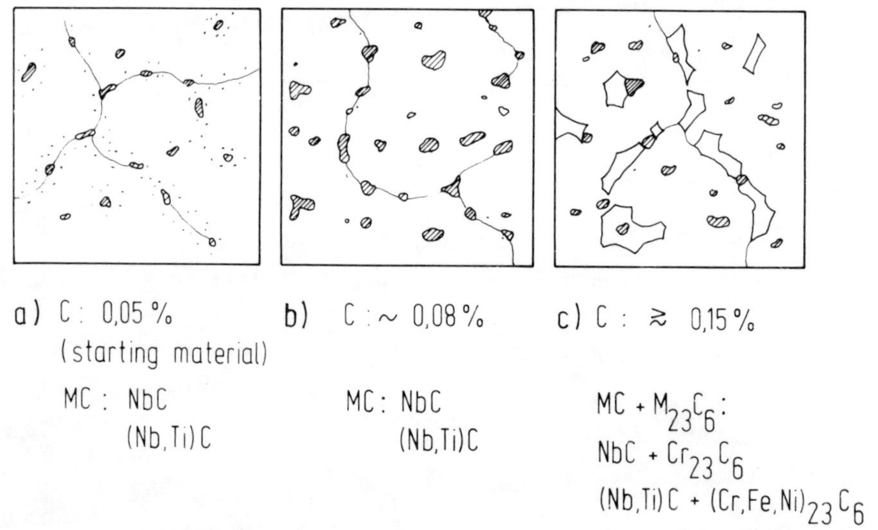

Fig. 2 Schematic representations of microstructures, showing carbide morphology and distribution in Alloy 800 with Nb-additions
a) after solution anneal at 1150°C
b) after first stage of carbusization: NbC formation
c) after second stage of carburization: NbC and $M_{23}C_6$ in regions of high local strain.

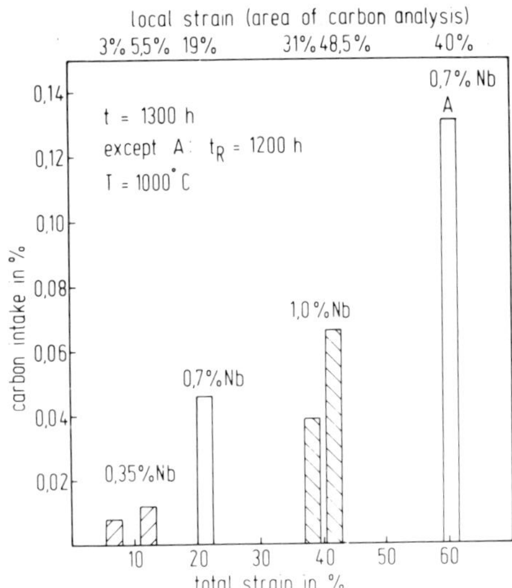

Fig. 3　Carbon content of Alloy 800 with different Nb-additions, after 1300 h creep test in oxidizing and carburizing atmosphere, in dependence on the total and local elongation.

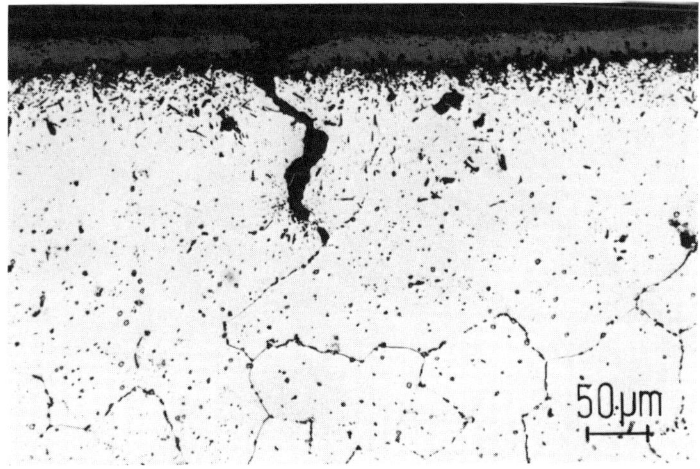

Fig. 4　Microstructure of the surface zone of Alloy 800 with O.35 % Nb as observed after creep test in $CO-H_2O-H_2$ at 1000°C for 1300 h.

DISCUSSION

H.W. Grünling: You mentioned chromium-rich carbides in the highly
strained portions of your specimens. Is the formation of such
carbides related to strain or strain rate or to time since it was
a surface effect?

J. Hemptenmacher: Our creep experiments show the effect of creep
rate and strain on the carburization. Since all experiments with
specimens of different Nb-content were stopped after the same time,
the time of exposure to the oxidizing and carburizing atmosphere
is not decisive for the different state of carburization. However,
the different specimens showed different creep rates and different
strains are obtained in the experiments. The higher the strain the
more often cracking of the oxide scale will have occurred - and the
cracking of the oxide scale is the principal cause of carburization
and formation of chromium carbides in the highly strained parts of
the specimens.

THE CREEP BEHAVIOUR OF HK 40 AND ALLOY 800 H IN A CARBURIZING ENVIRONMENT

V. Guttmann and R. Bürgel

Commission of the European Communities

Joint Research Centre, Petten (The Netherlands)

INTRODUCTION

The mechanical behaviour of metallic materials at elevated temperatures is strongly dependent on their environmental resistance. A corrosive attack often causes a degradation of properties although sometimes also a favourable influence can be observed (1, 2, 3). The mechanisms of the environmental-mechanical properties interactions can be of a very complicated nature and as a result their understanding is by far incomplete.

This paper presents some preliminary results gained on the creep behaviour of the two high temperature steels HK 40 and Alloy 800 H in a purely carburizing environment. The results are compared to those obtained in air. The choice of these alloys has been made because of a) their current use in the high temperature petro-chemical industry and b) their consideration as candidate materials to be applied in coal conversion processes. Atmospheres to be encountered in both of these application areas are oxidizing, carburizing and/ or sulphidizing in nature. Although most frequently a combined attack occurs, sometimes one reaction predominates as for instance carburization in the ethylene cracking process.

EXPERIMENTAL PROCEDURE

The chemical analyses of alloys used for the experiments are listed in table 1. The centrifugally cast alloy HK 40 has been investigated in the as received condition by machining creep samples from a 13 mm thick tube wall. The Alloy 800 H type material has been solution annealed at 1150°C for 1 hour followed by water quenching. Test pieces had a gauge length of 30 mm and a diameter of 6 mm. The test temperature for all experiments was 1000°C.

For exposure under carburizing conditions a CH_4/H_2 mixture has been used containing about 1-vol.% CH_4 which gave a carbon activity, $a_c = 0,8$ at a gas pressure of about 1,3 bar.

RESULTS

An example of the results gained on the creep behaviour of HK 40 is shown in Fig. 1 where creep rate is plotted against time. After a few hours of testing samples exposed to the carburizing environment exhibit a higher creep rate than those in air. More-over the curve is characterized by the occurance of a second relative minimum creep rate.

The results gained on rupture strength are shown in Fig. 2. Rupture time in a carburizing atmosphere is shorter than in air although the difference is small and only by a factor of about 2.

Ductility properties are presented using a plot of elongation to fracture against time to rupture in Fig. 3. Under carburizing conditions elongation to fracture is remarkably improved when the life time exceeds about 100 h. This increase of ductility is more pronounced at longer life time (lower stresses) i.e. at longer exposure times.

It seems worthy to note that the severity of the deleterious effect due to carburisation on creep resistance of HK 40 becomes obvious only if the creep strain-time relation is considered. As indicated in Fig. 1, under the testing conditions employed, 1% elongation is reached after about 1100 h in air whereas only 100 h are needed for the same elongation under carburizing conditions. From this result one reaches a significantly different conclusion than from the rupture strength which is only weakly influenced by carburisation. The latter of course reflects an integral creep behaviour and for the carburized material condition it includes a strain range which exceeds clearly that of technical interest.

Concerning the creep behaviour of Alloy 800 H an example of the creep rate-time relationship is shown in Fig. 4. Again remarkable differences between creep tests conducted in air and in carburizing atmoshpere are observed. In a similar fashion to HK 40, the creep rate under carburizing conditions is higher than in air at the beginning of the test. However, for Alloy 800 H the situation reverses with increasing time, and the carburized samples then show the lower creep rate. The fraction of total life time during which this lower creep rate is observed extends with decreasing stress. Similar to the observations on HK 40 but more strongly marked, a second relative minimum occurs in the creep rate-time curve for samples tested in the carburizing environment.

Results of the rupture strength behaviour of Alloy 800 H are shown in Fig. 5. In contrast to HK 40 an improvement of rupture strength occurs which is pronounced at lower stresses (about 4 times longer life times).

It has to be emphasized that this improvement of creep rupture strength in a carburising environment does not imply a general increase in creep strength. For instance, as for HK 40, the time to reach 1% strain is shorter under carburizing conditions than in air. (see indication in Fig. 4).

Concerning the ductility of Alloy 800 H, nearly no change has been observed when tested under carburizing condition. This is in clear contrast to the observations made on HK 40.

DISCUSSION

The present investigations have shown that the creep behaviour of the two steels HK 40 and Alloy 800 H exhibit a different response to the change from air to carburizing atmosphere. The results on HK 40 can be expected from the creep properties determined for samples fully precarburized before testing in carburising gas. Preliminary results on this material have shown the same features for creep strength and creep ductility (4). The lowering of creep resistance during carburization can be interpreted by comparing the microstructure of carburized and uncarburized samples as shown in Fig. 6a/b. Much coarser carbides are found after carburization which will be less effective in restricting creep deformation than are the finer and more densely distributed carbide particles of the uncarburized air samples. These disappear during carburization in favour of the larger carbides. The effectiveness of the small carbides to block dislocations can be concluded from the TEM investigations, Fig. 7, showing the dislocation precipitate reactions. Moreover, although probably of less importance, carburization will cause chromium depletion in the matrix thus reducing the amount of chromium available as a solid solution strengthener.

No simple explanation can be given for the increase of ductility occuring during carburization. One possible interpretation, based on a theoretical consideration, takes into account the fact that the cast HK 40 tested in air atmosphere shows a very low ductility of about only 3% elongation to fracture whereas the wrought version of the same material exhibits values of about 25% (5). A logic assumption should be that an impurity induced intergranular embrittlement occurs for the cast material. This means that impurities which are concentrated at grain boundaries as a

consequence of the casting process will enable smaller creep
cavities to become stable and cracks to advance faster as already
reported for different metals and alloys (6, 7). Considering now
the structural consequence of exposure under carburizing conditions,
an increased area of interface is produced by the carbide
formation at grain boundaries. Hence the impurities will be re-
distributed and due to their lower concentration the danger of
embrittlement should be reduced.

Another possible explanation for the relatively high ductility of
carburized HK 40 has been derived from further analysis of the
most recent investigations. It is based, in principal, on the
characteristic feature that in a partially carburized sample
damage always starts in the uncarburized core section, as shown
for example in Fig. 8a. This observation is consistent with those
made on the failure mode of ethylene cracking tubes in service
where first cracks habe been found to be generated in the uncarburized outside surface (8). With progressing damage the
$M_{23}C_6$-carbide containing layer of a sample (nearest layer to the
core) becomes included in the damaged zone but, as shown for
example in Fig. 8b, all cracks are blocked at the $M_{23}C_6$-M_7C_3
boundary line and no damage occurs in the outside layer containing
the M_7C_3-carbides (beside the crack finally causing failure).

This sequence of creep damage development can be explained by
differences in ductility between the different parts of the sample
cross section. The uncarburized (or only weakly carburized) middle
zone withstands only a small strain, i.e. its ductility is low
compared to those parts containing the $M_{23}C_6$- and the M_7C_3-
carbides, respectively, the latter obviously exhibiting the
highest ductility.

Because in all cases creep damage is characterised by cavities and
cracks associated with carbide particles situated along grain
boundaries (see Fig. 9a), it is worthwhile to consider in detail
the amount and the morphology of the boundary carbides. Their
density increases from the core to the M_7C_3 containing outside
layer. Carburization occurs faster along the grain boundaries
than in the matrix and the higher the carbide content the more
advanced is the ripening process. The M_7C_3-layer exhibits a specific
structural feature compared to the other parts of the sample cross
section, Fig. 9a/b. In this part of the sample individual grain
boundary carbides are no more observed. Instead of this, band-like
carbide layers have been formed caracterized by a smooth interface
with the adjacent matrix.

On the basis of these observations it is probable that the differences in ductility observed for the various parts of a sample under carburizing exposure can be explained in the following way. Concerning the core section the danger of creep damage will be high because these large and often irregularly shaped precipitates act as sources of cavity and crack formation, Fig. 9a. A high impurity concentration will of course reinforce this deleterious effect. Regarding the $M_{23}C_6$ zone one could expect a higher probability of creep damage due to an increase in carbide density. However the latter will cause a decrease of grain boundary sliding. This beneficial effect can finally overcompensate the deleterious one, leading to an improved ductility.

The superior resistance to cavitation and cracking which is exhibited by the M_7C_3-layer can be explained by the smooth interface between the grain boundary carbide layer and the matrix. Although the contribution of grain boundary sliding probably increases accommodation processes become easier and cavity and crack formation will be reduced, i.e. the ductility will increase.

Specific features have been observed for the creep damage of HK 40 at low stress, i.e. when exposure times are relatively long and specimens become thoroughly carburized. If the stresses are still sufficiently high to cause first damage to occur before the state of full carburization is reached, cavities and cracks are found to be concentrated in the core of the sample. Compared to higher stress tests the delay in sample failure allows the carburization front to pass regions where first damage has already occured. As a consequence cavities and cracks become surrounded by carbide phases as indicated in Fig. 10. This process probably stops the propagation of these first defects as can be postulated from the following observation. Firstly, cavities situated closer to the outside regions are found to be smaller than those nearer to the core. Moreover they are rarely interlinked to cracks. Secondly, relatively long cracks which are assumed to be nearly at the threshold for propagation through the whole specimen cross section exhibit a carbide phase border on one side only, indicating that these cracks have been created in association with carbide-matrix interfaces after full carburization has taken place, i.e. at a late state.

The improved creep rupture strength of Alloy 800 H under carburizing conditions can be explained by the retention of a finer carbide distribution. This is due to the fact that in Alloy 800 H TiC-type precipitates are still available after carburization, the amount of them being increased during carburization due to an excess in titanium in the virgin 800 H alloy. Because of their small size and uniform distribution these carbides appear to be particularly effective for creep resistance. In the HK 40 material the small precipitates of the Cr-carbide type disappear and finally only the extremly coarse particles exist when tested in the CH_4/H_2 atmosphere.

The observation made on Alloy 800 H, that the time to reach for
instance 1% strain, is shorter under carburizing conditions than in
air, while the reverse situation occurs when creep rupture strength
is considered, is caused by the fact that temporarly the total stress
can exceed the applied stress. This is due to the carbide precipitation process, which leads to an increase in the specific
volume of the outside layer of a sample, thus increasing the tensile
stress on the sample core. On the other hand, as mentioned before,
carburization of Alloy 800 H leads to strengthening. This effect
overcompensates the former as obvious for instance in Fig. 4
after about 100 h.

SUMMARY

Investigations have been performed on the creep behaviour of two
high temperature steels HK 40 and Alloy 800 H in a carburizing
atmosphere. Compared to air testing, HK 40 exhibits a loss of
creep rupture strength whereas creep ductility increases
considerably. In contrast Alloy 800 H shows a remarkable increase
of rupture strength and the creep ductility remains nearly
unchanged.

The results can be explained mainly by the differences in carbide
morphology and distribution in the matrix and at the grain
boundaries, respectively.

ACKNOWLEDGEMENT

The authors would like to acknowledge valuable discussion with
Dr. J.B. Marriott (JRC Petten) and Prof. Dr. Ing. F. Erdmann-
Jesnitzer (University Hannover, FRG). The structural investigations
have been assisted by Mr. P. Tambuyser and Mr. P. Helbach. The
creep experiments on HK 40 were conducted by Mr. H. Schönherr.
The authors also acknowledge, with thanks, the permission to
publish this paper given by Mr. P.J. van Westen and Dr. M. Van de
Voorde, Director and Division Head respectively at JRC Petten.

REFERENCES

1. R.H. Cooke and R.P. Skelton, Int. Met. Rev., 1974, Vol. 19,
  p. 199

2. H.W. Grünling, B. Ilschner, S. Leistikow, A. Rahmel and
  M. Schmidt, Werkstoffe und Korrosion, 1978, Vol. 29, p. 691

3. V. Guttmann and J.B. Marriott, Proc. of the Conference on
  "Environmental Degradation of High Temperature Materials",
  Isle of Man (U.K.) 31st. March to 3rd April, 1980

4. V. Guttmann, Joint Research Centre, Petten (The Netherlands)
  unpublished research, 1980

5. W. Steinkusch, Pose-Marre Edelstahlwerk GmbH (4006 Erkrath bei Düsseldorf), private communication

6. H.R. Tipler and D. McLean, Metal Sci. J., 1970, Vol. 4, p. 103

7. C.L. Briant and S.K. Banerji, Int. Met. Rev., 1978, Nr. 4, p. 164

8. W. Steinkusch, Werkstoffe und Korrosion, 1979, Vol. 30, p. 837

TABLE I

HK 40

C^*	Cr	Ni	Si	Mn	P	S
0.30	24.05	20.50	1.36	0.35	0.017	< 0.010 %

$*$ The carbon content is somewhat lower than required by the specification (0.35 - 0.45 % C).

Incoloy 800 H

C	Si	Mn	Cr	Ni	Al	Ti	N
0.06	0.37	0.66	20.05	31.35	0.28	0.48	0.01 %

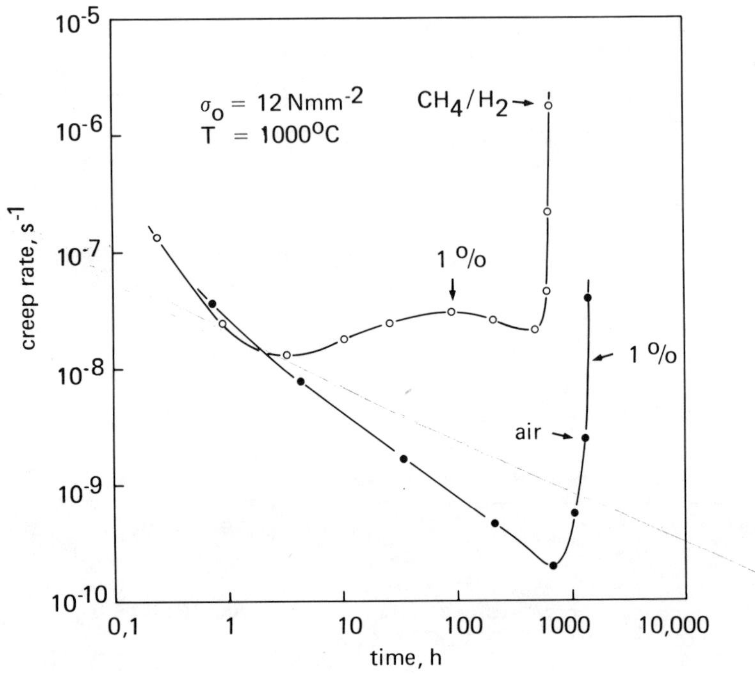

Fig. 1   Creep rate versus time for HK40 in different environments

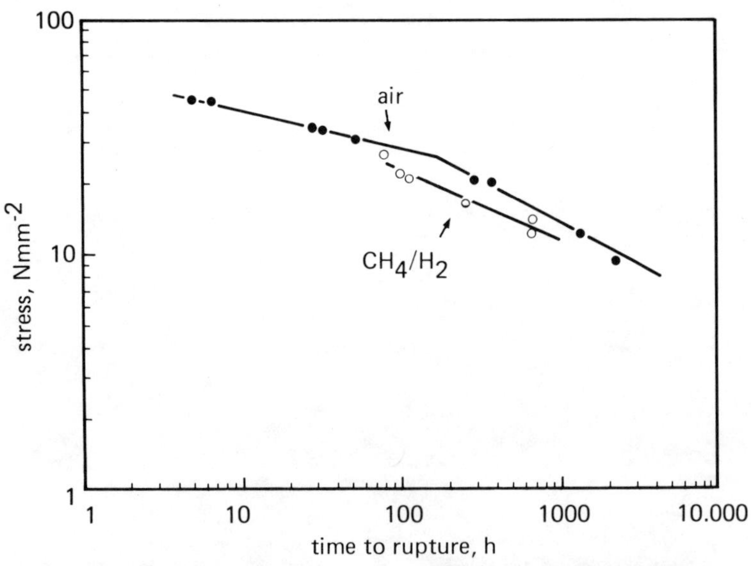

Fig. 2   Creep rupture strength for HK40 in different environments
$T = 1000^{\circ}C$

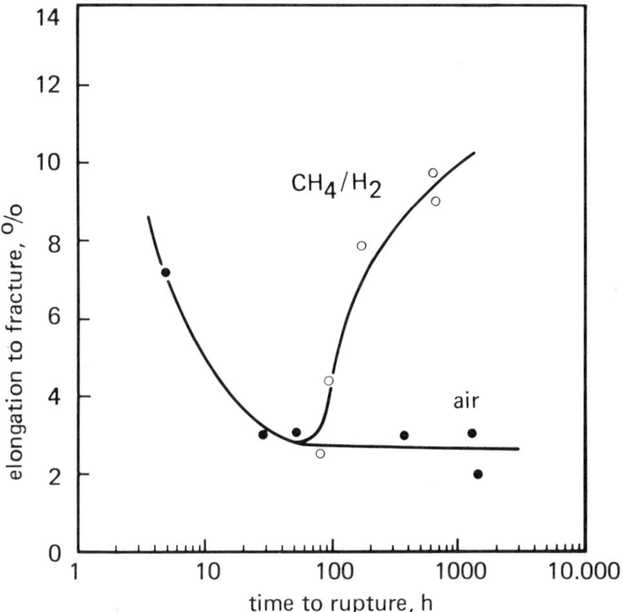

Fig. 3 Elongation to fracture versus time to rupture for HK40 in different environments T = 1000°C

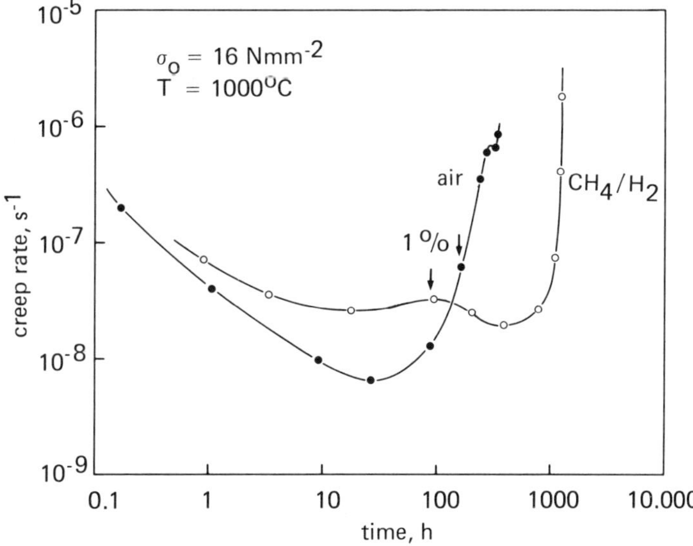

Fig. 4 Creep rate versus time for Alloy 800H in different environments

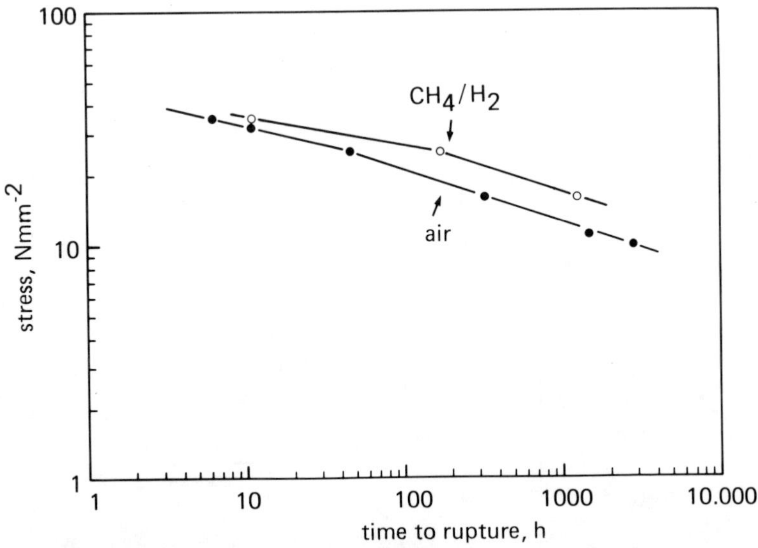

Fig. 5   Creep rupture strength for Alloy 800 H in different environments
                T = 1000°C.

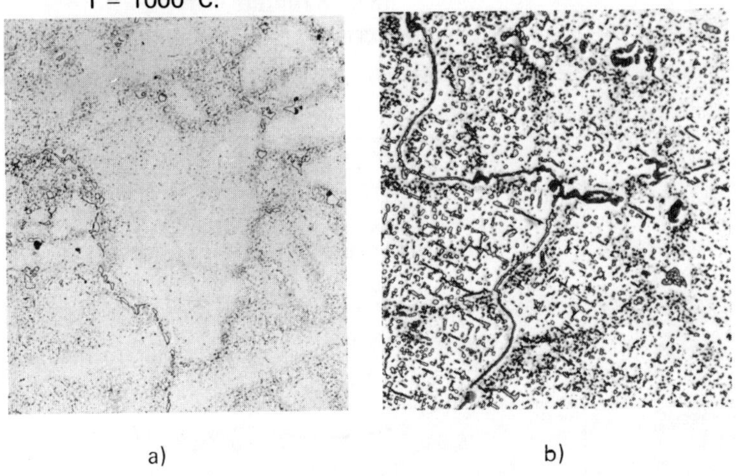

a)   b)

50 µm

Fig. 6

CARBIDE DISTRIBUTION IN HK40

a)   AIR TEST; T = 1000°C; t = 1360 h
b) CH_4/H_2 TEST; T = 1000°C; t = 82 h

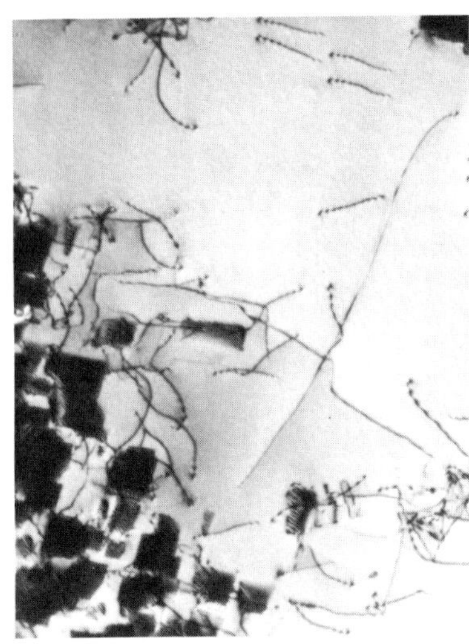

Fig. 7

Microstructure of HK40 air test
$T = 1000^\circ C$, $t = 75h$; $\sigma = 20$ Nmm^{-2}

0.5 μm

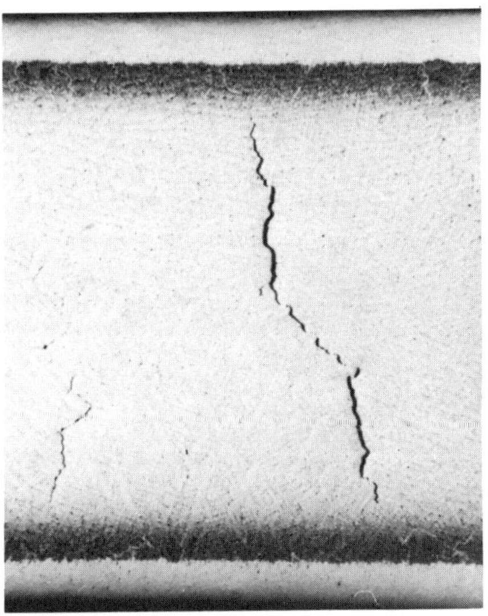

a)

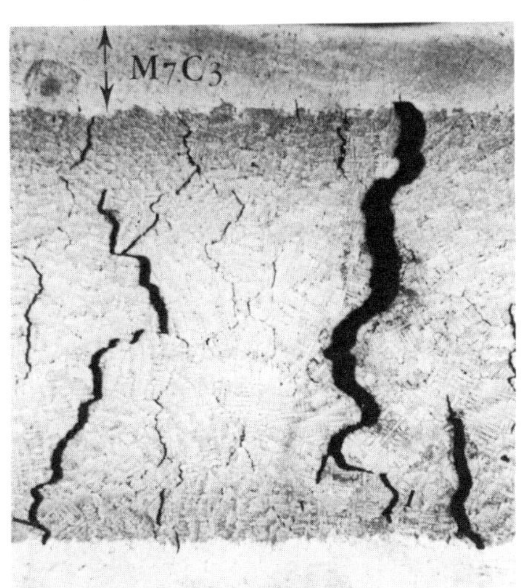

b)

Fig. 8

Creep damage in HK40. CH$_4$/H$_2$ test
$T = 1000^\circ C$; $\sigma = 20$ Nmm^{-2}
a) Onset of damage in the uncarburised core (t = 52h)
b) Stop of crack propagation at
M$_7$C$_3$ zone (t = 97h)

1 mm

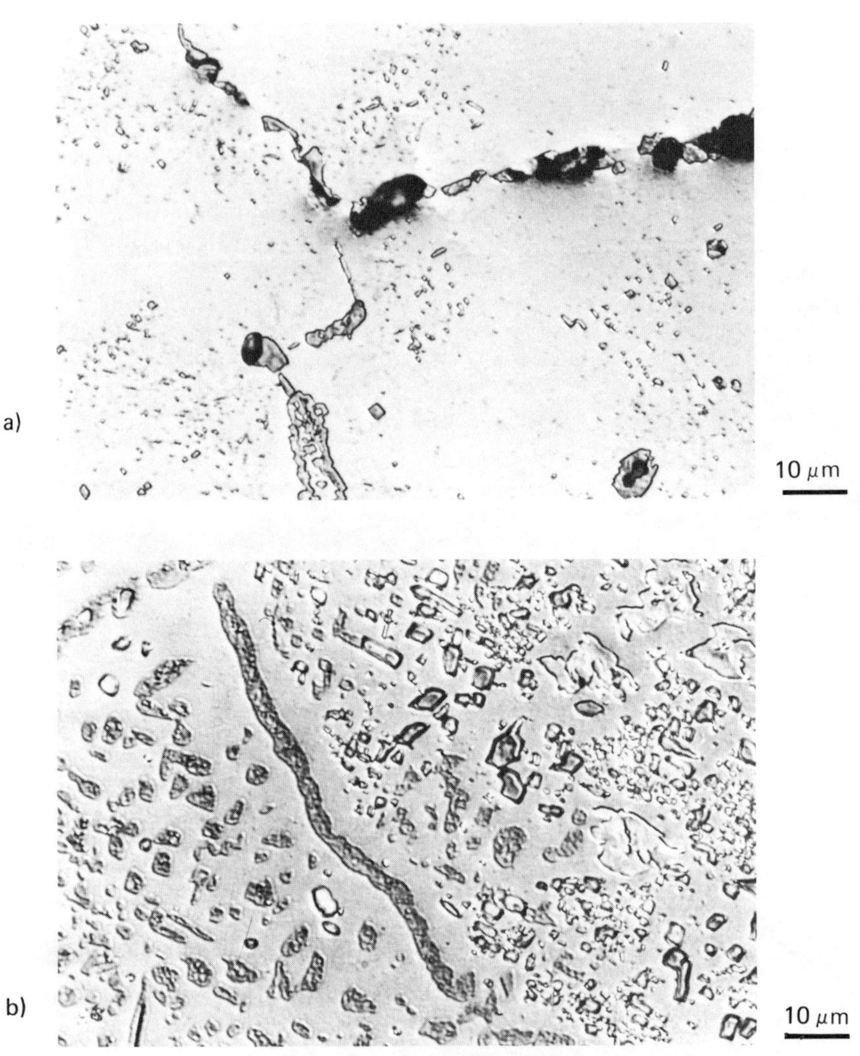

Fig. 9   Grain boundary morphology in HK40 , CH_4/H_2 test
         $T = 1000^{\circ}C$; $\sigma = 20\ Nmm^{-2}$; $t = 97h$
         a)   nearly uncarburised core zone with single carbides and voids
         b) M_7C_3-zone with continuous carbide layer at a grain boundary

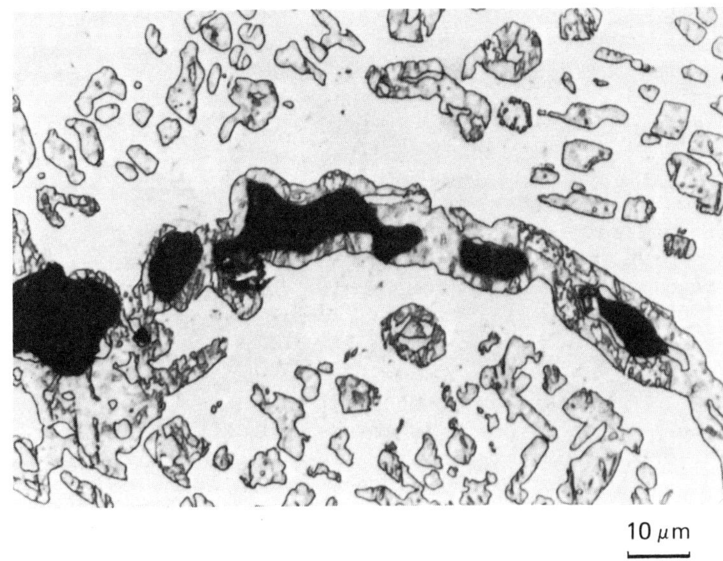

Fig. 10   Voids surrounded by carbides in carburised HK40
$T = 1000^{\circ}C$; $\sigma = 12\ \text{Nmm}^{-2}$; $t = 660h$

DISCUSSION

<u>H.D. Marsch</u>: The investigation conveys the impression that by
carburisation, HK40 is weakened and the properties of Incoloy 800
are improved. This conclusion is quite confusing compared to
practical experiences with steam reformers and other structures.

Isn't it advisable to consider transient stresses caused by
changing temperatures across a wall or caused by thermal expansion
imposing bending stresses? Such stresses are unavoidable in any
structure and sometimes reach the yield point.

Another influence seems to be of great importance in practical
service, i.e. intermittent variations in carburising and oxidising
conditions.

Testing time seems to be too short to draw final conclusions.

V. Guttmann: You mean that confusion is arising because usually creep strength is improved when the carbon content in heat-resistant steels is increased. However, in the present case we are concerned with extremely high carbon contents (about 4% max.). Hence, the situation has to be reconsidered. The decrease in creep strength of HK 40 can be explained, as already noted, by the lack of fine carbides in the carburised material (see Fig. 6) and by the reduction in the chromium content of the matrix.

Concerning the second part of your contribution, I think we all agree that creep testing conditions should be very similar to service conditions. Unfortunately this requires very often unrealistic high financial efforts, as can be immediately concluded from your examples.

H.J. Grabke: The increase in ductility which V. Guttmann observed after carburisation of HK40, is somewhat puzzling. Usually, carburisation of high temperature alloys causes a decrease in ductility. In previous work with A. Schnaas, we did tensile tests at 950°C and at room temperature with Alloy 800, carburised to different carbon contents. For the high temperature, the strength and ductility of the material stayed nearly unchanged, but at room temperature, the ductility was strongly decreased by carburisation. Would V. Guttmann please comment on how to explain the increase in ductility observed for HK40 and Alloy 800H in his studies?

V. Guttmann: I would first like to make it clear that our investigations have shown that carburisation causes an increase in ductility of HK 40, whereas that of Alloy 800 H remains virtually unchanged. The latter finding obviously corresponds to your observation made for tensile tests,

Concerning the room temperature behaviour, we do not have exact data although some bending tests made in order to check qualitatively the ductility of the carburised material have shown that it behaves in a completely brittle manner.

Concerning a possible explanation for the ductility improvement of HK 40 in the carburised state I can only repeat that one can assume that the redistribution of impurities at grain boundaries during carburisation could cause an increase in ductility. Due to some recent investigations it seems probable that the change in the amount and the morphology of the grain boundary carbides will also be of some influence.

O. Götzmann: This HK40 seems to be a particular alloy. Do you have any information on the nature of the precipitates in the central region of the cast material prior to carburisation? What you referred to as 'impurities'. Also, do you have any information

of the grain size of the cast material as compared with the wrought
material?

V. Guttmann: Investigations concerning the type of impurities at
grain boundaries of HK 40 are in progress. However, without the
experimental evidence it is to be expected that, as usual for iron-
based casts, an increased content of sulphur and phosphorus should
be present at the grain boundaries.

Concerning the grain size this was about 140 µm (5) and about
300-700 µm in the wrought and in the cast material, respectively.

O. Götzmann: Have you verified that the precipitates on the grain
boundaries contained sulphur and phosphorous?

V. Guttmann: No, I have not yet verified this assumption.

T. Boniszewski: Carburisation at 1000°C and the formation of
carbides may cause re-crystallisation of the cast HK40 structure.
New crystallographic orientations may be formed by local strains
and epitaxial growth on carbides. This would be equivalent to hot
working, which is known to improve the ductility of HK40. X-ray
studies could show this: Perhaps preferred orientations formed
during casting are eliminated.

V. Guttmann: Examinations carried out by means of light microscopy
have revealed no evidence of recrystallisation in HK 40 due to
carburisation. Therefore we believe that a recrystallisation effect
cannot explain the increase in ductility.

B. Ilschner: Just in answering to the question by Dr. Boniszewski and
what Dr. Guttmann said, I would also not expect recrystalisation due
to internal strain raised by carburisation because these internal
strains, which will certainly exist, will be released by recovery.
The time scale and the stress level which will be reached by carbon
diffusion is of a nature that will be released by relaxation or
recovery rather than by recrystalisation.

R. Rolls: Is there any evidence that the inhomogeneities of the
as-cast structure have been acting as effective sources/sinks for
dislocations which could account for the high ductility observed?

V. Guttmann: Certainly strong interactions will occur between
dislocations and carbide precipitates. These have been analysed
in the present work for instance in the uncarburised material by
means of transmission electron microscopy. As would be expected,
carbides strongly hinder moving dislocations and dislocations are
emitted from the carbide matrix interface. However, independent
on the type of interaction, I cannot see how it could be possible

that these interactions could increase the ductility under the present conditions, i.e. high temperatures and low strain rates. Moreover all these interactions occur in the same way in both, the carburised and the uncarburised alloy.

O. Götzmann: Since the cast material after ageing does not behave any better than the untreated cast material, I think the increase in ductility after carburisation is a result of the interaction between carbon and the impurites concentrated in the grain boundaries of the central region of the cast material. Carbon forms multinary carbides with those possible impurites which adopt carbide properties. Carbides do not reduce the strength of grain boundaries as other precipitates do, so it is quite conceivable that the transformation of the original precipitates on the grain boundaries into multinary carbides increased the strength of the grain boundaries of the cast HK material.

Corrosion Behaviour of Incoloy 802 Creep-Rupture-
Specimens during Exposure in Air and Helium Environment of 1073 K to 1273 K

J. Ebberink, K. Krompholz, E. te Heesen

INTERATOM, Internationale Atomreaktorbau GmbH

5060 Bergisch Gladbach 1, FRG

<u>ABSTRACT</u>

The corrosion behaviour of Incoloy 802 specimens after
creep-rupture tests in PNP^+-Helium and air up to
10,000 h is described. The helium temperature lies
between 800 and $1000^{\circ}C$ and the stresses applied between
13 to 81 MPa. Thermodynamic considerations show, that
gas impurities in the helium (i.e. H_2, CH_4, H_2O, CO
and N_2) can interact with the surface of the specimens
and are able to build up stable oxides.

Experimental examinations are compatible with the
thermodynamic considerations, which predict the presence of oxide layers.

The microprobe examinations confirm the formation of a
thin oxide layer of Cr-, Mn- and Si-oxides. Chemical
bulk analyses show decarburization of all creep-rupture
specimens, except one, after exposure. Comparative
metallurgic and microprobe examinations of specimen
sections with different stresses show no influence of
stress on the thickness and the composition of the
surface layer.

+) PNP: Prototype Nuclear Processheat

1. INTRODUCTION

Helium as the coolant in the primary systems of a HTGR contains small impurity levels of H_2, H_2O, CO, CO_2, CH_4, and N_2 which originate from the graphite core and the walls of the metallic components as a result of outgassing, desorption, and in-leakage processes. Structural metallic materials are exposed to this helium at different temperatures between $300^{\circ}C$ and $1000^{\circ}C$, the latter one being the maximum temperature expected to be at the outlet of the reactor core. Materials are affected by the reactions like oxidation, carburization, and/or decarburization; these reactions can occur simultaneously along the wall of e.g. a helium heat-exchanger depending on the temperature gradient along the length of the tubes. Many investigations on the corrosion and on the creep behaviour of the structural materials in helium containing these impurities at various levels have been reported [1-11].

The interaction between gas impurities and metals depends on the impurity levels in helium and the metallic components involved in the system. So far it is assumed that the oxidation potential of the gas is mainly controlled by the pH_2O/pH_2-ratio, whereas the carbon activity is determined by the p_{CH_4} pressure.

In this paper the theoretical aspects of the decarburization/carburization of the alloy X 35 NiCrTiAl 33 21 (Incoloy 802) in the specific PNP standard-atmosphere at different temperatures are described. For confirmative reasons the surface layers of creep-rupture specimens with rupture-times of up to 10,000 h were examined by means of metallography and microprobe technique.

2. THERMODYNAMIC CONSIDERATIONS

The impurity levels of the PNP atmosphere are shown in Table 1. In this highly diluted system gas-gas reactions can be neglected as the authors have shown experimentally within inert systems. Impurity depletions and/or enrichments preceed according to catalytic reactions on metallic surfaces.

The carburization potential of the atmosphere is given by

$$CH_4 = \underline{C} + 2 H_2 , \qquad\qquad (1)$$

PNP Standard Helium

– System Pressure 1.8 bar

– Gas Impurity Levels / μbar

H_2	CH_4	H_2O	CO	N_2
500 ± 50	20 ± 5	1.5 ± 1.0	15 ± 5	≤ 5

PNP	BF GHT HRB KFA RBW	Table 1
	Gas Impurity Levels for the PNP Helium Atmosphere	

$$CO = \underline{C} + 1/2\ O_2 \ , \tag{2}$$

while hydrogen and oxygen combine to give water

$$H_2 + 1/2\ O_2 = H_2O \ . \tag{3}$$

On the other hand the oxygen potential of the atmosphere is mainly controlled by the reaction

$$H_2O = H_2 + 1/2\ O_2 \ , \tag{4}$$

and with chromium the reaction

$$\frac{2}{3}\ Cr + H_2O = \frac{1}{3}\ Cr_2O_3 + H_2$$

may occur.

Calculation gives

$$\log \frac{p_{H_2O}}{p_{H_2}} = -\frac{\Delta G_T^\circ}{2.303\ R\ T} - \frac{2}{3}\ \log a_{Cr} \ , \tag{5}$$

where ΔG_T° is the standard free energy of the reaction, R the gas constant, T the absolute temperature, p_{H_2O} and p_{H_2} the partial pressures of the impurities and a_{Cr} the activity of chromium. In Fig. 1 equation (5) gives the horizontal oxidation equilibrium line. The combination of equations (1) - (3) gives

$$\Delta G_T^{\,o} = - R\,T\,\log\,\frac{p_{H_2O}}{p_{H_2}}\,\frac{p_{H_2}^{\,2}}{p_{CH_4}\,p_{CO}}\,, \qquad (6)$$

and a simple rearrangement gives

$$\log\,\frac{p_{H_2O}}{p_{H_2}} = -\frac{\Delta G_T^{\,o}}{2.303\,R\,T} + \log\,\frac{p_{CH_4}\,p_{CO}}{p_{H_2}^{\,2}} \qquad (7)$$

which is plotted also in Fig. 1. The thermodynamic data for these estimations have been taken from literature[12].

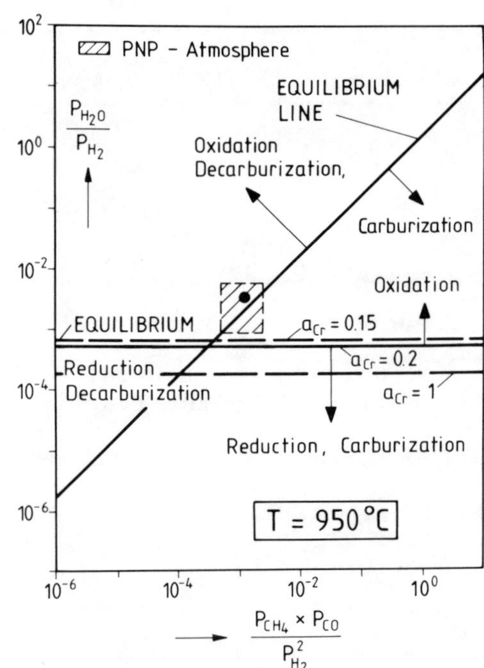

Fig. 1

PNP	BF GHT HRB KFA RBW
	Fields of Carburization and Oxidation in Relation to Partial Pressure Products of Impurities

The plot clearly shows that the boundaries of the PNP-standard-helium allow carburization and decarburization of alloys.

With respect to the oxide film an estimation was made
for the thermodynamic stability of different oxides
according to the reaction

$$x \, Me + y \, H_2O = Me_xO_y + y \, H_2 \, , \tag{8}$$

in which the activity of metallic components in solid
solution is assumed to be equal to its mol percent
content (Fig. 2).

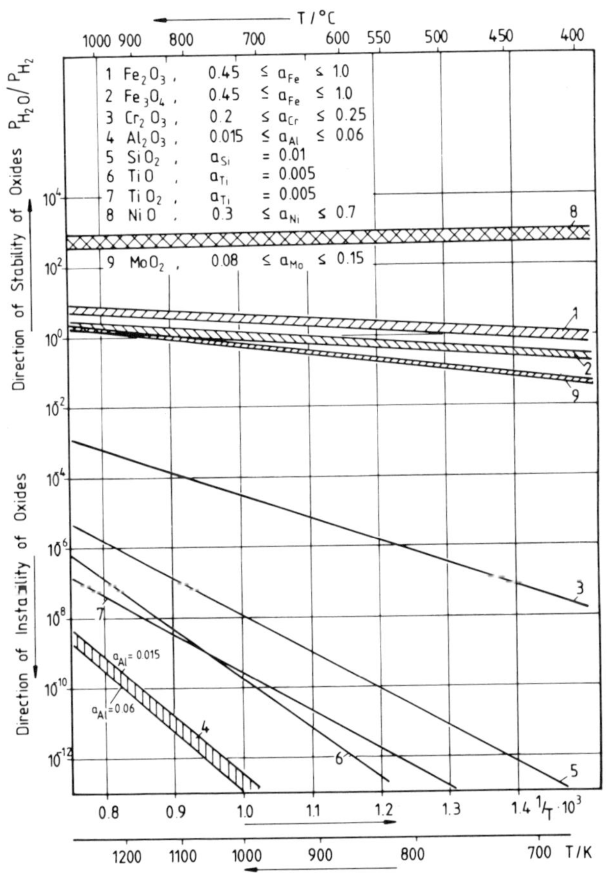

Equilibrium Lines for Different Types of Oxide Scales

Fig. 2

If one assumes that

$$\frac{P_{H_2O}}{P_{H_2}} \approx 10^{-3} \ldots 5 \cdot 10^{-3}$$

one sees that even Cr_2O_3 can be stable up to $1000^\circ C$. Only NiO, MoO_2, and Fe_2O_3/Fe_3O_4 do not form in the regarded temperature range within this atmosphere.

It should be mentioned that the stress to which the creep specimens are subjected to contributes to the thermodynamic data. But the change in the surface tension is small so that the contribution in this approximation can be neglected.

3. <u>EXPERIMENTAL EXAMINATIONS OF SURFACE LAYER OF CREEP-RUPTURE SPECIMENS</u>

3.1 <u>Experimental Conditions</u>

The creep-to-rupture specimens were taken from a tube with a 132 mm outer diameter and 14 mm wall thickness in a solution annealed condition.

The chemical composition (wt.%) of the material Incoloy 802 is given below (Table 2):

C	0.32
Si	0.30
Mn	0.84
P	0.017
S	<0.003
Cr	20.12
Mo	0.28
Ni	31.0
Al	0.17
Ti	0.77

The specimens were creep tested in air and in PNP-helium at temperatures between 800 and $1000^\circ C$, respectively. Fig. 3 shows the stress-rupture properties of the tested material.

The gas composition during the creep rupture tests was controlled by a gas chromatograph at the inlet and at the outlet of the furnaces. To demonstrate the changes in the gas composition the gas atmosphere in passing one specimen ($950^\circ C$, 1354.9 h) is recorded in Fig. 4.

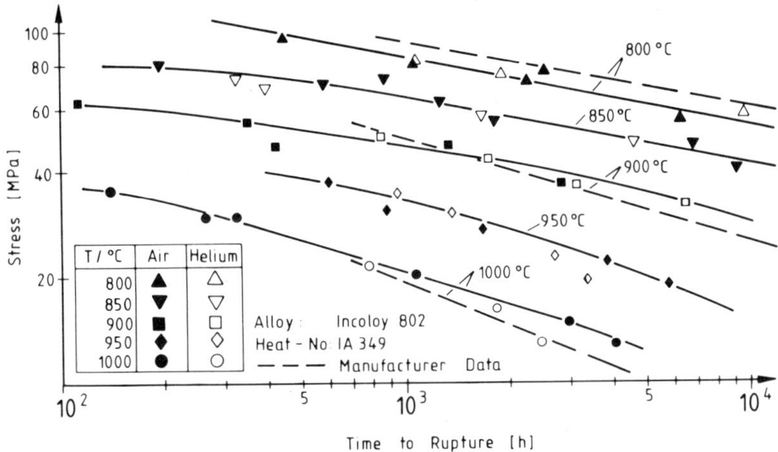

PNP	BF	GHT	HRB	KFA	RBW	
	Stress Rupture Properties of Incoloy 802					

Fig. 3

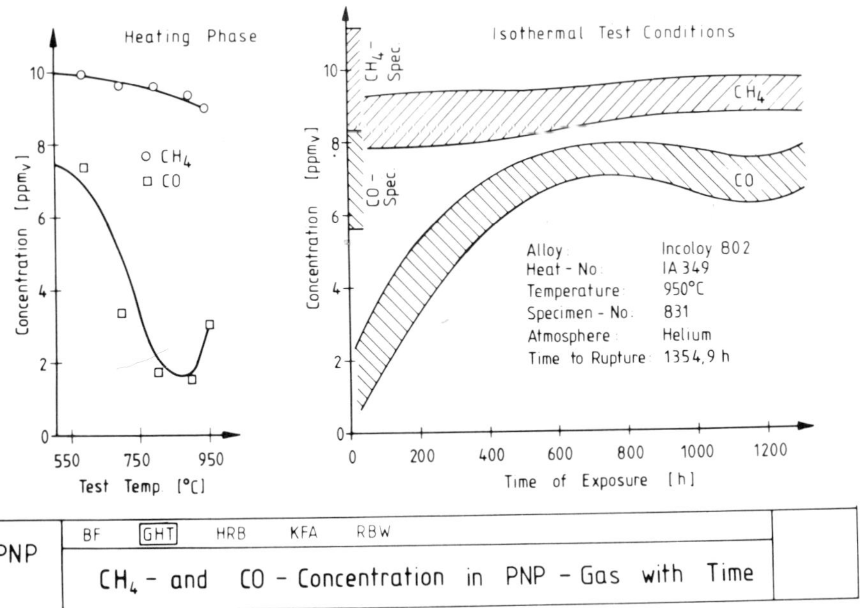

PNP	BF	GHT	HRB	KFA	RBW	
	CH₄ - and CO - Concentration in PNP - Gas with Time					

Fig. 4

3.2 <u>Post-Test Examinations</u>

<u>Surface layer microprobe analysis</u>

Fig. 5 and 6 show the carbon and oxygen profiles from microprobe examinations of three specimens which were exposed to PNP-helium at different temperatures (900, 950 and 1000°C, respectively).

The profiles were measured on a transverse microsection of the gauge length in a region, representative of the total surface-layer.

The direction of measurement was from the surface to a depth of about 500 µm into the specimen.

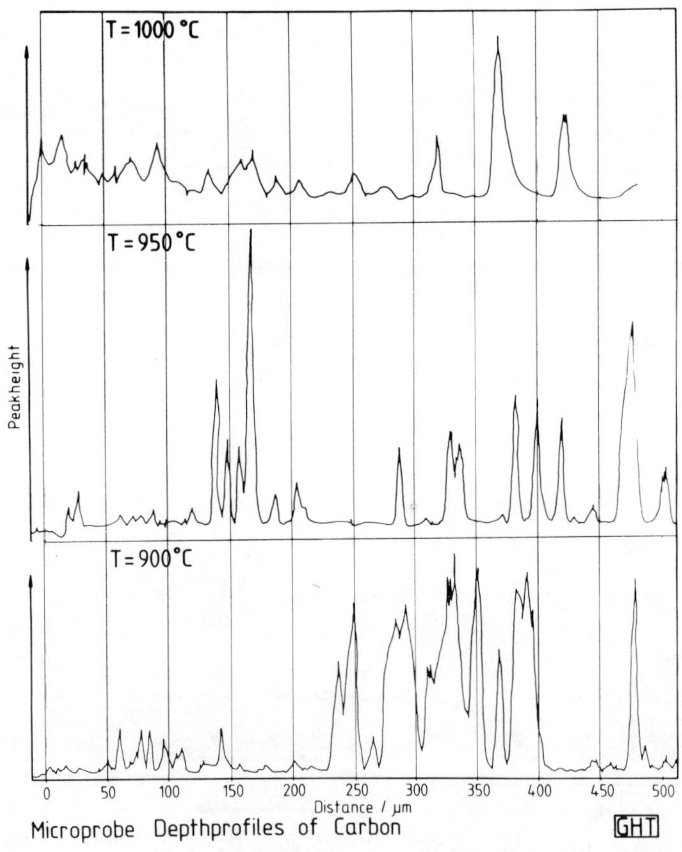

Microprobe Depthprofiles of Carbon

Fig. 5

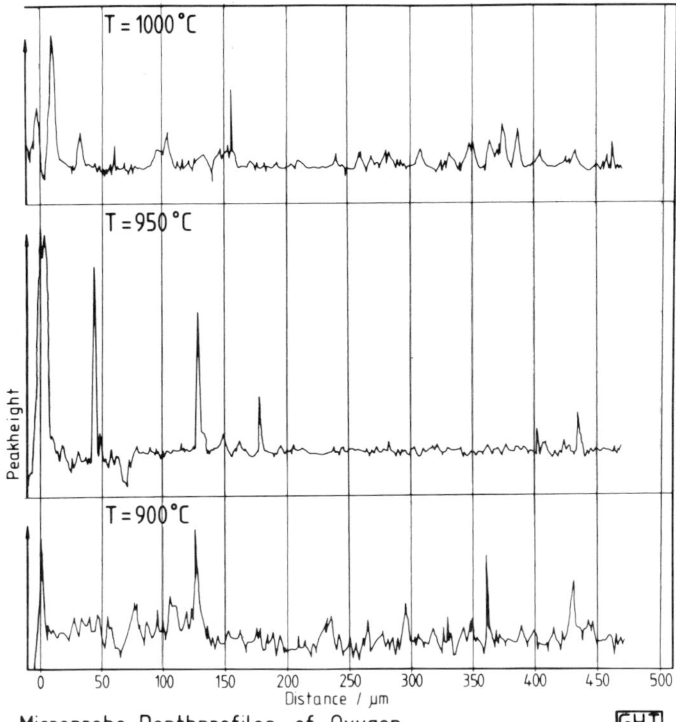

Fig. 6

Chemical bulk analysis

Bulk carbon analyses were made from 22 specimens,
tested in helium (17 specimens) and air (5 specimens),
respectively; the results of the carbon analyses as
compared to the as received value are shown in Fig. 7
and Fig. 8.

Impact of stress on surface layer

The stress to which the creep specimens were subjected
to is assumed to cause only a second order effect to
the structure and thickness of the surface layer. To
demonstrate this, a longitudinal cut through the gauge
length and the head of a specimen (900°C, 6.276 h) was
made. Both sections had a different diameter, thus
corresponding to different stresses. The stresses of
the gauge length and the head of the specimen was
32.1 MPa and 14.25 MPa, respectively.

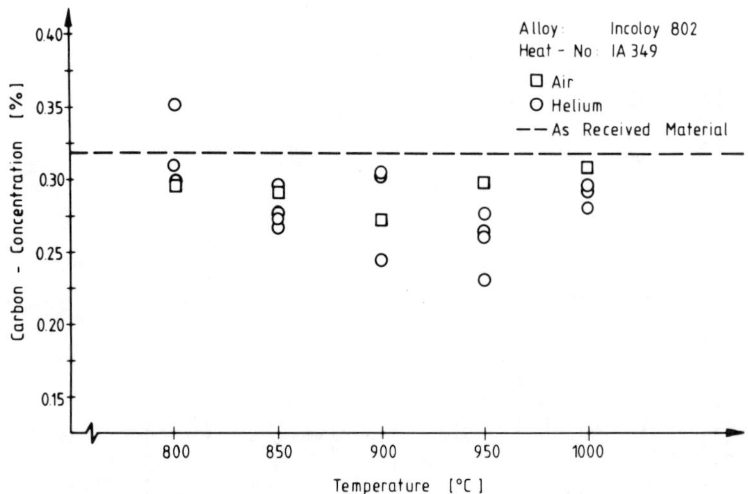

Fig. 7

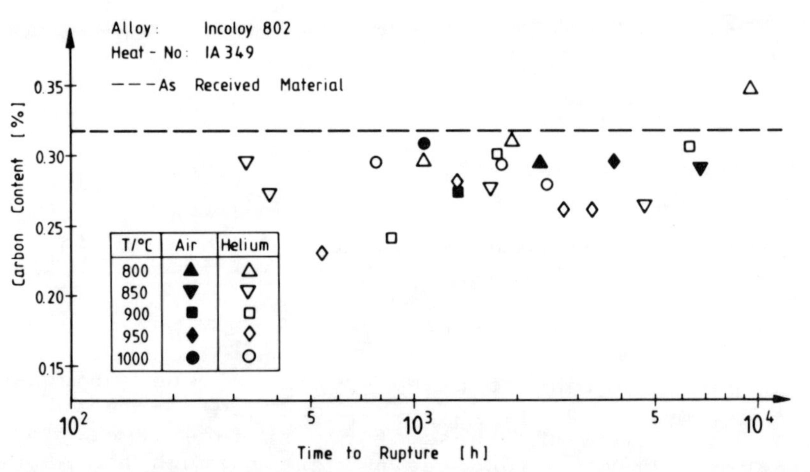

Fig. 8

From the two sections a metallographic documentation
as well as microprobe analyses with respect to carbon
and oxygen were made (Fig. 9 and Fig. 10).

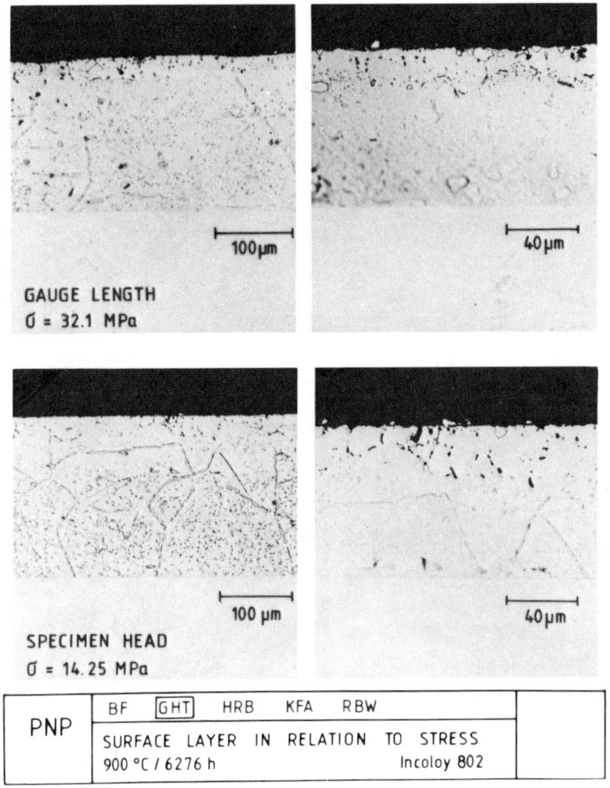

Fig. 9

4. DISCUSSION

The creep-rupture test results (Fig. 3) indicate about
the same stress-to-rupture properties up to 10,000 h
test time in the low temperature region for both atmospheres, air, and helium.

At the higher temperatures, 950 and 1000°C helium seems
to affect the stress-rupture properties negatively
after 2,000 h test time.

The gas depletion of CO during the heating phase of the
test furnaces can be seen in Fig. 4. During the isothermal test period the gas atmosphere - especially the
CO content - is recovered to the nominal composition
of the PNP-specification. Significant changes in the

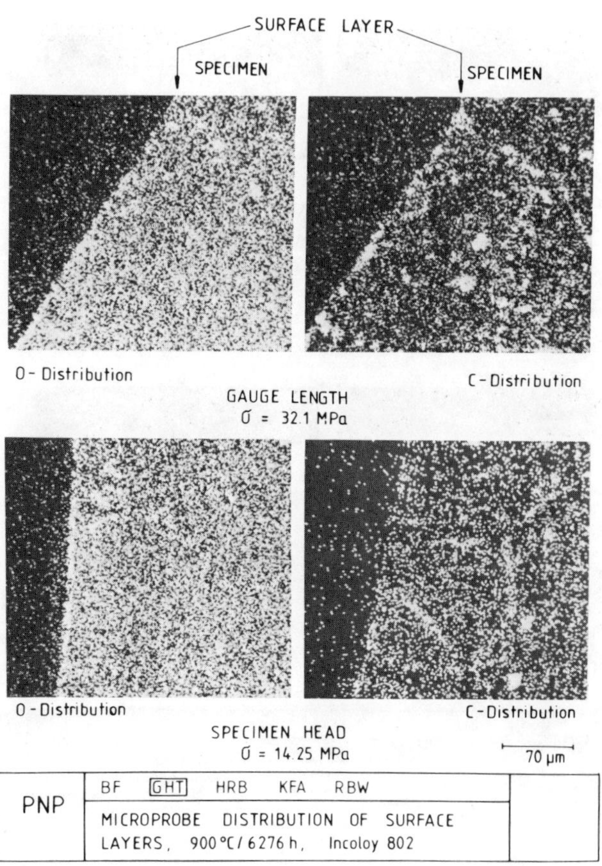

Fig. 10

CH_4-concentration cannot be observed. It is assumed
that the CO depletion is caused by the formation of a
thin oxide layer on the specimen during the heating
phase and during the first period of the isothermal
part of the experiments; the oxides resulting from the
long term experiments are mainly due to H_2O depletion
under simultaneous dissolution of carbides or carbon
from the metallic surface. Microprobe profiles for
oxygen (Fig. 6) show the oxide layer on specimens at
900, 950 and 1000°C. The thickness of the surface layer
is about 10 µm. Oxygen enrichments could be also detected in the grain. Mn, Cr and Si could be detected
in the oxide layer with an energy dispersive system.
It is concluded that the layer contains Mn-, Cr- and
Si-oxides in accordance with the thermodynamic considerations.

The examinations in relation to carbon show the absence
of a carbide surface layer (Fig. 5). Only local carbides in the grain and at the grain boundaries could
be detected. Chemical bulk analyses show the decarburization of the creep specimens (Fig. 7 and 8).
Only one specimen had an elevated carbon concentration
in relation to the as received condition. This specimen
had the longest exposure time of all 22 specimens after
an exposure to helium of $800^{\circ}C$.

A correlation between time-to-rupture and carbon concentration cannot be confirmed, as demonstrated in
Fig. 8. The maximum carbon depletion is about 28 %
at a stress to 47.2 MPa, 857.2 h time-to-rupture and
$950^{\circ}C$.

No influence of stress on thickness and composition of
the surface layer was observed; Fig. 9 shows the surface-layer of two specimen sections with different
stresses; the higher stress of the gauge length had
no influence on the thickness of the surface-layer.

The oxygen and carbon distributions with the microprobe in Fig. 10 show also no stress effects. Carbide
precipitations do occur in the material, but no carbide
surface-layer could be identified. A thin oxide layer,
independent on stress, exists.

5. CONCLUSIONS

Thermodynamic calculations, which show that oxide
layers can be stable in PNP gas atmosphere up to $1000^{\circ}C$,
could be confirmed experimentally.

A CO depletion, which can provide the oxygen together
with H_2O was measured. Microprobe analysis of the surface-layer, chemical bulk analysis and metallographic
investigations were performed which verified the presence of a thin oxide layer as well as a decarburization.

No influence of stress on the oxide layer thickness and
composition was observed.

ACKNOWLEDGEMENT

The authors gratefully acknowledge experimental work of J. B. Pierick, F. Sprotte, and R. Ehrmann.

The work has been performed under the terms of the cooperative agreement between Bergbau-Forschung GmbH, Gesellschaft für Hochtemperaturreaktor-Technik mbH, Hochtemperatur-Reaktorbau GmbH, Kernforschungsanlage Jülich GmbH and Rheinische Braunkohlenwerke AG dealing with the development of processes for the conversion of solid fossil fuels with heat from high temperature reactors, with the assistance of the Federal Minister of Research and Technology and the State of Nordrhein-Westfalen.

REFERENCES

1. R. G. Shepheard and M. J. Donachie, Jr., Trans. ASM 1962, 45-56
2. J. Board, J. Brit. Nucl. Energy Soc. 1970, 101-112
3. H. Willermoz, L. Berry, J. Dixmier et Ph. Olivier, 5ème Congrès Européen de la Corrosion, Paris, 24 - 28 Sept. 1973
4. J. Dixmier, H. Willermoz, and R. Roche, BNES Int. Conf. Inst. Civ. Eng., London, 26 - 28 Nov. 1974, No. 41, 41.1 - 41.9
5. R. H. Cook and R. P. Skelton, International Metallurgical Reviews 1974, 19, 199-222
6. R. J. Pearce, A Status Report of Alloy 800, BNES Conf., Reading, UK, 1974, 129-149
7. R. H. Cook, A Status Report of Alloy 800, BNES Conf., Reading, UK, 1974, 150-169
8. F. N. Mazandarany and P. L. Rittenhouse, Nuclear Technology 1976, 28, 406-423
9. H. G. A. Bates, W. Betteridge, R. H. Cook, L. W. Graham, and D. F. Lupton, Nuclear Technology 1976, 28, 424-440
10. R. H. Cook, J. Nuclear Materials 1977, 66, 257-262
11. T. Hirano, M. Okada, H. Yoshida, and R. Watanabe, J. Nuclear Materials 1978, 75, 304-308
12. O. Kubaschewski, E. L. L. Evans, and C. B. Alcock, Metallurgical Thermochemistry, Pergamon Press, Oxford, 4th Edition, 1967

DISCUSSION

H.J. Grabke: (Comment)   It is very surprising that in two papers
of this session, the same environment - the so-called PNP - helium,
acts carburising in studies at one institution (KFA Jülich) and
decarburising in investigations of the other institution (Interatom).
Usually, it is tried to explain the corrosion phenomena in the test
helium atmospheres, which are not in thermodynamic equilibrium, by
thermodynamic calculations and considerations. However, the state
which shall be simulated by the tests, high temperature alloys in a
fast flowing atmosphere with constant composition, is a steady state
and not a thermodynamic equilibrium. A steady state   carbon and
oxygen activity is established on the solid surface, which depends
on the relative rate of carburising and decarburizing reactions.
Since the decarburisation by water vapour is a very fast reaction
this steady state will strongly depend on the water vapour content
of the flowing gas atmosphere, if it can be controlled well enough
and can be kept constant.
Concerning this, my question to E. Ebberink is: did you control
the water vapour content of your test atmosphere at the inlet
and outlet of the furnaces?

J. Ebberink: Under the used test conditions of our experiments a
fast impurity depletion, especially water, between inlet and outlet
was observed (Fig. 4). In our experiment the samples were just at
the inlet of the single specimen test cell. The samples showed no
carbon gradient over the length. As Fig. 1 shows the carburization
potential of the atmosphere can easily shift from decarburization
to carburization by depletion of the impurities within the specified
tolerance.

P. Kofstad: I would like to further emphasize some of the comments
by H.J. Grabke.
One should be very careful in using thermodynamic calculations in
predicting alloy behaviour in these gases. The gaseous impurities
are not in equilibrium, and the reactions which will take place
will, to a large extent, depend on reaction kinetics. As to the
fact that some laboratories observe decarburisation and carburisation, respectively, I can add that in Oslo, we found that the
type of reaction behaviour observed was dependent on the gas flow
in the furnaces and the location of the specimen in the furnace.
Decarburisation was under such conditions observed at the inlet
side of the furnace (effect of H_2O), while carburisation was
observed at the outlet side where the gas was depleted in H_2O. It
is important to continuously monitor the gas composition at both
the inlet and the outlet side of the furnace.

J. Ebberink: Yes, not only thermodynamics   determined the material
behaviour. It was interesting to see in our experiments, that the
gas composition did not change when no metallic surface was available when the gas passed through a high temperature zone. Your observation of carburization and decarburization is a result of the
depletion of H_2O in your furnace. The thermodynamic calculations
show that you may have shifted from the decarburization at the
inlet to carburization at the outlet as shown in Fig. 1.

B. Ilschner: I would like to draw attention to the
fact that the metal (or oxide, or carbide) surface will exhibit
a high catalytic activity, not only with respect to the gas/solid
reaction kinetics, but also with respect to equilibrium within the
gas mixture. Thus, the gas composition within the adhering
boundary layer may be quite different from the bulk composition
which is monitored either at the inlet or the outlet. Just this
composition is, however, responsible for the carbon and oxygen
activity in respect to the alloy.

J. Ebberink: We hope to overcome the problem when we have found
an alloy forming inert protective oxide layers.

D.M. Ward: Nitrides are formed during creep testing in air and in
service in oxidising conditions in HK Alloys (25Cr/20 Ni steel)
and HP alloys (25Cr/35N steels). Microstructures showing these
chromium nitride particles are given in ASTM book on Microstructures of High Temperature Alloys.

J. Ebberink: We observed no nitridization.

The Mechanical Properties of Alloys in High
Temperature Reactor Environments

P. J. Ennis and H. Schuster

Kernforschungsanlage Jülich GmbH

D-5170 Jülich, FRG

SYNOPSIS

The results of creep-rupture tests with test durations of up to
10 000 hours show no differences in the creep strength of a number
of high temperature alloys measured in air, in helium environments,
and in methane reforming process gas. The rupture ductility is
however generally lower in helium than in air or process gas.
Metallographic examination of ruptured specimens show significant
differences in the corrosive attack of the three environments. The
effect of carburisation on the room-temperature tensile properties
has been determined. Maximum tolerable carburisation rates for high
temperature reactor components, assuming a minimum room temperature
ductility throughout service life of a component have been derived.

1. Introduction

Because of the high core outlet temperatures which can be
achieved (ca. 950 ^{O}C), the helium-cooled, high temperature
reactor (HTR) has the potential capability of providing heat for
chemical processes, the most important of which is at the present time the gasification of coal. To meet the high standards
of safety and reliability required for components of a nuclear
process heat plant, a thorough understanding of the behaviour
of materials under all service conditions is essential. The particular problems are the effects of the various reactor environments on the mechanical behaviour of components, especially in
consideration of the very long service times required due to the
heavy economic penalties and technical difficulties involved in
the replacement of components.

In this report we are considering mainly the effect of the reactor
primary circuit coolant, which is helium containing various
impurities, on the creep-rupture properties and the effect of
carburisation on the short-time tensile properties of some high-
temperature alloys, including INCOLOY alloy 800 H (Fe - 20 Cr -
32 Ni - 0.08 C) and INCONEL alloy 617 (Ni - 22 Cr - 9 Mo -
12,5 Co - 1.5 Al). First results obtained from creep-rupture
tests of INCOLOY alloy 800 H in the process gas of the methane
reforming process will also be mentioned.

2. HTR Environments

2.1 Primary Circuit Coolant

The helium coolant of the reactor primary circuit contains very
low levels of impurities, the most important of which are hydrogen, water, carbon monoxide and methane. The sources of these
impurities in a process heat reactor are

H_2 -   hydrogen from the process gases by permeation through
                the tube walls of the heat exchangers plus hydrogen
                split from water which is desorbed from fresh fuel ele-
                ments;

H_2O -   residue of water from the reaction mentioned above;

CH_4 -   radiolytic reaction of hydrogen with carbon in the core;

CO -   product of the water-gas reaction in the core.

Various "standard" test atmospheres have been defined based on
data from steam-generating HTRs and on predictions of the impurity concentrations in a process heat reactor. Table 1 summarises
the helium test atmospheres considered in this paper.

In the test atmospheres, the levels of impurities are far from
the thermodynamic equilibrium. Therefore, the reactions between
these gases and the alloys cannot be described by a single oxidising or carburising potential; rather the individual reactions
which can lead to oxidation or carburisation and their competitive kinetics must be considered. The various roles played by
water, methane and carbon monoxide in the formation of surface
scales on a number of high temperature alloys exposed in HTR
helium environments have been discussed by Lupton[1]. It is shown
that complex kinetics determine the growth of the surface scale,
which in turn governs the extent of carburisation which will
occur in an alloy exposed in HTR-helium.

In considering the effects of HTR-helium on the mechanical properties of high temperature alloys, we must therefore take
account of oxidation (surface oxidation and internal oxidation)
and more particularly of carburisation.

2.2 Secondary Circuit Helium

In some designs of gasification plant, for example, the steam
gasification of hard coal, an intermediate circuit will be

necessary to transfer heat from the reactor primary circuit to the
coal gasification fluidised bed. This environment presents relatively few problems as the coolant may be doped with controlled
additions of oxidising species to provide protective oxide films.

2.3 Process Gas Environments

Hydrogasification of brown coal and steam gasification of hard
coal are the key processes which will be coupled to an HTR. At
KFA we are concerned with the steam reforming of methane, the
heat consuming process for the hydrogasification. The reformer
gas mixture is similar to that found in conventionally fired steam
reformers, and a test gas composition of 35 vol. % H_2, 50 vol. %
H_2O, 5 vol. % CO, 5 vol. % CO_2 and 5 vol. % CH_4 has been selected
for the evaluation of candidate reformer tube alloys.

3. Possible Effects of HTR Environments of Mechanical Properties

3.1 Creep-Rupture Properties

The possible effects of aggressive environments on the time-de-
pendent mechanical properties of metallic materials have been
recently reviewed by Guttmann and Marriott[2]. A number of mechanisms exist both for strengthening and for weakening of alloys
by corrosion under creep loads. Of relevance to HTR-helium environments are the strengthening effects due to precipitation of
carbide phases, especially at the grain boundaries, during carburisation, and the weakening effects arising from

-   the loss of solid solution strengthening elements, such as
    Cr and Mo, from the matrix to form carbide phases;
-   the internal stresses developed by the volume changes which
    occur as carbon diffuses into the structure and precipita-
    tes as carbides;

- the formation of brittle phases which can act as notches;
- the reduction of load-bearing cross-section resulting from
  internal oxidation.

In process gas environments similar effects could be expected
with the emphasis on the weakening effects of internal oxidation
and the possibility of strengthening by the blunting and bridging
of cracks by oxides.

3.2 Tensile Properties

The carbon content of high temperature alloys is carefully specified and controlled by alloy manufacturers to ensure that the
materials maintain good tensile properties, especially sufficient
ductility during service operation. The effect of carbon content
on short-time mechanical properties is usually most marked at
temperatures ranging up to 600 - 700 $^{\circ}$C, in which range the ductility is very sensitive to carbon content. Because construction
and subsequent maintenance operations are carried out at ambient
temperatures, room temperature should also be regarded as a service temperature. The room temperature properties are therefore
important. We can reasonably expect that if carburisation occurs
in HTR-helium or in the process gases, the associated increase in
carbon content and precipitation of carbide phases will lead to
a decrease in the ductility of the alloy at temperatures from
room temperature up to around 500 - 700 $^{\circ}$C.

4.   Results

4.1 Creep-Rupture Strength

Creep-rupture tests have been carried out in HTR-helium environments, in air, and in the reformer gas, up to test durations of
around 10 000 hours. To date, no detrimental effects of environ-

ment on the 1 % strain limit and on the rupture strength have been
observed.

For INCOLOY alloy 800 H the creep-rupture properties in the three
types of environment are shown in Figure 1. Metallographic examination of the broken creep test pieces showed very significant
microstructural differences between specimens tested in air, in
HTR-helium and in reformer gas. Microstructures of test pieces
which had rupture lives of between 2300 and 4300 hours at 900 $^{\circ}$C
in the three environments are compared in Figure 2. The following
features should be noted:

-   in air, a thick oxide scale containing metallic inclusions
    has formed and filled surface cracks. An acicular precipi-
    tate can be seen in the grains, and extensive grain-boundary
    cracking has occurred; oxide films can be seen on the inter-
    nal crack surfaces.
-   in HTR-helium (in this case, PNP-standard), extensive inter-
    granular surface cracking has occurred throughout the gauge
    length of the specimen which had rupture elongation of
    23.8 %. Cavities have formed at triple points but there is
    less grain-boundary separation compared with the air-tested
    specimen and no indication of enhanced cracking in the car-
    burised zone. The average carbon content of the specimen
    after testing was twice the original carbon content.
-   in reformer gas, we have extensive intergranular cracking,
    with wide cracks in the central part of the cross-section.
    Surface cracks at grain boundaries were filled with oxide,
    and a thick oxide scale was found on the specimen surface.
    No carburisation was observed.

In spite of considerable changes in the microstructures as a
result of reaction with the different test environments, there
are no differences in the creep strengths in the different
atmospheres, at least for test durations of up to 10 000 hours.

We have mentioned that the formation of brittle phases could lead
to a reduction in creep strength. However, in the creep specimens
examined to date, the presence of carbides in the structure does
not seem to have influenced to any great extent the cracking
processes. Figure 3 shows the microstructure of an INCONEL alloy
617 specimen creep tested at 850 ^{O}C. The test was terminated at
3400 hours with a strain of 11 %. There are surface cracks at
nearly every transverse grain boundary, some of which must have
formed very early in the test as there is considerable internal
oxidation extending into the alloy from the crack surfaces. The
interference contrast photomicrograph (Figure 3) shows that
although the specimen had a strain of ca. 11 %, there was no sign
of cracking or cavitation associated with the extensive carbide
precipitation, and the surface cracks showed no tendency to follow
the carbide particles.

4.2 Creep Rupture Ductility

Although no significant differences in the creep strength in
HTR-helium, reformer gas and air have been observed, rupture
ductility in HTR-helium appears to be lower than that in air.
Figure 4 shows the rupture elongations of INCOLOY alloy 800 H and
NIMONIC alloy 86 (Ni - 25 Cr - 10 Mo) as a function of rupture
time. This points to a decreased strain which can be achieved in
the tertiary creep stage, possibly due to the absence of oxide
to blunt or bridge over the creep cracks. The trend to lower
ductility in HTR-helium seems to apply to all of the high-temperature, wrought alloys which we have tested (INCOLOY alloys 800 H
and 802, NIMONIC alloy 86, INCONEL alloy 617 and HASTELLOY alloys
S and X).

4.3 Room Temperature Tensile Properties

Some results of our investigations into the relationship between
carburisation and ductility loss were presented at the last
Petten conference[3]. It was shown that twice as much carbon could
be tolerated in INCOLOY alloy 800 H than in INCONEL alloy 617 before the room temperature ductility dropped below 10 % elongation.
The ductility results together with the tensile strengths and
0.2 % proof stresses are shown in Figure 5.

On the basis of the ductility-carbon content relationships and
assuming that a minimum ductility will be required for components
throughout their service lives, it is possible to define the maximum tolerable carburisation rates. Using the ductility curves
shown in Figure 5 and assuming a service life requirement of
140 000 h and a linear carburisation rate (i.e. the carburisation
is surface-controlled), the end-of-life, room temperature ductility can be plotted against carburisation rate for given component
thicknesses, as shown in Figure 6.

If we now specify a minimum ductility requirement - say 5 % elongation, which is roughly equivalent to an average ductility of
10 % as plotted in Figure 6 - the maximum carburisation rates can
be derived. Thus for a reformer tube of INCOLOY alloy 800 H, a
relatively high carburisation rate of $3 . 10^{-7}$ g.cm^{-2} h^{-1} can be
tolerated for the alloy to retain a minimum ductility of 5 % after
140 000 h service. This rate is equivalent to 0.3 wt. % increase
in the carbon content of a 5 mm diameter specimen exposed for
10 000 hours. Similar carburisation rates have been observed for
a large number of alloys in the HTR-helium environment I in
Table 1[4].

In contrast, for a thin-walled tube, a typical example being an
intermediate heat exchanger tube, the maximum tolerable carburisation rates for the retention of a minimum room temperature elon-

gation of 5 % are 0.5 . 10^{-7} g cm^{-2} h^{-1} for INCOLOY 800 H and
0.2 . 10^{-7} g cm^{-2} h^{-1} for INCONEL alloy 617. These maximum tolerable rates are equivalent to increases in carbon content of
5 mm diameter specimens of 0.05 and 0.02 wt. % respectively after
10 000 h. The importance of carrying out long-term environmental
testing (greater than 10 000 h) to ascertain whether carburisation
rates below the ductility loss limited, maximum rates can be
achieved is clear from these calculations.

5.   Summary

The results available to date which have reached around 10 000 h
have shown no detrimental effects of HTR-helium environments or
a process gas environment on the creep strength of several high
temperature alloys, compared with the properties obtained in air.
Metallographic examination and carbon analysis of the broken creep
specimens tested in HTR-helium environments showed that considerable carburisation can occur. In the reformer gas, thick oxide
scales were found which filled the large number of surface cracks
present. The creep strength appears to be insensitive to the
corrosive attack of HTR-helium and reformer gas, at least to
times of up to 10 000 h. However, a marked trend towards lower
rupture ductility in HTR-helium atmospheres than in air was found.

The room temperature tensile properties were significantly affected by carburisation, and it has been shown that carburisation
rates must be limited to very low levels in order to ensure
reasonable end-of-life ductilities for HTR components.

6.   Future Work

Because of the sensitivity of the short-time tensile properties,
especially ductility, to carburisation effects, a large programme
has been undertaken in which alloys are being exposed under stress

        P.J. ENNIS and H. SCHUSTER

at 850 ^{O}C in air and in two HTR-helium environments (III and IV
in Table 1). Stresses were selected to give about 1 % strain in
10 000 h, and after exposure for 1000, 3000 or 10 000 h, the
specimens are tensile tested principally at room temperature and
at 850 ^{O}C. Additional post-exposure tensile tests at intermediate
temperatures will be carried out to determine the temperature
dependence of the strength and ductility of exposed specimens.

In a second programme, the effect of carbide morphology on the
tensile properties of carburised alloys is being determined, using
a series of simple ternary alloys.

7.   Acknowledgements

We wish to thank Dr. U. Bruch who made available the results
obtained in reformer gas.

The work has been performed under the terms of the cooperative
agreement between

> Bergbau-Forschung GmbH
> Gesellschaft für Hochtemperaturreaktor-Technik mbH
> Kernforschungsanlage Jülich GmbH
> Rheinische Braunkohlenwerke AG

dealing with the development of processes for the conversion of
solid fossil fuels with heat from high temperature reactors, with
the financial support of the Federal Minister for Research and
Technology and the State of North Rhine-Westphalia.

8. References

1. D. F. Lupton
    KFA-Report JÜL-1639, January 1980

2. V. Guttmann and J. B. Marriott
    Conf. on Environmental Degregation of High Temperature
    Materials, Isle of Man, April 1980

3. P. J. Ennis and D. F. Lupton
    Int. Conf. on the Behaviour of High Temperature Alloys in
    Aggressive Environments, Petten, Oct. 1979: Paper 60

4. A. V. Dean
    ibid; Paper 13

Table 1: Compositions of HTR-Helium Environments

Atmosphere		Nominal Composition (μbar)			
		H_2	H_2O	CO	CH_4
I	CIIR Oslo, Phase 4	500	1,5	30-50	50
II	ZEMAK IV "Start up"	1800	1	70	100
III	PNP-Standard (Process-heat reactor)	450-550	1,5	10-20	15-25
IV	HHT-Standard (direct-cycle helium turbine reactor)	50	5	50	5

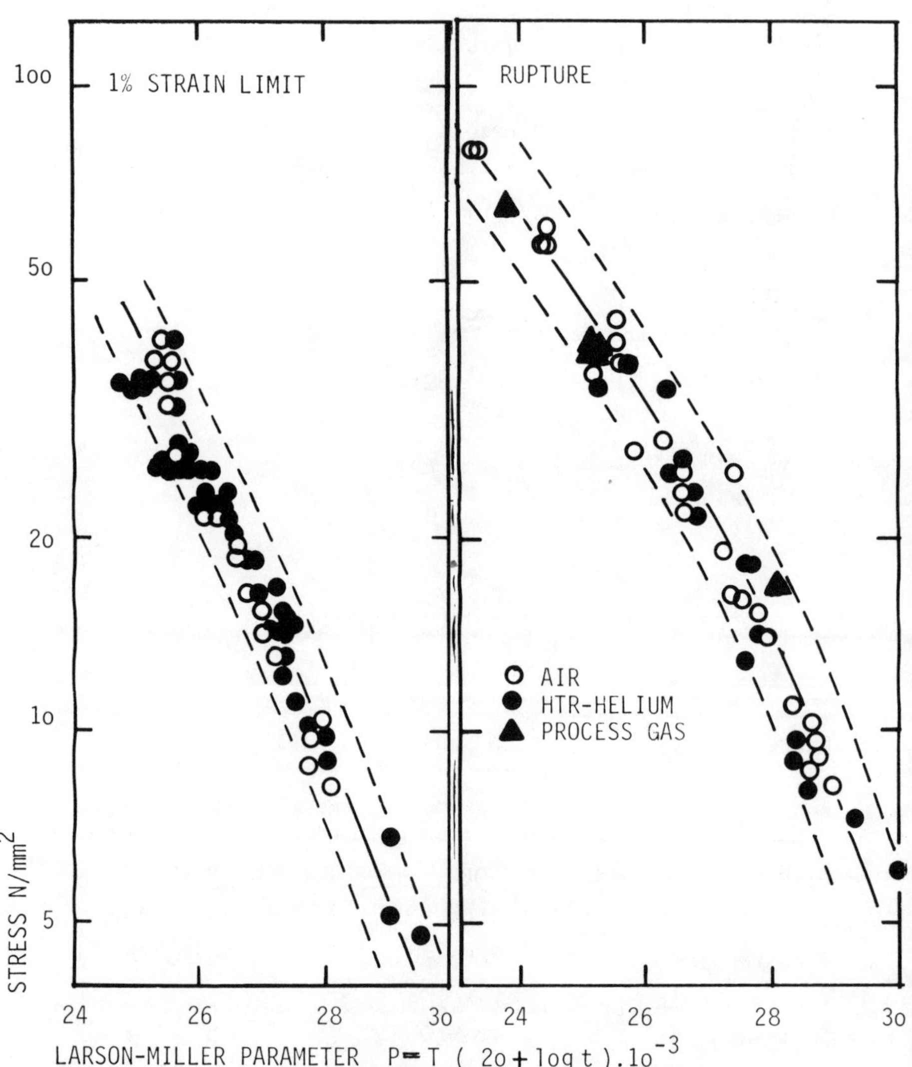

Figure 1

Creep-rupture Properties of INCOLOY alloy 800H in Air, HTR-Helium and Process Gas

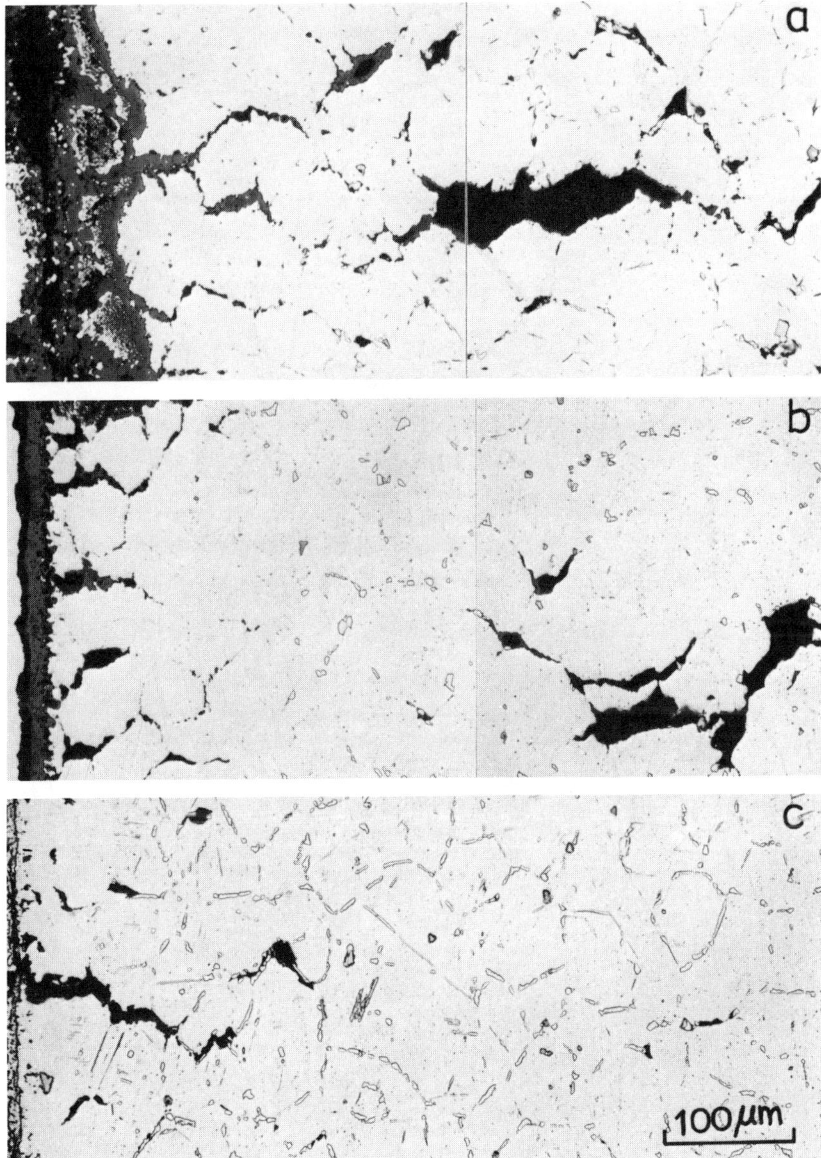

Figure 2 Microstructures of INCOLOY 800H creep specimens tested at 900ºC: a- air test, 21 N/mm^2, rupture time 2300 h, elong. 28% : b- RSO process gas, 15 N/mm^2, 4260 h, 29,4% : c- PNP helium, 18 N/mm^2, 3500 h, 23,8%.

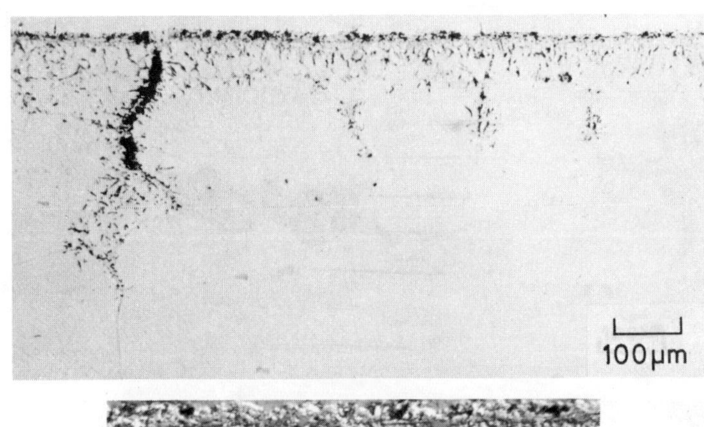

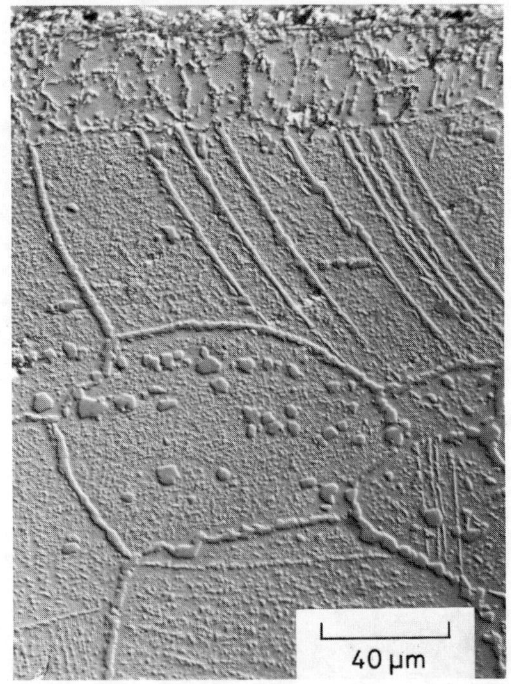

Figure 3

Microstructure of INCONEL 617 Creep Specimen Tested at 850°C, Stress 65 N/mm^2; Test Discontinued After 3400 Hours, Creep Strain 11,5%.

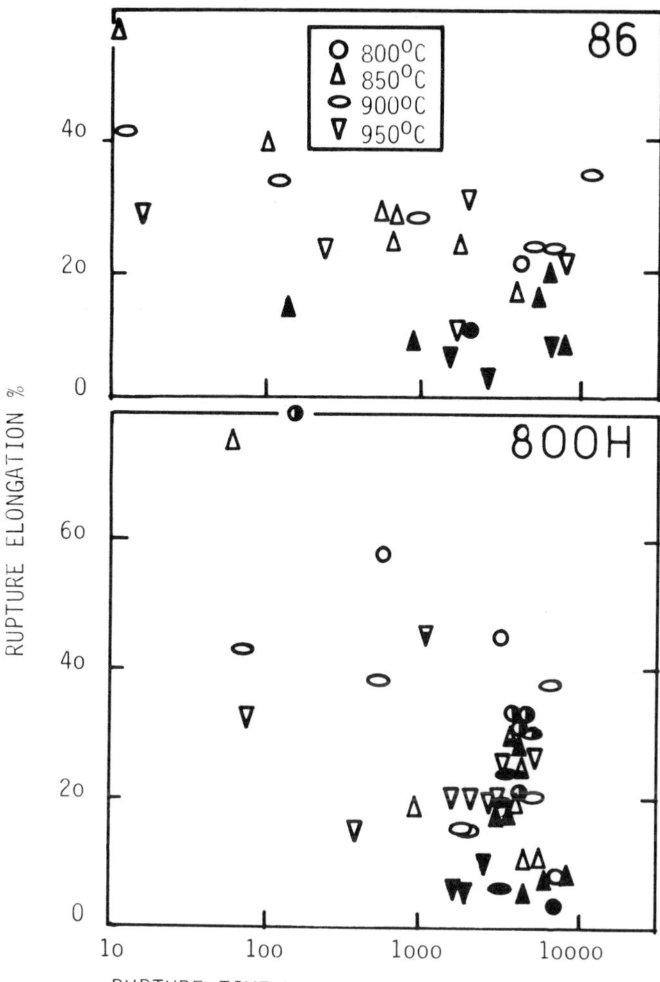

Figure 4

Rupture Ductilities of NIMONIC 86 and
INCOLOY 800H in Air (open symbols), in
HTR-Helium (filled symbols) and in RSO-
Process Gas (half-open)

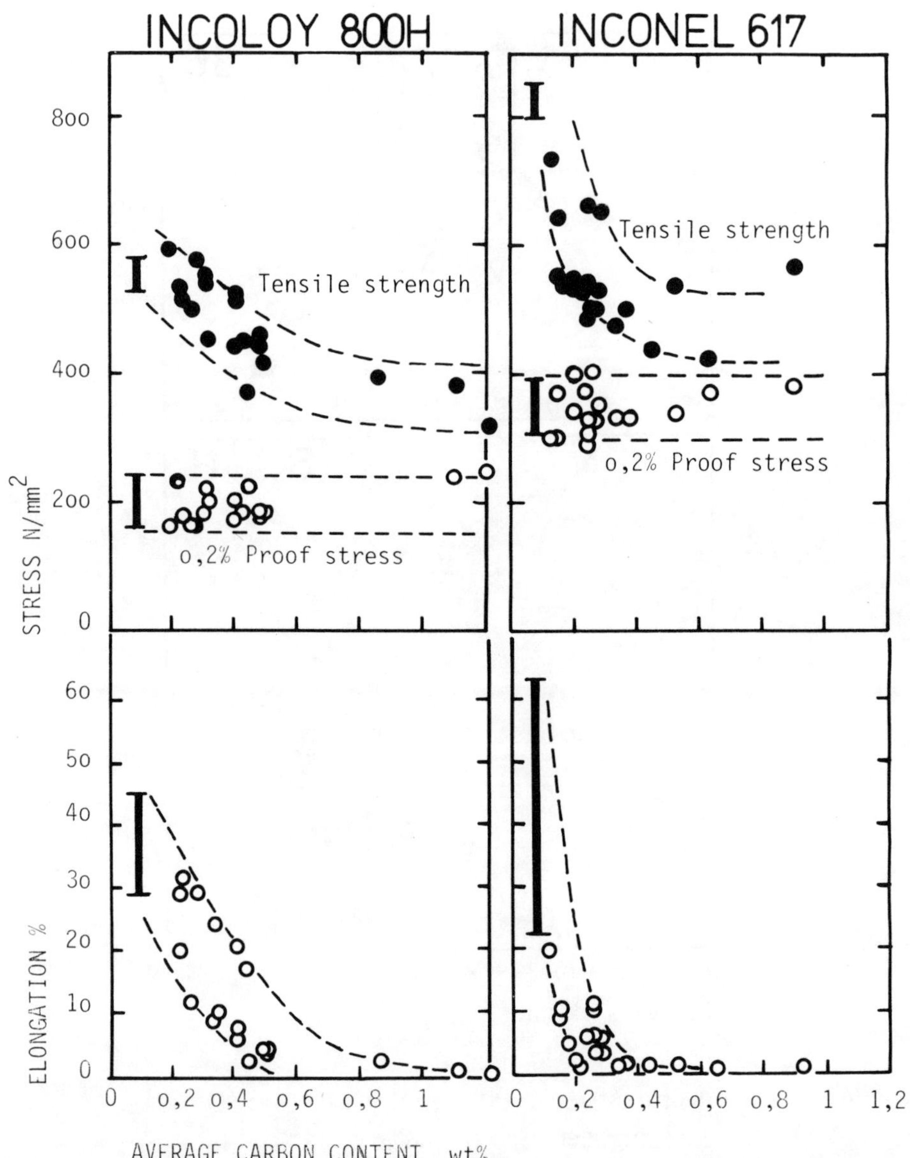

Figure 5

Relationship Between Room Temperature Ductility and Carbon Content
for INCOLOY 800H and INCONEL 617 Carburised at 950°C.

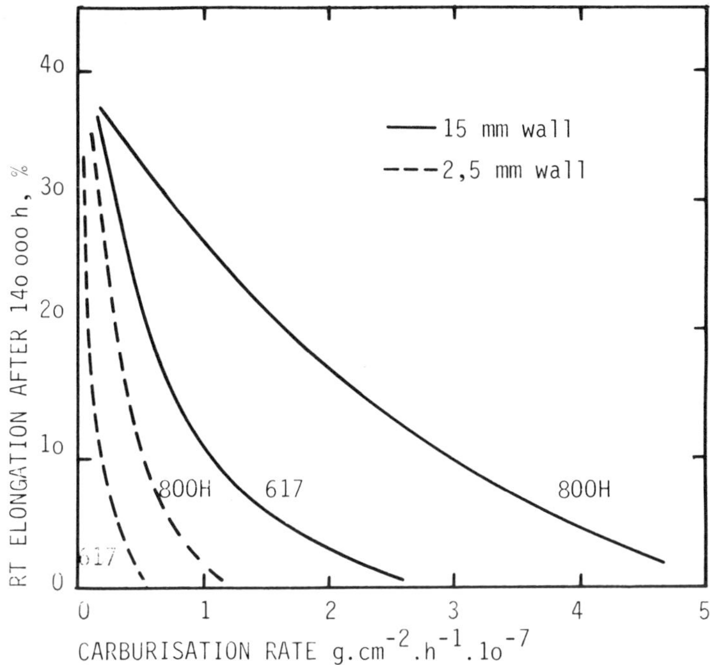

Figure 6

Predicted Average Room Temperature Ductility of INCOLOY 800H and
INCONEL 617 Components After 140 000 Hours Exposure as a Function
of Mean Linear Carburisation Rate.

DISCUSSION

<u>D.B. Meadowcroft:</u> On the question of choice of the atmoshere, I
believe the various compositions were choosen to be the bulk
compositions because the high flow rates could give an average
level round the circuit. Therefore should not the flow rate in
autoclaves be fast enough so that this is achieved. The only
alternative is that if, as mentioned by another speaker, the metal
surface causes catalytically local equilibriation of the gas
composition. Then, if kinetic effects are not critical (which
they probably are because of the very low concentrations present)
the thermodynamic equilibriated gas composition could be used in
experiments. Have these two possibilities been compared experimentally.

<u>P.J. Ennis</u>: I do not think that increasing the flow rate to avoid
impurity depletion in the test rigs is the whole answer. Although
in the reactor itself primary coolant flow rates are extremely high,
there is also a very large metallic surface in the heat exchangers.
What may be more important than flow rate is the actual amount of
each of the impurities related to the area of metal which can react.
This ratio of available impurities to reactive metal areas gives
a clue to the observed variations in the response of materials to
HTR helium. For instance, Japanese workers in the helium corrosion
field have worked with helium flow rates of 6 $Nl/cm^2/h$, providing
16o x $1o^{-8}$ $gH_2O/cm^2/h$, and found strong decarburization of the
test materials; in our facilities and in the CIIR Oslo creep rigs,
the helium flow rates are around 6o - 1oo Nl/h, which provides 1,5
- 2,o x $1o^{-8}$ $gH_2O/cm^2/h$, and carburization is observed. Such considerations provide an explanation for the contradictory results reported in this conference for very similar helium atmospheres.

The big question is which test conditions simulate more closely
conditions in a reactor. We consider that the "drier" atmospheres
may turn out to be more representative of reactor conditions.

Regarding the use of thermodynamically equilibrated gas mixtures,
again we do not think that we would be simulating reactor conditions. For example under equilibrium conditions methane completely
decomposes at the temperatures of interest (8oo - 95o$^{\circ}$C) so that
the equilibrated gas would not contain methane.

<u>J.F. Norton</u>: In your presentation, you referred to a metallic
phase present within the surface oxide. We have observed perhaps
a similar phenomenon on laboratory-prepared specimens which, with
the help of a collaborating establishment, were exposed to the
carburising/oxidising process gas environment of a reformer plant.
Have you established the composition of the 'metallic' phase and
secondly, what mechanisms do you propose to account for it being
there.

<u>P.J. Ennis</u>: We have looked at the composition of the metallic
particles embedded in the oxide scale formed on the specimens
exposed to methane reforming gas. If I remember correctly, the
composition was similar to that of the matrix but we slightly
lower Cr and Al contents.

In our specimens these oxides lay below the orginal specimen
surface and I think the particles were left behind as the
oxidation front moved into the bulk material. The other possibility that the oxide has been reduced in some way to reconstitute
metallic particles can be discounted in our case, as we found
similar embedded metallic particles in the oxides on air-tested
specimens, and I cannot envisage any reduction mechanism being
available in such circumstances.

I.R. McLauchlin: I would like to comment on P.J.
Ennis' statement concerning the variability in the effects of
carburisation and also on his observation of the presence of
nitride precipitates in some of his specimens.   In work on
9Cr-1Mo steel, I have found different behaviour between tests
in air and CO_2 at different stresses. At high stresses, where
exposure time is short, creep rates in air were lower than those
in CO_2 , suggesting perhaps an air strengthening effect. Nitride
precipitation in air tests was suggested as an explanation. At
lower stresses, and longer exposures, strain rates in CO_2 became
lower. Carburisation strengthening thus appears to be greater than
nitride strengthening after longer exposure times and the effect
of exposure time may be relevant when tests in air, rather than in
vacuum, are compared with those in a carburising environment.

Is Air a Suitable Environment for Simulation of Zircaloy/Steam-

High Temperature-Oxidation within Engineering Experiments?

S. Leistikow, R. Kraft, E. Pott

Institute for Material and Solid State Research
Nuclear Research Center Karlsruhe, FR Germany

Synopsis

The typical fuel cladding material of nuclear pressurized water
reactors Zircaloy 4 was tested by creep and creep-rupture experiments at 800 and 1000°C to find out whether steam as oxidizing environment under loss-of-coolant accidental (LOCA) conditions can
be substituted by air within out-of-pile engineering experiments.

At 800°C similar creep curves were measured in steam and air. At
and above 1000°C different creep-rupture properties were measured
because of an up-coming variation in the air oxidation mechanism.
While in steam a compact and adherent structure of oxygen-containing phases, which surrounded the tubing like a compact shell, of
increased strength and decreased ductility was formed, a highly
defective scale of considerable thickness was formed in air, which
degradated under creep by the combined action of scale growth and
applied stresses without any strengthening effect, as observed in
steam. Therefore it was concluded that - under restriction to exposures up to 15 min - steam should not be substituted by air above
a temperature limit of about 800°C.

Introduction

The mode and extent of oxidation of compact zirconium metal in
oxidizing gases at high temperatures depend —as reported by literature [1,2] — among others on their chemical composition, and in
case of gas mixtures like air, on the concentration of the constituents. One experiment was reproduced in our laboratory in uncon-
sciousness of existing warnings [1,2]: In pure nitrogen thin continuous ZrN-layers of about 10 μm thickness were formed at 1150°C. [3]
But when a superficially nitrided specimen was subsequently exposed
to pure oxygen at 1000°C, it inflammed instantly under high heat
production, locally up to 2000°C, causing melt-down of the specimen and destruction of the quartz test section. This behaviour proved the reactivity of nitrogen containing zirconium either as a
continuous ZrN-layer or/and dissolved in the metallic matrix.

Later experiments [3] have compared the oxidation behaviour of Zircaloy 4 tube sections, at 1150°C separately exposed to steam, oxygen,
and air. They resulted in relatively similar extent of oxidation
below and at 1000°C for all gases. Above this temperature limit
pronounced differences became obvious (Fig. 1). Air was shown to
be the most aggressive environment, followed by oxygen and steam
(Fig. 2). Within 15 minutes air penetrated the whole metallic wall
thickness under formation of ZrO_2 (ZrN) and O_2-stabilized α-phase,
respectively. The specimens in oxygen and steam were attacked to
a comparatively lower extent.

Within out-of-pile engineering experiments under hypothetical
LWR-loss-of-coolant conditions, the question was raised whether
steam can be substituted by air, thus providing an equally oxidizing and more conveniently applicable environment to the test
procedure. Our decisive experiments were expanded therefore to
even more severe conditions in superposing mechanical stresses
to the oxidation reaction. Thus, surplus attack of the metallic
matrix by creep deformation and vice versa creep by scale growth
and applied tangential stresses was promoted. Comparative creep
and creep-rupture tests were performed at 800° and 1000°C in
steam and air.

Experiments

Experimental Procedure of Creep-Rupture and Creep Testing

Short sections of typical pressurized water reactor Zircaloy 4
cladding material (l = 50 mm, OD = 10.75 mm, ID = 9.3 mm) were
closed by welding on one side with an end plug, on the other side
they were connected with a tube of lower diameter for pressure supply. After external cleaning in a mixture of nitric and fluoric
acid, the capsules were washed in hot distilled water and dried.

The specimens were creep tested under defined internally pressurized conditions at the central location of a test - section in a
resistance heated furnace. This test-section could be purged either
by purified argon or - in case that the environment should be one
of the oxidizing gases - by steam or air. For creep-rupture tests
in argon, the capsules were heated up to the required test temperature and then brought up to the required internal pressure. When
the tests had to be performed in steam or air, the capsules were
heated equally in argon which was then substituted by a slow steam
or air flow. The beginning of the oxidation was indicated by a
short temperature peak, the tube capsules were then immediately internally pressurized. The specimens remained under constant internal pressure through the whole duration of the experiment, the creep
deformation was compensated automatically by pressure adjustment.
The experiments were continued until either the desired time was
reached to interrupt the experiment or rupture occurred.

The results obtained so far were plotted either as initial tangential stress versus time-to-rupture in double-logarithmic scale
(creep-rupture functions) or as maximum circumferential creep deformation versus time of exposure (creep curves) for each test temperature.

Pictures of the capsules were taken before cutting them into pieces
for post-test investigations, like metallography and scanning electron microscopy.

2.2 Experimental Results

2.2.1 Creep Experiments at 800°C

Comparative creep experiments in steam and air at 800°C showed in
principal a similar behaviour. In case of exposure to air a slight

trend towards longer rupture-life and higher creep deformation
(Fig. 3) was observed. The appearance of the deformed oxidized surfaces was equally very similar after steam and air exposure. This
kind of behaviour is in agreement with our fore-running experiments
in which no remarkable differences in weight gain and scale morphology of unstressed tubing could be detected.

Creep-Rupture Experiments at 1000°C

Comparative creep-rupture experiments in steam and air at 1000°C
however, came to quite different results (Fig. 4). The slight
trend towards longer rupture-life, which was observed at 800°C
in air, turned to shorter time-to-rupture accompanied by relatively high capsule creep deformation. Thus, the creep-rupture function gained in air was more similar to that in argon than in steam.
Only the rupture strain was influenced by dissolution of gases causing consequently matrix hardening of moderate extent (Fig. 5).

On the other hand, the results gained in steam showed that those
applied low internal pressures, which did not rupture the capsules
within about 600 s, did not burst the capsules at all. The superimposed oxidation strengthened the capsules (Fig. 6) by formation
of a multi-layer of three oxygen-containing phases so much that
very high burst pressures (around 50 bars) were necessary to produce rupture in an independent approach. This effect becomes even
more obvious when preoxidized capsules were inserted into the experiments (Fig. 4). Combined with a further gain in strength, a
sharp drop in ductility was measured.

Extent of Oxidation at 1000°C

By application of planimetric methods to the post-test evaluation
of the oxidized specimens it could be stated that the extent of
oxidation by air of ruptured tube capsulses which is plotted in
Fig. 7 as function of (rupture-) time was, compared to our results
in steam, much higher. This indicates also, that the intial tangential stress was raised significantly by the interaction of oxidation and creep straining so that the true effective tangential
stress is much higher, at least during tertiary creep.

Discussion

The oxidation of Zircaloy 4 cladding material in the temperature
range of 900 - 1300°C is a relatively well understood process which

can be described at least during the first 15 minutes duration by a parabolic rate law. This indicates that a protective and adherent oxide scale is built up which controls the progress of the reaction by oxygen ion inward diffusion and contributes to an increase of mechanical stability, measured as gain of strength and loss of ductility. Furthermore, in case of stress-induced oxide scale cracking quick metal oxidation at the tip suppresses a fast local crack growth and penetration.

On the other hand, the oxidation in air seems to proceed at $800\,^{\circ}C$ according to a similar mechanism. But at and above $1000\,^{\circ}C$ a different mode of oxidation becomes effective. Dissolution of nitrogen in the metallic matrix and/or reaction is going to form individual phases (layers) or particles of nitride[1] which are known as volume-expansive, brittle, and highly reactive. In this way, scale growth stresses are built up, oxide scales are cracked and the competitive oxidizing and nitriding attack is proceeding much faster than the only oxidizing one in steam. This trend is strongly accelerated by superimposed tensile stresses which by creep are forming the typical appearance of oxidized deformed surfaces: thick defect scales (as "islands" distributed on relatively similar distances) containing metallic particles, very thin α-layers of dissolved gases below them, indicating a quick oxide forming process accompanied by a slower gas dissolution. In fact, these defect scales do not act as strengthening "sandwich"-structures like the ones created in steam. Therefore it can be understood that the mechanical properties as measured within these experiments are similar to those in argon.

A very pronounced effect of stress-induced accelerated oxidation was found in air at $1200\,^{\circ}C$. Local cracking of the oxide scale could not be stopped by a healing process, it rather caused the penetration of a special form of corrosion attack under fast consumption of the oxygen (and nitrogen) containing brittle sublayers (Fig. 8).

This process is a self-accelerating one, since scale growth and applied stresses are causing creep and are accelerating oxidation, which create stresses and so on. Thus the typical behaviour of a at this temperature non oxidation-resistant material was revealed: the material cannot withstand its own oxide growth stresses and is exposing by creep more and more of its unoxidized matrix to the oxidizing environment.

Conclusion
____

At $800\,^{\circ}C$ a nearly similar oxidation and consequently mechanical behaviour was found for Zircaloy 4 capsules when tested in steam and

air. At and above 1000°C pronounced differences became obvious.
While steam oxidation formed an adherent "sandwich-structure" of
three oxygen-containing phases which slowed down oxidation and
strengthened the matrix, oxidation in air, supported by applied
stresses, went on much faster. Defective scales of different composition appeared to be the reason that no influence of corrosion
protection and strengthening was exerted , as observed in steam.
Scale growth and applied stresses are degradating the relatively
thick scale so heavily that it does not resist against further
creep and consequently oxidation. These results are indicating
that above 800°C steam should not be substituted by air for
LOCA-test purposes.

References

1
    J. Bénard
    "L'oxydation des métaux"
    1. Edition, p. 313/343 (1964), Paris Gauthier-Villars Editeur

2
    E.B. Evans et al.
    "Critical role of nitrogen during high temperature scaling of
    zirconium"
    Proc. Conf. Met. Soc. Aime, Boston, Mass. 1972, p. 248/281

3
    S. Leistikow et al.
    KfK 2700 (1978) p. 4200-47/63

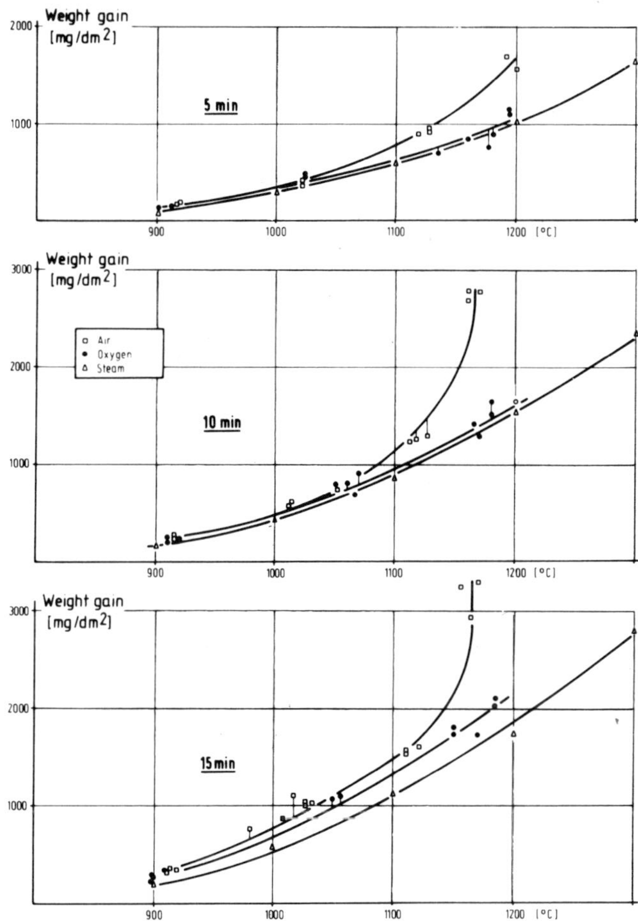

Fig. 1 Zircaloy 4/High Temperature Oxidation by Steam, Oxygen and Air as Function of Temperature and Time of Exposure (5-15 min, 900-1300°C, 1 bar)

 S. LEISTIKOW et al.

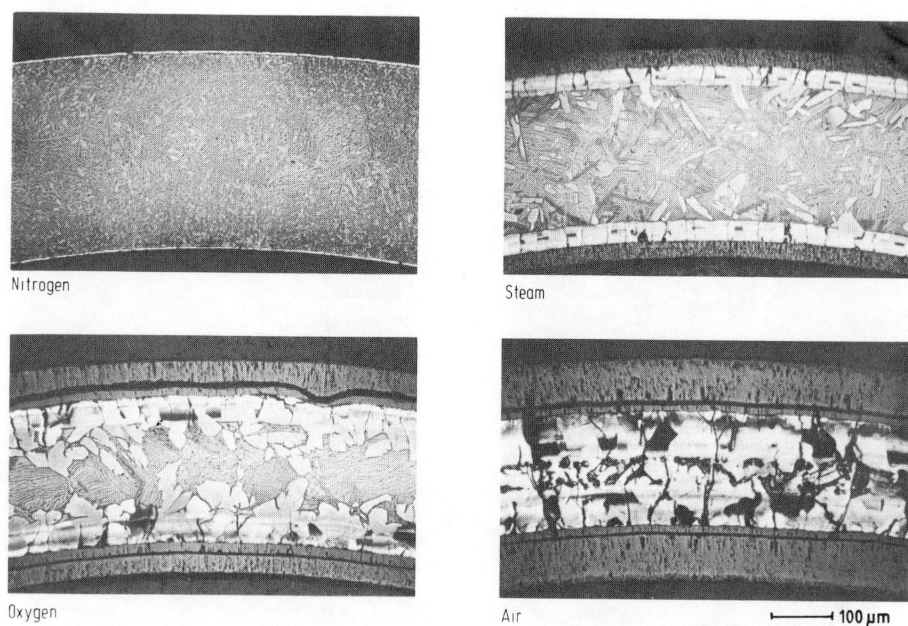

Fig. 2   Comparative Isothermal Tests of Zircaloy 4 in Nitrogen, Steam, Oxygen, and Air (15 min, 1150°C)

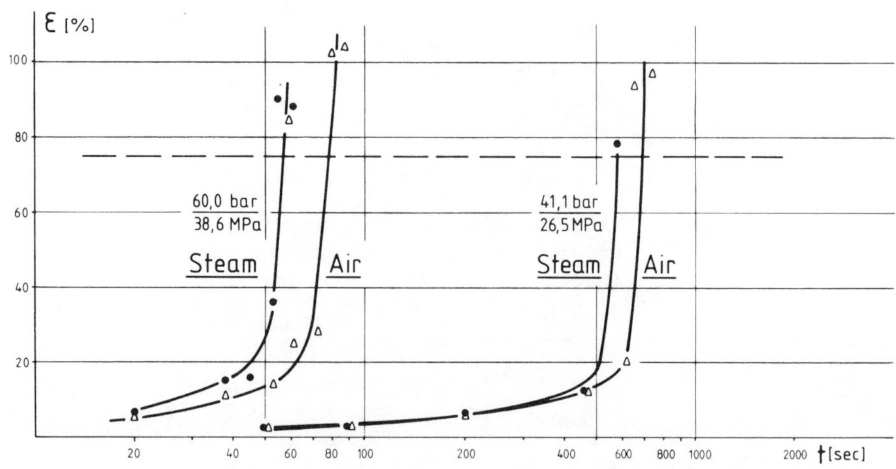

Fig. 3   Creep-Curves of Zircaloy 4 in Steam and Air at 800°C

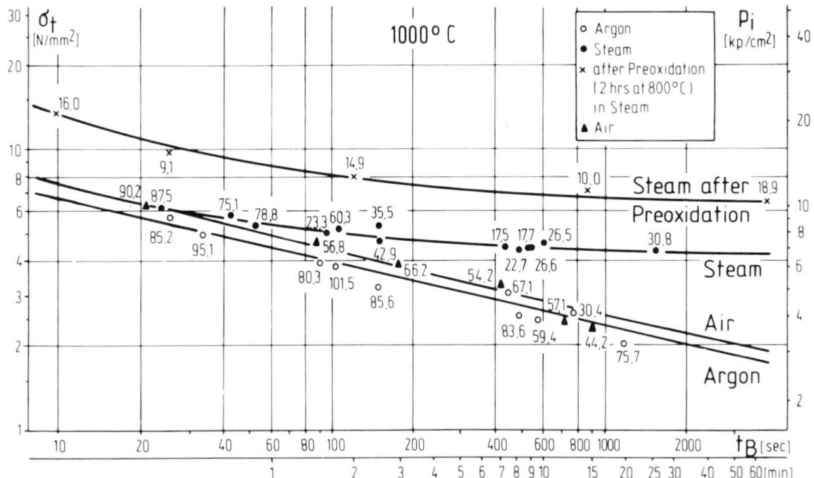

Fig. 4   Creep-Rupture Functions of Zircaloy Tube Capsules
tested in Argon and Steam without and with Preoxidation (2 hrs, 800°C) at 1000°C
(Numbers: Rupture Strain)

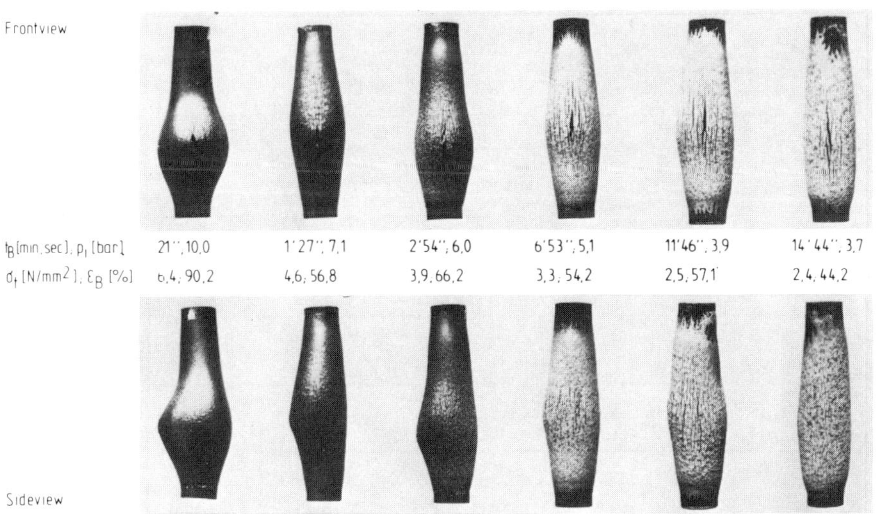

Fig. 5   Zircaloy 4 Tube Capsules after Creep-Rupture Testing
in Air at 1000°C

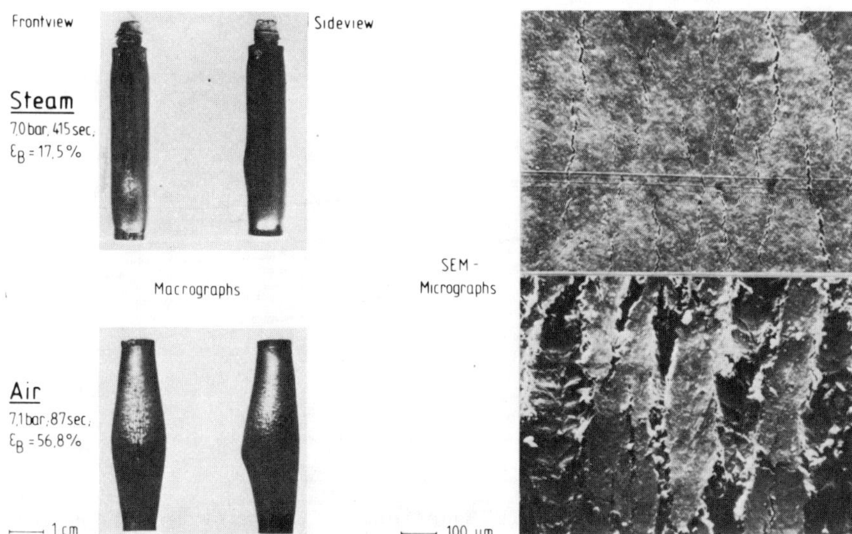

Fig. 6   Macro- and Micrographs of Zircaloy 4 Tube Capsules after Creep-Rupture Testing in Steam and Air at 1000°C

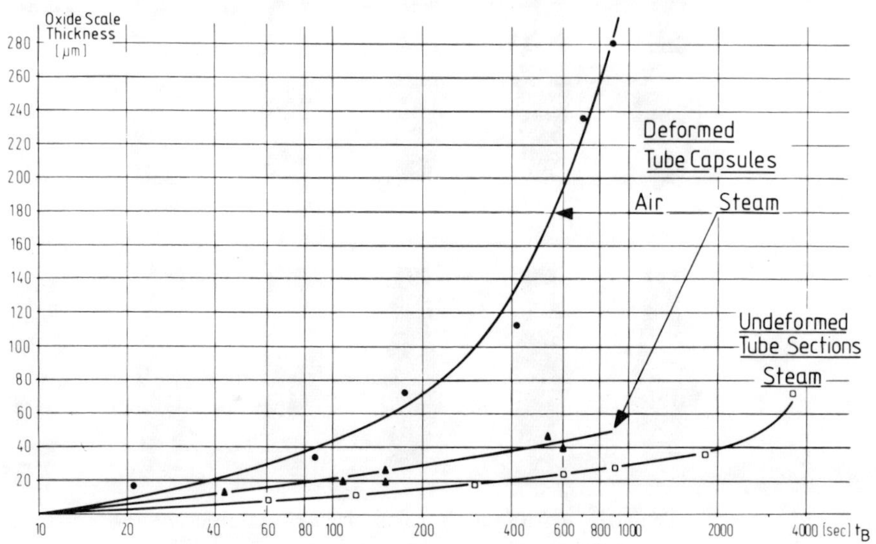

Fig. 7   Oxide Scale Thickness of Undeformed Zircaloy 4 Tube Sections and Deformed Tube Capsules after Exposure to Steam and Air at 1000°C.

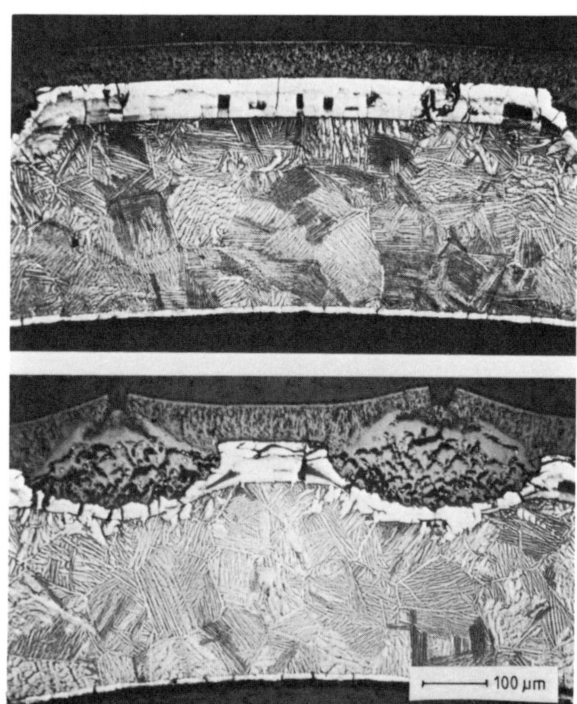

Fig. 8   Creep-Rupture Tests of Zircaloy 4
Cladding Tubes in Air (1 min 24 sec,
1200°C, 5 bar)

DISCUSSION

A. van der Linde: The Zircaloy-4 tubes used for the creep rupture
tests had a length of only 5 cm and were closed at both ends with
a welded cap. Are the results, in particular the extent of bulging,
not influenced by the presence of the relatively ductile weld
zones at the ends of the test capsule?

In view of the short duration of the tests at 1000°C, one can
think of a pre-conditioning of the capsules in steam before the
creep rupture test in air. Thus, the strengthening and ductility
loss normally occurring during testing in 1000°C steam may be
partly reproduced.

S. Leistikow: The Zircaloy 4 capsules were fabricated by closing
a tube section on one side with an end plug and by connecting the
other side to a tube for pressure supply; all done by welding. The
observed burst strain proved not to be influenced by the narrow
weld seams or the restriction in length, when compared to capsules
tested in double length or to short and full size rods.

We did no consecutive testing in steam and air. I suppose that
preconditioning in steam can moderate the subsequent attack by
air. The gain in strength depends highly on the integrity of the
oxide scale.

CORROSION INDUCED FAILURE OF INTERNALLY PRESSURIZED ZIRCALOY TUBES AT HIGH TEMPERATURES

P. Hofmann

Kernforschungszentrum Karlsruhe
Institut für Material- und Festkörperforschung
Postfach 3640, D-7500 Karlsruhe
Federal Republic of Germany

Abstract

The fuel elements of nuclear reactors are exposed to a variety of chemical and mechanical loads. In the course of nuclear fission about 30 different fission product elements are generated in the fuel. In addition during nuclear fission of the oxide fuel (UO_2) oxygen is released which likewise can react with a Zircaloy cladding material. Besides the chemical attack the cladding material has to carry mechanical load. Therefore, with respect to the behavior of fuel elements during reactor transients and accidents, it is important to know the influence of the complex chemical environment on the mechanical properties of the cladding material at rather high temperatures. For this reason, extensive creep rupture and burst tests of Zircaloy cladding tube specimens containing different simulated fission product elements or compounds, respectively, were conducted between 700 and 1100°C. It appeared that, in particular, iodine exerts a high influence on the mechanical properties of Zircaloy (service life, burst strain). While the burst strain of the cladding tubes is reduced by the effect of iodine below 850°C only, the service life is always shortened for all temperatures applied in the creep rupture tests conducted.

1. Introduction

In the course of burn-up of reactor fuel rods a considerable reaction potential builds up in the fuel elements since in addition
to the oxygen released during nuclear fission of the UO_2 fuel
about 30 different fission product elements are generated (atomic
number 34-63) [1]. The oxygen released during nuclear fission ($UO_2 \rightarrow$
2 f.p. elements + O_2) is bound by the fission products formed, the
fuel, and the cladding material. It depends on the thermodynamic
stability of the oxides, the concentration of the oxygen binding
fission products, and the initial O/U-ratio of the fuel to which
extent the oxygen will react with the different substances [2]. The
greater part of the fission product elements react with each other,
the fuel and the Zircaloy cladding material, the remainder stays in
elemental form in the fuel rod. The extent of the inner corrosion of
the Zircaloy tubing depends, besides on the temperature and time,
substantially on the burnup condition of the fuel rod, since the
burnup determines the amount of oxygen and fission products available [2].

The cladding material of Light Water Reactor (LWR) fuel rods
(Zircaloy-4; chemical composition in weight-%; Sn : 1.57, Fe : 0.22,
Cr : 0.10, Ni : < 0.035, O : 0.145, H : 0.0018, N : 0.045, Zr : remainder) is neither thermodynamically stable with respect to the
cooling medium (water or steam) and to the oxide fuel (UO_2) nor
with respect to some of the fission products. However, under normal
reactor operating conditions, the temperature of the Zircaloy cladding is too low for significant chemical interaction of the cladding material. However, this cannot be assumed a priori for much
higher temperatures, even for a short term exposure.

Both outside and inside corrosion entail a change of the mechanical
properties of the Zircaloy cladding material. Therefore, it is important to investigate the influence of the complex chemical environment in the mechanical properties of the Zircaloy in LWR transients.

2. PROBLEM DEFINITION

During a fast startup of a LWR or in case of rapid power increases
cladding tube damage may occur especially after extended reactor
operation at low or medium power levels for fuel rods with medium
and high burnups. Today, it is generally accepted that the Zircaloy cladding tube will fail due to Stress Corrosion Cracking (SCC).
However, the prerequisite is the presence of chemically reactive
fission products such as iodine, in addition to local strain concentrations in the Zircaloy tubing due to mechanical interactions
between the fuel and the cladding material. By slow power increases
and hence reduced mechanical interactions damage to the fuel rod
can be prevented [8].

For LWR transients and accidents such as ANTICIPATED TRANSIENTS
WITHOUT SCRAM (ATWS) and LOSS OF COOLANT ACCIDENTS (LOCA) where
the cladding material temperature exceeds the normal operating
temperature ($\approx 350^{\circ}$C) the question arises if a cladding tube failure has to be expected as a result of SCC. Whilst in ATWS the tensile hoop stress of the cladding tube is due to mechanical interactions between the fuel and the cladding (different thermal expansion coefficients), in LOCAs it is due to the decreasing coolant
pressure which will be exceeded by the gas pressure (filling gas,
fission gas) in the fuel rod. In both types of transients, the increase in fuel temperature also leads to a higher release of volatile fission products.

For safety considerations, it is important to know whether and how
the mechanical properties (service life, burst strain) of the Zircaloy tubing are influenced in these transients (ATWS, LOCA) by
fission products. Therefore, in the frame of the Nuclear Safety
Project, comprehensive out-of-pile experiments investigating the
stress corrosion cracking behavior of Zircaloy-4 cladding tubes
under simulated LWR transients were performed. They serve to determine the critical conditions of cladding tube failure (stress,
stress intensity factor, deformation, iodine concentration) as a
function of temperature.

3. EXPERIMENTAL PROCEDURE

The out-of-pile experiments were conducted with unirradiated Zir-
caloy-4 cladding tube specimens (10.76x0.72 mm; 60-400 mm long)
mainly using inert gas as outside atmosphere. The mechanical load
was imposed to the cladding tube sections with helium or argon gas
pressures via a capillary tube. Most of the tubular specimens were
heated by thermal radiation. Some experiments were performed with
fuel rod simulators with internal heaters in steam atmosphere. The
test designs and the experimental procedures are described in [3,4,5].

Isothermal - isobaric creep-rupture as well as temperature and
pressure transient experiments were performed between 700 and
1100°C. During the test, the internal pressure and the cladding
temperature were measured continuously. In addition the cladding
tube deformation was recorded in some tests [5]. The Zircaloy tube
material was mainly used in the as-received condition. Filling of
the specimens with simulated fission product elements or compounds,
gas-tight sealing and welding were made in gloveboxes under high-
purity inert gas.

The following substances simulating fission products have been
examined:

- Se, Mo, Cd, Sn, Sb, Te, I, Cs
- TeO_2, TeI_4, $ZrTe_2$, ZrI_4, I_2O_5, Cs_2O, Cs_2Te, CsJ

- Cs_2ZrO_3, Cs_2MoO_4.

The simulated fission product concentration varied between 0.01 and 0.50 g/cm^3 (1 cm^3 ≙ 4.3 cm^2 cladding tube inside surface). These concentrations corresponds to high burnup and/or release rates. But, one objective of these investigations was to determine if the examined fission product elements and compounds have an influence on the mechanical properties of Zircaloy tubing exposed to transient conditions at all. In a later step the critical concentrations are being determined; first results are described in [2,6].

4. EXPERIMENTAL RESULTS

As appears from the results of the out-of-pile experiments, all examined simulated fission product elements and compounds caused some change of the burst strain of the Zircaloy-4 cladding tubes compared with reference specimens filled with helium only (fig.1). However, only the test specimens containing elemental iodine or iodine compounds (TeI$_4$, ZrI$_4$, I$_2$O$_5$), except CsI, exert a strong influence on the burst strain. With respect to the time-to-failure only in the creep rupture tests iodine also influences the service life of the tube specimens. Whereas for the as-received specimens only a minor shortening of the time-to failure occurs for internally notched or precracked specimens a considerable reduction of the time-to-failure was determined due to the action of iodine (fig.2).

Since only iodine or volatile iodine compounds have a strong influence on the burst strain and partly also on the service life of the Zircaloy tubing and since the action of the iodine compounds (except CsI) is comparable to that of iodine the following statements are restricted solely to the influence of iodine on the mechanical properties of Zircaloy. The influence of iodine on the burst strain and time-to-rupture of internally oxidized or notched Zircaloy tube specimens as well as on specimens containing UO$_2$ is described in [3,4].

4.1 TEMPERATURE AND PRESSURE TRANSIENT BURST EXPERIMENTS

The results of transient tests measuring burst temperature as a function of pressure indicated no significant differences between specimens containing iodine and the reference specimens free of iodine (fig.3). Also with respect to time-to-failure (t_{20^oC} → trupture: 70-300 s) significant differences cannot be noticed [3,4].

The burst strains, however, of the as-received Zircaloy cladding specimens, tested in inert gas atmosphere were substantially smaller for the specimens containing iodine than for the iodine-free reference specimens for burst temperatures below 850°C. While at low burst pressures (< 40 bar) and associated high burst temperatures (> 850°C), practically no or only small differences occur in the maximum circumferential burst strains, significant differences exist at burst temperatures below 850°C associated with burst pressures above 45 bar (fig.4). The burst strains of the specimens containing iodine vary between 10 and 40% as compared to about 60 to 100% for the reference specimens free of iodine.

The hoop stress at rupture of the Zircaloy cladding tubes is defined taking into account the burst strain.For the specimens containing iodine it can be recognized that this rupture stress below 850°C is smaller than for the iodine-free reference specimens (fig.5). The hoop stress so determined might be the critical stress in the cladding material which, in the presence of corrosive media, such as iodine, results in SCC and hence in the failure of the cladding tube. At temperatures above 850°C the presence of iodine exerts practically no influence on the deformation and rupture behavior of the Zircaloy cladding tubes. Most probably the critical stresses in the cladding tube required for failure are no longer obtained at those elevated temperatures, since the stresses are relaxed by plastic deformation before reaching a critical value. The extent to which the β-phase formation above 850°C influences the rupture has not yet been clarified.

4.2 ISOTHERMAL-ISOBARIC CREEP RUPTURE TESTS

In addition to the influence of iodine below 850°C on the burst strain and stress at rupture the creep-rupture tests revealed an influence of iodine on the service life in the whole temperature range investigated (700-1000°C).

With respect to the burst strains the test results found already in the transient tests have been confirmed, i.e. below 850°C the burst strains of tubular specimens containing iodine are markedly smaller than those of iodine-free reference specimens (fig.6). For the reference specimens the burst strains ranged from 85 to 115 % at temperatures of 700 to 800°C, whereas the burst strains of specimens filled with iodine varied between 15 and 50% at the same temperatures (fig.7).

Metallographic investigations of the burst specimens clearly show that at 700 and 800°C the cladding tube will fail in the presence of iodine without local necking (fig.8). Moreover, the micrographs show that the rupture is intergranular. Other intergranular incipient cracks were also observed at the internal circumference of the cladding tube. Above 800°C, however, local plastic deformations first takes place at the circumference of the cladding tube before rupture occurs.

For the hoop stress at rupture defined in the previous section
similar results are obtained for the isothermal-isobaric tests as
for the temperature and pressure transient tests. However, in the
creep rupture tests the critical hoop stresses causing SCC cladding failure in the presence of iodine are markedly lower than in
the transient burst tests [3,4].

Creep rupture tests with various initial iodine concentrations
(0.01-100 mg/cm^3) in the tube specimens reveal that a low ductility failure of the Zircaloy tubing as a result of SCC occurs only
after a temperature dependent critical iodine concentration has
been exceeded. The critical initial iodine concentration amounts
to about 0.3 mg/cm^3 (0.07 mg/cm^2)* at 700°C (fig.9). Above the
critical iodine concentration the burst strains of the as-received
Zircaloy cladding tubes will,practically independent of the iodine
concentration, be strongly reduced to low values (fig.9). By contrast, at temperatures of 900°C and above the influence of iodine
with any concentration on the deformation and rupture behavior of
the cladding tubes can be neglected.

5. SCANNING ELECTRON MICROSCOPY OF ZIRCALOY BURST SPECIMENS

In order to determine the mechanisms causing cladding failure the
burst specimens containing iodine as well as the iodine-free reference specimens were examined systematically by means of a
Scanning Electron Microscope (SEM).

Evidently the presence of iodine will lead to quite a number of incipient cracks on the cladding tube inner surface at temperatures
below 850°C, even with specimens not pre-cracked nor subjected to
internal pre-oxidation (figs. 8 and 10). A representative inner
surface and fracture surface of a Zry-4 cladding tube containing
iodine after rapid failure due to pressurization are shown in
fig.10. Numerous crack initiation sites are observed on the internal cladding surface. The fracture has nucleated at the surface
exposed to the corrosive environment. The first step of the attack
is always intergranular cracking, presumably because less energy
is required than for transgranular cleavage. There are two well defined types of fracture features, namely intergranular cracking,
and ductile shear. Sometimes a transition range exists in the fracture surface where the crack path gradually changes from intergranular to transgranular before ductile shear occurs. This change

---

*   An iodine concentration of 0.07 mg/cm^2 corresponds to the fission iodine generated in a LWR fuel rod after a burnup of a-
   bout 0.5 at.% ($\approx$ 4500 MWd/t), assuming total release of iodine
   from UO_2 in elemental form.

of fracture mode takes place because the thickness of the remaining tube wall decreases but still has to sustain the internal load. The stress finally exceeds the tensile strength and the remaining cladding wall instantaneously fails in a ductile mode (fig.10).

In absence of iodine the fractography shows an entirely ductile mode of failure independent of burst temperature. The same holds for burst specimens containing iodine for temperatures above 850°C [3,4].

6. DISCUSSION AND CONCLUSIONS

Both the temperature and pressure transient burst tests and the isobaric-isothermal creep-rupture tests unambiguously show that up to about 850°C Zircaloy-4 tubing can fail due to SCC by iodine. The burst strains of as-received tube specimens containing a sufficient amount iodine are much lower than those of iodine-free reference specimens. The burst strain of iodine bearing specimens decreases with decreasing temperature and increasing burst pressure. Above 850°C, however, practically no influence of iodine is observed on the burst strain of the Zircaloy tubing, even at very high iodine concentrations.

In the transient tests, no significant differences could be found in the time-to-failure, between specimens containing iodine and iodine-free specimens. However, in the isothermal-isobaric tests the presence of iodine accelerated the failure of the Zircaloy due to SCC. Obviously, in the creep-rupture tests Zircaloy was strongly attacked by iodine during the pressureless heatup period which, under the subsequently applied stress, considerably shortended the period of crack formation (incubation period); but in the transient burst experiments these processes take place simultaneously.

Below 850°C the Zircaloy-4 cladding tubes fail in the presence of iodine as a result of SCC at much lower tensile hoop stresses compared to reference specimens free of iodine. Correspondingly, internally notched specimens containing iodine suffer from SCC at much lower nominal stresses and at much shorter time-to-failures than specimens without notches.

The appearence of SCC rupture is almost identical with that of brittle fracture. To interpret SCC of Zircaloy by iodine, it is assumed that the major role of iodine is to reduce the energy of the atomic bonds at the crack tip [7]. The reduction in binding strength between the atoms is a result of chemical sorption of iodine. This process allows brittle crack propagation and leads to failure of the material at a reduced stress. Fig.5 clearly shows that below approximately 850°C the rupture stress of the cladding material in the presence of iodine is markedly lower than in iodine-free reference specimens.

The safety related significance of these results is that through the action of the fission product iodine the burst strain of the Zircaloy cladding material during a LOCA transient may be strongly reduced at temperatures up to about 800°C. As a consequence the possibility of reduced coolability of the reactor core by the energency core cooling system may be diminished. However, this is probably effective for medium and high burnup fuel rods only.

7. ACKNOWLEDGEMENTS

I wish to thank Professor Dr. W. Dienst and Dr. S. Dagbjartsson for their critical suggestions during the preparation of the manuscript. I also gratefully acknowledge the assistance of Mr. H. Metzger in performing the experiments and the metallographic investigations, and Mr. J. Burbach for the SEM-investigations.

The paper was sponsored by the "Projekt Nukleare Sicherheit", KfK.

8. REFERENCES

1. P. Hofmann,   KfK-Ext. 6/70-2 (1970)

2. P. Hofmann,   KfK-2785 (1979)

3. P. Hofmann, Journ. of Nucl. Mat. (1979), Vol.87, No.1, pp.49-69

4. P. Hofmann,   KfK-2661 (1978)

5. H. Lehning, K. Müller, D. Piel, L. Schmidt,   Jahrestagung "Kerntechnick 1980", Berlin, 25-27th March (1980), pp.231-234

6. P. Hofmann, 5th International Conference on "SMiRT", Berlin, 13-17th August (1979)

7. H.J. Engell, M.O. Speidel,   Korrosion 22, Verlag Chemie, Weinheim (1969)

8. ANS Topical Meeting on "Water Reactor Fuel Performance", 9-11th May (1977), St. Charles, Illinois, USA

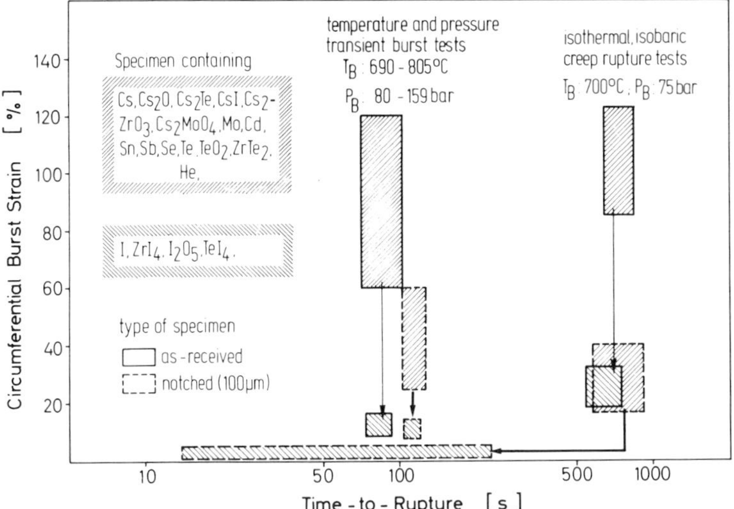

Fig.1: Effect of simulated fission products on the burst strain
and time-to-rupture of as-received and notched Zircaloy-
4 tubing in He. In the presence of I, ZrI_4, I_2O_5 or TeI_4
the burst strain of the cladding tubes is strongly reduced. In the case of internally notched specimens also
the time-to-failure is markedly shortened under creep
rupture test conditions.

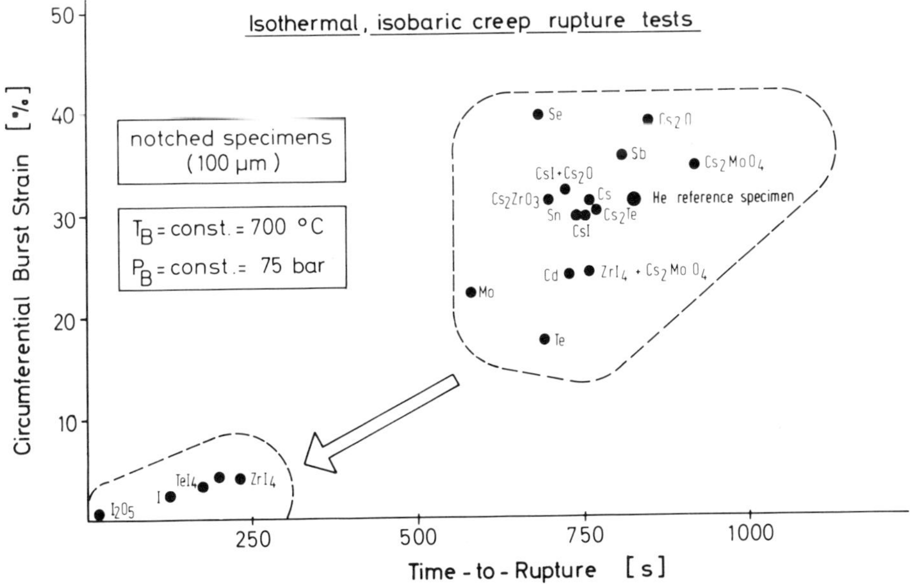

Fig.2: Effect of simulated fission products on the burst strain
versus time-to-rupture behavior of internally notched
Zircaloy-4 tubing at 700°C in He.

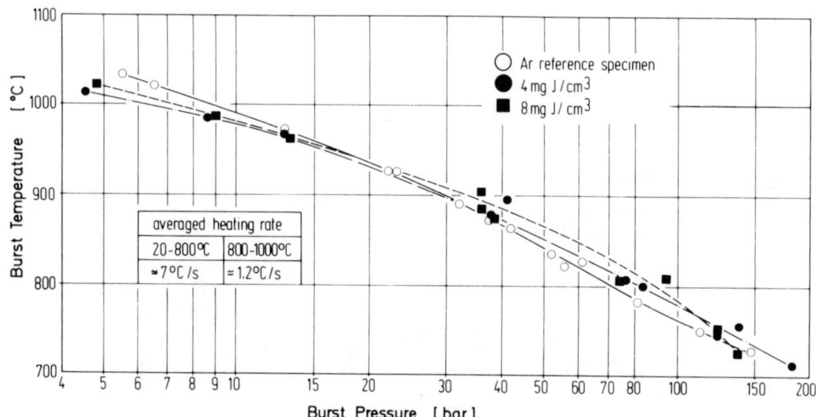

Fig.3:   Effect of iodine on the burst temperature versus burst
           pressure behavior of as-received Zircaloy-4 cladding tube
           specimens (temperature and pressure transient burst tests).

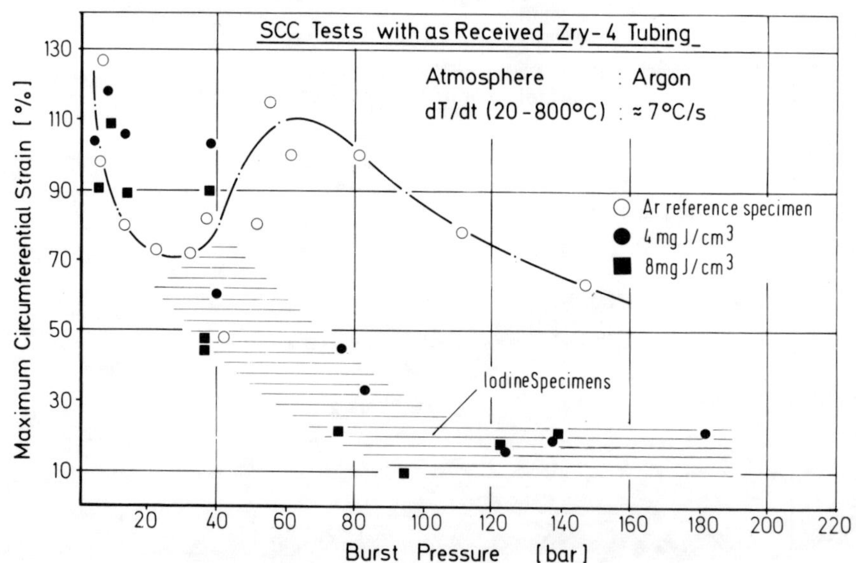

Fig.4:   Circumferential burst strain as a function of burst pres-
           sure of as-received Zircaloy-4 tubing; influence of io-
           dine on the burst strain (temperature and pressure tran-
           sient burst tests).

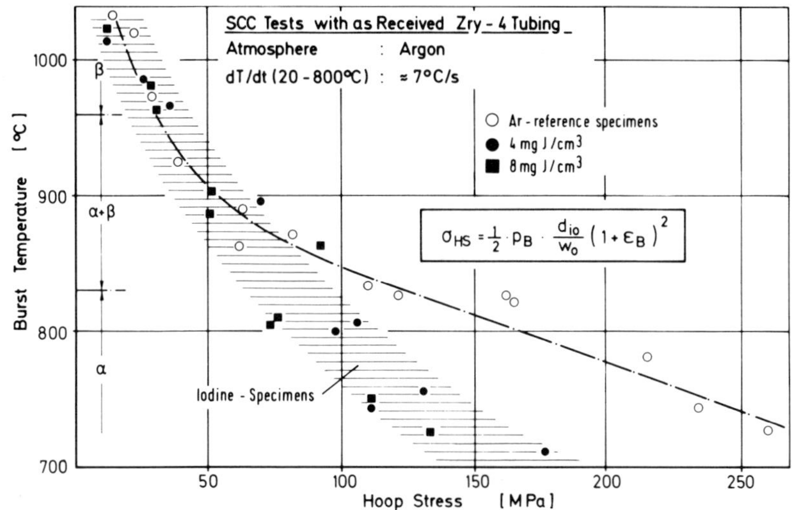

$$\sigma_{HS} = \frac{1}{2} \cdot p_B \cdot \frac{d_{io}}{w_o} \left(1 + \epsilon_B\right)^2$$

Fig.5: Effective hoop stress at rupture versus burst temperature for as-received Zircaloy-4 tubing; influence of iodine on the stress at rupture (temperature and pressure transient experiments).

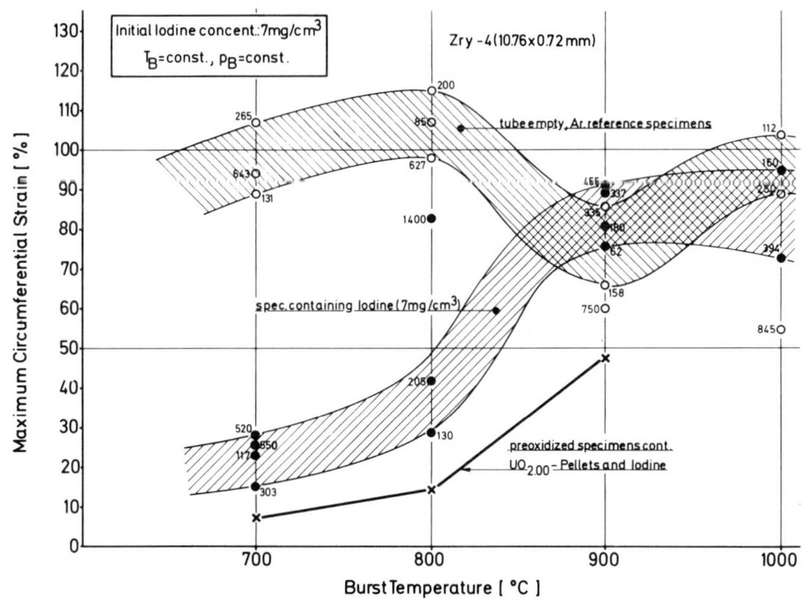

Fig.7: Circumferential burst strain versus temperature of as-received Zircaloy-4 tubing internally pressurized with argon or argon-iodine-gas mixtures (with the indication of the endurance times).

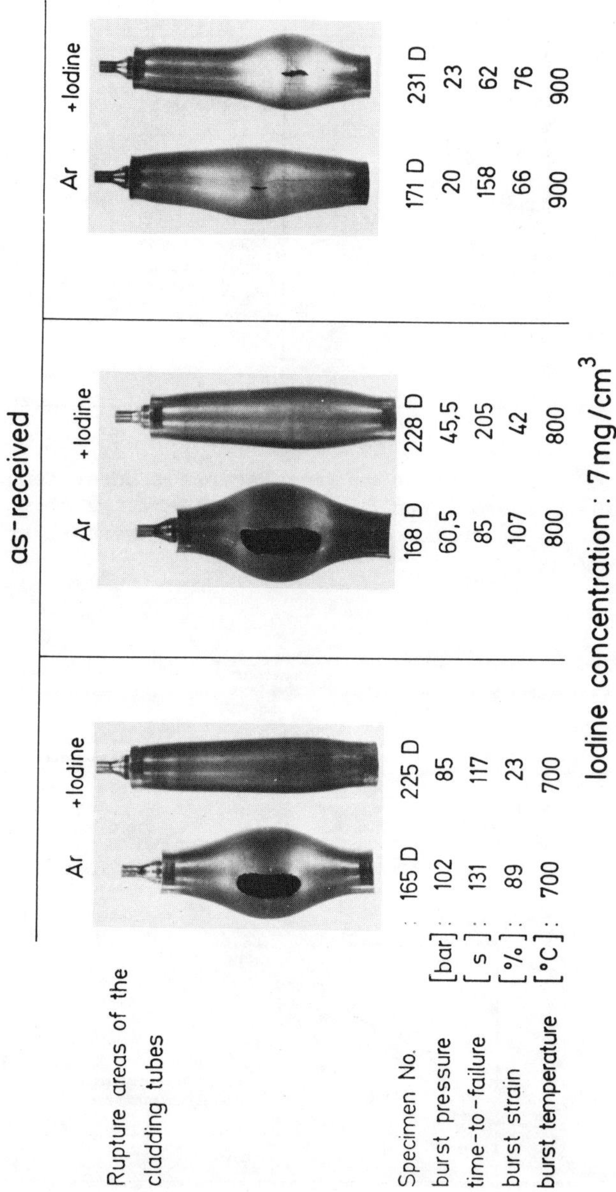

Fig.6: As-received Zircaloy-4 cladding tube specimens after creep-rupture tests in Argon at 700, 800 and 900°C. Influence of iodine on the burst strain and time-to-failure.

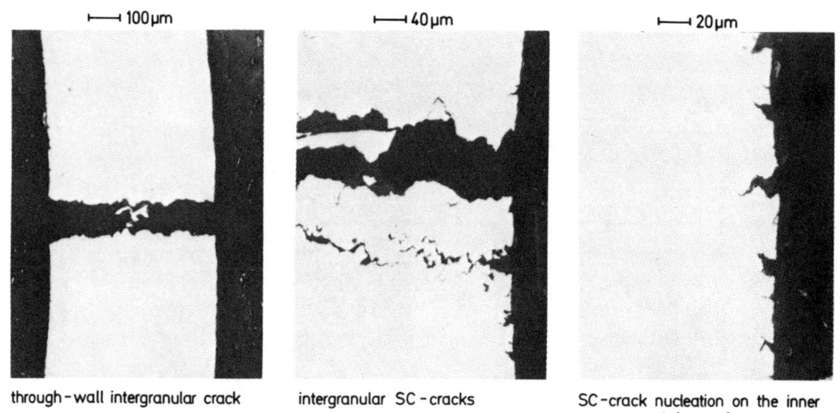

229 D : $T_B = 800°C$, $p_B = 50,7\,bar$, $t_B = 130\,s$, $\varepsilon_B = 29\%$

iodine concentration: $7\,mg/cm^3$

Fig.8:   Stress-corrosion cracks in as-received Zircaloy-4 tubing failed under argon-iodine-gas pressurization. No local necking takes place at the point of rupture (intergranular SCC path).

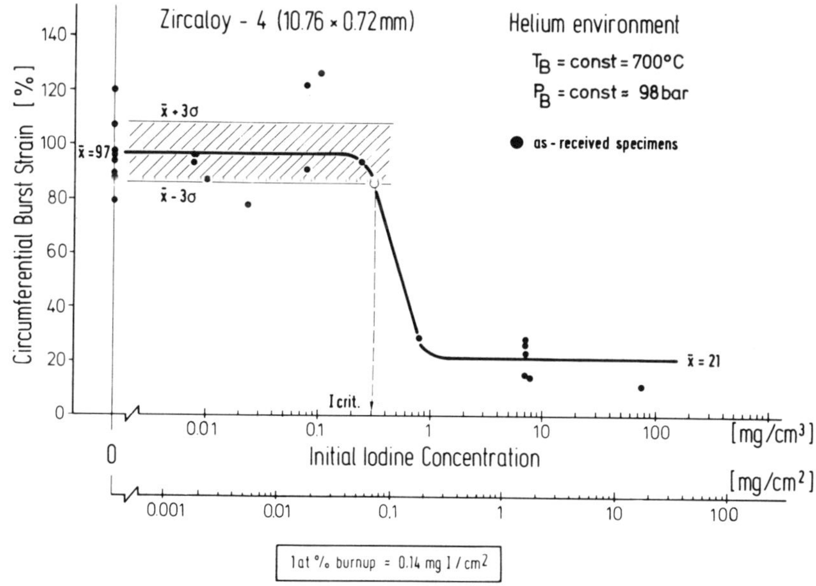

Fig.9:   Influence of the initial iodine concentration on the burst strain of as-received Zircaloy-4 tubing at 700°C in He. Above the critical iodine concentration the burst strain of the tubing will undergo a strong reduction to low values.

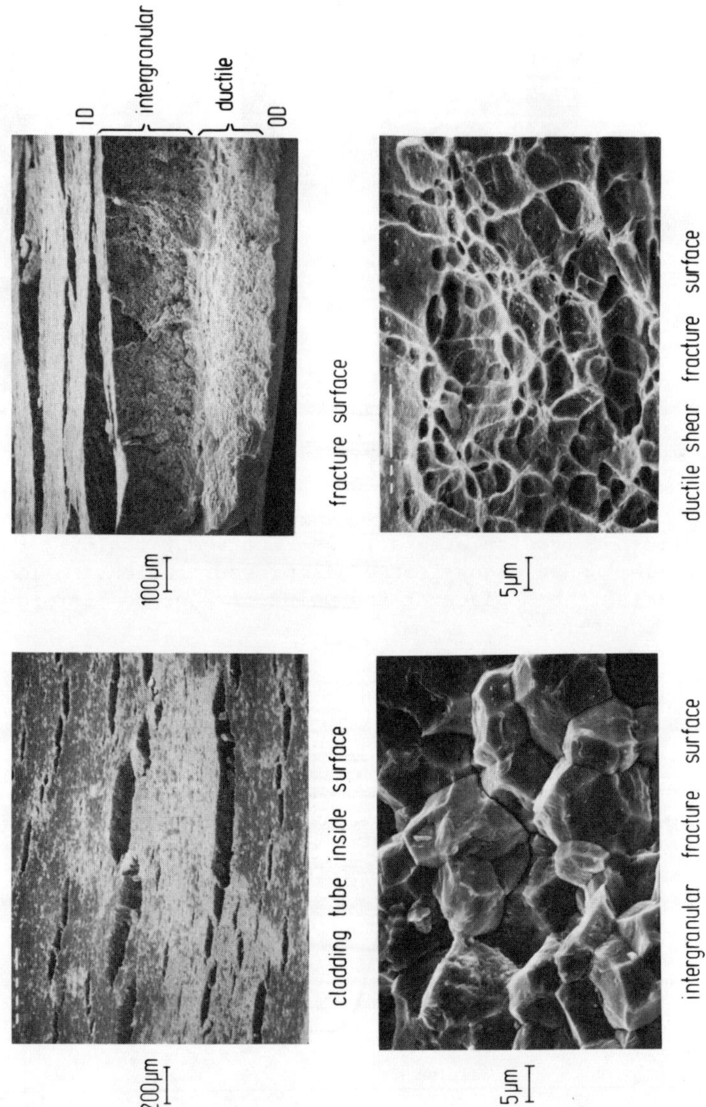

Fig.10:   Inside surface and fracture surface of as-received Zir-
            caloy-4 tubing after failure under argon-iodine-gas pres-
            surization. The cracks on the inside surface form during
            the creep rupture test.

DISCUSSION

T. Boniszewski: The disappearance of the detrimental effect of
iodine at 900°C as opposed to the embrittling effect at $700\text{-}800^{\circ}$C
on Zircaloy-4 can be interpreted purely in chemical terms rather
than the effect of temperature on metal ductility. An aggressive
species will help the crack growth only if it can react to form
an atomic compound at the crack tip and hence to 'break' the metal
in this way. This is what happens in hydrogen embrittlement up to
$200\text{-}300^{\circ}$C in steel. Above a certain temperature, the 'compound'
which would result from the reaction between the metal and the
aggressive species becomes unstable. Therefore, at 900°C, iodine
ceases to be detrimental.

P. Hofmann: In addition the following effects can be responsible
for the disappearance of the influence of iodine on the deformation and burst behavior of Zircaloy-4 tubing at temperatures above
850°C:

a) the critical stresses at the crack tip required for crack
   growth and hence failure of the tubing are no longer attained
   at the elevated temperatures, since they are reduced by plastic
   deformation (SEM studies of the internal surface of the tubing
   after rupture revealed many small cracks which did not grow;
   the tubing failed due to local necking);

b) the phase change of the Zircaloy-4 (α-phases $\rightarrow$ β-phase) above
   850°C has possibly also an influence.

B. Ilschner: With relation to the question concerning the mechanism of interaction between the corrosive medium and Zircaloy,
it is my impression that the results presented by Hoffmann are in
good agreement with results obtained in environment-assisted
subcritical crack growth. Probably, the halogen assists transport
flow between the highly stressed areas very close to the crack tip
towards the stress-free crack surfaces further away. This would
increase crack growth.

Even high temperature plasticity would not essentially change
this picture, since there will always be a flow stress present at
the crack tip.

The Influence of Oxidation on the High-Temperature Creep Behaviour of an Fe-1Si Alloy

by

R. Rolls and M.H. Shahhosseini

Joint University/UMIST Metallurgy Department,
Manchester, U.K.

SYNOPSIS

The influence of preoxidation and continuous oxidation on the creep behaviour of an Fe-1Si alloy at 823 and 853K at constant tensile stresses of 4 and 8 MN m^{-2} has been studied. Preoxidation and additional oxidation during tests was found to increase the transient creep rate and to decrease the steady-state creep rate. At higher temperatures (973 to 1073K) and higher stresses (14 to 21 MN m^{-2}) the steady-state creep rate increased with increasing applied stress, temperature and oxygen partial pressure (10^{-7} to 10^{-2} N m^{-2}). At 998K with higher oxygen pressures (6 to 10 KN m^{-2}) creep rates decreased with increasing oxygen pressure. The weakening effects of oxidation were primarily a consequence of oxidation-induced vacancies assisting dislocation movement in the substrate. Strengthening by oxidation was due to the combined effects of dislocation pinning at internal oxides, blocking of dislocation egress by the oxide scale and the mechanical constraint imposed by the strongly adherent scale, which comprised layers of wustite and fayalite.

1.0   INTRODUCTION

The influence of oxidation on the creep behaviour of metals and alloys can have striking and unexpected results. In some cases a strengthening effect is observed when an adherent oxide film acts as a barrier to dislocation egress from a substrate, and also when the film imposes a mechanical constraint to substrate deformation. In other instances a weakening effect may result if an adherent film facilitates oxidation-induced vacancy injection into the substrate or when film growth stresses augment an applied

tensile stress in the substrate. Furthermore, in systems under-
going internal oxidation, interactions may arise between strengthening particles and mobile defects like vacancies and dislocations.
The outcome of such interactions may of course depend upon whether
the predominant deformation mechanism is diffusional or dislocation
creep. Many of these aspects of the creep of metals in corrosive
environments have been critically considered in several recent
reviews.[1,2,3]

The present work forms a continuing part of a detailed study
of the influence of oxidation on the creep behaviour of iron-base,
binary alloys. It sets out to establish the specific contributions
of simple and complex oxide films and scales, and internal oxides,
which are associated with different alloying elements, to particular
creep regimes (diffusional and dislocation). Earlier studies[4]
indicated that at 823K for an applied tensile stress of 4 MN m^{-2},
α-iron exhibited an increased creep rate with increasing oxygen
pressure from 10^{-6} to 133 N m^{-2} when a surface film of magnetite
was present. At higher temperatures (923 to 1123K), α-iron showed
a decreased creep rate with increasing oxygen pressure with a surface film of wustite present. Further work[5] showed that for constant stress tests up to 16 MN m^{-2} in the 823 to 1223K range, the
steady-state tensile creep behaviour of high-purity iron followed
a linear relationship of the form:

$$\log \dot{\varepsilon}_{ss} = a \log p_{O_2} + b$$

where a and b are constants depending on temperature, oxide film
thickness and composition. In general, the steady-state creep rate
($\dot{\varepsilon}_{ss}$) increased with increasing oxygen pressure (p_{O_2}).. The effect
of oxidation enhanced the influence of temperature on ε_{ss}. This
seemed to be true irrespective of the creep deformation mechanism,
although Tien and Davidson[3] have expressed the view that one would
not expect a surface environment to result in a drastic change of
the thermally activated processes of dislocation motion, which are
usually rate-controlled by vacancy self-diffusion. Nevertheless
our values[5] for the stress exponent n and for the activation energy
Q in relationships of the general form for power-law creep:

$$\dot{\varepsilon} = A \sigma^n \exp (-Q/RT)$$

also showed a dependence on oxygen partial pressure.

It was anticipated that for an Fe-1Si alloy, oxide film composition, internal oxidation and solute strengthening effects would
promote some differences in creep behaviour when compared with high
-purity iron. This knowledge would enable an assessment to be made
of the extent to which the elasticity and/or plasticity and
consequently the adhesion of films of magnetite or wustite (as
modified by silica or fayalite) significantly weakened or

strengthened the substrate in a particular creep regime. Both
preoxidation before testing and continuous oxidation during testing were examined with a view to isolating the separate effects
associated with oxidation products and oxidation mechanisms respectively.

2.0   EXPERIMENTAL PROCEDURE

Details of the creep testing facility for ultra-high vacuum
(UHV) and controlled atmosphere conditions (Fig. 1) used in this
investigation have been described elsewhere[6] and therefore only a
summary of the principal steps of the procedure is now presented.

Creep test specimens of Fe-1Si (machined to 25 mm gauge
length x 5 mm diameter) with electropolished surfaces were finally
cleaned in the test rig by adapting the UHV system to permit
hydrogen reduction of the air-formed surface oxide film before
commencing a test. The partial pressure of oxygen in the test
chamber was then adjusted to lie in the range 10^{-7} to 10^{-2} N m^{-2}
to form either a surface film of magnetite (below 843K) or wustite
(above 843K) on the specimens. Initially, surface oxide film
thickness of 2 to 3 µm were produced by varying the time of oxidation before creep testing.

A preoxidation period of 6 hours was used before creep tests
at 823 and 853K respectively under constant tensile stresses of 4
and 8 MN m^{-2}. Additional oxidation periods of 3 hours in the
transient and at the start of the steady-state creep stages were
also included. In each instance oxidation took place under zero
applied stress conditions. The minimum detectable creep strain
was 5 x 10^{-3} per cent.

Further studies are continuing at higher temperature (973 to
1073K), higher stresses (14 to 21 MN m^{-2}) and higher oxygen pressures (6 to 10 kN m^{-2}).

Sectioned test specimens were examined by conventional optical
and electron microscopy techniques. Optical metallographic specimens were etched in a 3 per cent Nital solution.

3.0   RESULTS

3.1.   Creep behaviour at 823 and 853K

The results indicated in Fig. 2 confirm the expected influence
of stress on creep at 853K during continuous oxidation at an
oxygen pressure of 10^{-5} N m^{-2}.

Pre-oxidation for 6 hours at 823 and 853K respectively,
before testing in continuously oxidizing conditions was found to

increase the transient creep rate and to decrease the steady-state
creep rate when compared with tests performed with no pre-oxidation
stage.

In continuously oxidizing conditions, the effect of an
additional 3 hour oxidation period following an interruption of
the transient creep stage (in tests performed at 823K under a
stress of 4 MN m^{-2}) is shown in Fig. 3. This indicates that an
increase in the transient creep rate occurred when testing was
resumed.

When, however, an additional 3 hour oxidation period was
introduced at the start of the steady-state creep stage at both
823 and 853K with constant stresses of 4 and 8 MN m^{-2} respectively,
a decrease in creep rate was found when the test was resumed, as
shown in Figs. 4 and 5.

Metallographic examination of tested specimens showed that
some recovery and recrystallization of the ferrite grains had
occurred. Both internal oxidation (SiO_2) and cavitation of the
substrate was pronounced after the lengthy preoxidation and
additional oxidation periods. Fig. 6 shows the transverse microstructure of a specimen tested at 8 MN m^{-2} at 853K with a strain
of 5.5 per cent in 30 hours. A longitudinal section showed an
identical structure of equiaxed ferrite grains with finer
recrystallized grains throughout the bulk section. In contrast,
for a specimen tested at the lower stress of 4 MN m^{-2} at 853K to a
strain of 4 per cent, elongated grains can be seen in Fig. 7
representing a longitudinal section of the bulk structure.

In order to establish qualitatively whether the density and
distribution of dislocations differed in the near-surface and bulk
centres respectively of specimens tested at 823K to a strain of 4
to 5 per cent, thin foils were prepared by chemical thinning and
examined by transmission electron microscopy. The electron micrographs (Figs. 8 and 9) indicate significant differences in the dislocation features. For example, Fig. 8 shows a greater number of
subgrain boundaries and finer dislocation networks in the near-
surface or substrate region compared to the bulk centre zone
depicted in Fig. 9, which shows relatively coarser grains and an
absence of dislocation networks. Some dislocation pinning by an
internal oxide particle can also be seen in Fig. 8.

3.2. Creep behaviour at 973 to 1073K

Initial experiments have indicated that the steady-state creep
rate at these higher temperatures of 973 to 1073K also showed an
increase during continuous oxidation when the oxygen pressure was
increased from 10^{-7} to 10^{-2} N m^{-2} at a constant stress of 16 MN m^{-2}.
For constant oxygen pressures of 10^{-7} and 10^{-3} N m^{-2} respectively,

with an applied stress of 16 MN m^{-2}, the steady-state creep rate
increased as the temperature was increased from 973 to 1073K.
From these data, which fitted the general equation for power-law
creep:

$$\dot{\varepsilon} \;=\; A\,\sigma^{n}\,\exp\,(-Q/RT)$$

activation energies of Q $\sim$ 357 kJ mol^{-1} (at 10^{-5} N m^{-2} oxygen
pressure) and Q $\sim$ 378 kJ mol^{-1} (at 10^{-3} oxygen pressure) were
calculated for the 973 to 1073K temperature range under an applied
stress of 16 MN m^{-2}.

For the latest tests performed at 998K at higher oxygen
pressures of 6 to 10 kN m^{-2}, evidence of strengthening effects has
been observed in decreased creep rates with increasing oxygen
pressure. From data obtained in tests at 998K with applied
stresses of 14 to 21 MN m^{-2} and an oxygen pressure of 10 kN m^{-2}, a
value of n $\sim$ 9 for the stress exponent was calculated.

3.3. Microstructural features from specimens tested at 998K

(i) Uncrept specimens which were oxidized in atmospheres with
oxygen pressures of 6 kN m^{-2} for 5 hours, showed a near-surface,
substrate ferrite grain size of $\sim$5 µm in a band of depth $\sim$10 to 15
µm (Fig. 10) compared to the remainder of the substrate and bulk
structure grain size of 39 µm together with some internal oxidation. The external scale comprised wustite and fayalite layers.

(ii) The above differences in grain size were not observed when
uncrept specimens were oxidized for similar times in atmospheres
with a considerably lower oxygen pressure of 10^{-7} N m^{-2}.

(iii) The extent of ferrite grain growth in crept specimens (from
37 µm before testing to 50 µm after testing) at a constant stress
of 16 MN m^{-2} and with an oxygen pressure of 6 kN m^{-2} was greater
than that for similarly oxidized but uncrept specimens (39 µm
grain size after oxidation).

(iv) For specimens crept in atmospheres with oxygen pressures of
10 kN m^{-2}, internal oxidation (primarily SiO$_2$) was apparent for a
depth of $\sim$10 µm below the scale/metal interface. The external
scale comprising magnetite, wustite and fayalite layers was from
20 to 50 µm thick (Fig. 11).

(v) In uncrept specimens, similarly oxidized, internal oxidation
extended over a greater depth ($\sim$30 µm) and was primarily fayalite
along grain boundaries. The external scale thickness was from 20
to 30 µm.

(vi) Pronounced subscale formation occurred during oxidation at

1023 and 1073K, although the external scale adhesion became weaker.

4.0   DISCUSSION

The present work has confirmed the widely held view that the influence of oxidation on steady-state creep behaviour can in some instances cause a strengthening and in other instances cause a weakening effect. It is the purpose in this section to consider the phenomena and probable mechanisms associated with these contrasting modes of behaviour for the Fe-1Si alloy.

In the lower temperature tests at 823 and 853K, preoxidation produced a strengthening effect, as did an additional period of oxidation at the start of the steady-state creep stage (Figs. 4 and 5). There is support[1,2,3] for the claim that a surface layer of oxide can act as a dislocation barrier and consequently reduce the creep rate. But Hales et al[7] found that during the oxidation of magnesium the introduction of a surface supersaturation of vacancies, thereby opposing the vacancy flux of diffusional creep, could also reduce creep rate. The possibility that oxidation-induced vacancy injection can cause strengthening at low strains by a cluster-hardening or restrict grain boundary movement by void pinning at higher temperatures has also been discussed.[8] Recent work[9] on aluminium has indicated that the strength of the oxide, if strongly adherent, can exert a restraining effect whether the deformation mechanism is diffusional creep or dislocation creep.

Microstructural evidence for the Fe-1Si alloy shows some cavitation in the substrate. It also is apparent from the micrographs that a partial dynamic recovery/recrystallization process has been operative (Fig. 6) and the electron micrograph (Fig. 8) confirms that there has been a greater degree of dislocation activity near the oxide/metal surface region than in the bulk centre of specimens (Fig. 9). For the applied stress/temperature regime of the present work, the predominant creep mechanism is believed to be dislocation creep. Stang et al.[10] have shown that the steady-state creep rate for Fe-3Si is diffusion controlled and exhibits a power law stress dependence in the temperature range 1393 to 1678K. Our own Fig. 2 confirms the stress influence on the Fe-1Si alloy.

There was ample evidence of general internal oxidation in all the oxidized test specimens. For scales of 2 to 3 µm thickness oxide adhesion was strong. It is probable therefore, that if a vacancy injection mechanism had operated during creep under continuous oxidation conditions, additional internal oxidation from the long preoxidation period may have provided sinks for the vacancy flux, thereby minimising void formation, or the vacancy flux could have enhanced dislocation climb. A critical examination

of vacancy injection has been recently made by Harris[11] which
suggests that alternative explanations, like stress-induced void
growth, can account for the observed phenomena. We conclude,
nevertheless, that the principal cause of the observed strengthening is the combined effect of the mechanical constraint offered by
the adherent layers of wustite and magnetite respectively (which
was noted in our earlier work[4]), the dislocation barrier imposed
by the oxide layers and possibly some void pinning of grain
boundaries.

Although our main interest has been to study steady-state
creep behaviour, it is interesting to note that at 823 and 853K
both preoxidation and additional oxidation each caused an increase
in the transient creep rates (Fig. 3). During the initial preoxidation stage, the two dominant effects are likely to have been
that:-

(i)   oxide growth stresses would induce a tensile strain in the
near surface of the substrate with associated dislocation movement.

(ii)   vacancy injection into the substrate would occur if the
adhesion of magnetite or wustite was strong.

This latter process would certainly have assisted dislocation climb
in the initial stage of transient creep following the oxidation
period. During the additional oxidation in the interrupted
transient creep, it is conceivable that static recovery and perhaps
some recrystallization may have occurred. These events would
enable an initially rapid creep rate to be attained once transient
creep was resumed after the oxidation period.

In contrast to the strengthening effect associated with preoxidation and additional oxidation at 823 and 853K, current studies
in the higher temperature range (973 to 1073K) have indicated that
continuous oxidation during creep at the higher stress of 16 MN m^{-2}
produces a weakening effect, viz. that steady-state creep rates
increase with increasing oxygen pressure (for the 10^{-7} to 10^{-2} N m^{-2}
range). There are many factors that might be contributing to such
a weakening influence:-

(i)   loss of adhesion through scale growth fractures, and therefore a removal of any mechanical constraint imposed by the scale.

(ii)   loss of strengthening solute elements (e.g. Si) through
selective oxidation.

(iii)   oxidation-induced vacancies which could produce void
formation and/or assist dislocation climb.

(iv)   an enhancement of grain boundary sliding which could arise from oxygen at the boundaries.

These and other possibilities have been recently discussed by Tien and Davidson[3]. It is clearly apparent that interaction or competition between some of these factors (e.g. (ii) and (iii)) and potential strengthening effects like internal oxidation may ultimately determine the creep behaviour. Our most recent experimental work (still to be completed) at higher oxygen partial pressures (6 to 10 kN m^{-2}) has shown that the near-surface structure after oxidation and before creep testing possesses a finer grain size than the bulk structure (Fig. 10). Some internal oxidation and a partially adherent outer scale layer were also found. After creep, however, no pronounced fine-grained region was observed. The crept specimens showed a relatively uniform grain size suggesting that recrystallization had occurred uniformly throughout the section of the specimen. Several instances of a fine grain size being associated with a strengthening effect have been cited[3] although other complicating features make the grain size contribution less clear. For example, it has not been possible in the present work to assess whether grain boundary sliding, if any, has assisted the creep deformation processes.

We have certainly observed that where strong scale adhesion and an intergranular penetration of the subscale oxide have persisted (Fig. 11), the steady-state creep rate has been reduced. The possibility of strengthening arising from such a microstructural feature has been embraced in the concept of an environmental dependent back stress term in the power-law creep equation discussed by Tien and Davidson[3]. They imply that a positive back-stress would augment the effective stress to give an environmental weakening and a negative back-stress would be appropriate in the case of environmental strengthening.

For the present work it seems reasonable to conclude, therefore, that in the 973 to 1073K range where oxidation-weakening was observed at the low oxygen pressures of 10^{-7} to 10^{-2} N m^{-2}, oxidation-induced vacancies have promoted easier dislocation climb. The calculated activation energies of 357 and 378 kJ mol^{-1} are of the order expected for diffusion-controlled dislocation climb processes in iron. At the much higher oxygen pressure of 10 kN m^{-2} at 998K where oxidation-strengthening was found, the combined effects of internal oxidation and strong scale adhesion (with some plasticity of the wustite/fayalite inner scale at this temperature) have opposed dislocation movements. The calculated stress exponent n value of 9 for these test conditions gives a supporting indication of the notional combined stress sensitivity of mobile dislocation density and dislocation mobility. For pure metals and many single phase alloys, n is usually about 4 to 6.

Throughout the work it has been generally apparent that the

extent of oxidation (as indicated by oxide scale thickness) has
been increased by creep conditions. This is not unexpected where
dislocation climb processes assisted by oxidation-induced vacancies
enhance the vacancy flux, which in turn maintains scale adhesion
(in the absence of interface voids) and assists outward cation
diffusion through the scale. A recent study[12] has shown again
that the oxide growth stresses during iron oxidation at 973 to
1173K are large enough to cause appreciable plastic flow of oxide,
thereby decreasing the number of interfacial voids. The
possibility that internal oxidation processes may be especially
promoted in grain boundary regions above certain critical creep
strain rates has been further discussed by Grunling et al.[2].

Additional work by the authors is continuing in order to
obtain similar data to the above for higher Si contents and for
other alloying elements with iron over wide differences in oxidation pressures.

5.0   CONCLUSIONS

1.   Preoxidation at 823 and 853K before creep in continuously
oxidising conditions caused an increase in the transient creep
rate and a decrease in the steady-state creep rate.

2.   Additional oxidation during an interruption of transient
creep and at the start of steady-state creep caused similar
behaviour to that associated with the above preoxidation effects.

3.   In the higher temperature range of 973 to 1073K at lower
oxygen pressures (10^{-7} to 10^{-2} N m^{-2}) the steady-state creep rate
increased with increasing oxygen pressure.

4.   At the higher temperatures and higher oxygen pressures, (6 to
10 kN m^{-2}) where thicker, adherent scales of wustite and fayalite
formed with internal and intergranular oxidation of the substrate,
a decreased creep rate was found as oxygen pressure increased.

5.   The observed oxidation-weakening was primarily a consequence
of oxidation-induced vacancies assisting dislocation movement in
the near-surface of the substrate.

6.   The oxidation-strengthening was primarily a consequence of
dislocation pinning at internal oxides, together with the
mechanical constraint imposed by the strongly adherent scales.

6.0   ACKNOWLEDGEMENTS

The authors wish to thank Professors K.M. Entwistle and
E. Smith for the provision of research facilities, their colleagues
for many helpful discussions and Mr. I. Brough for his expert

assistance with the electron metallography.

7.0   <u>REFERENCES</u>

1.   R.H. Cook and R.P. Skelton, Int.Met.Rev., 1974, 19, 187, 199-222.

2.   H.W. Grünling, B.Ilschner, S. Leistikow, A. Rahmel and M. Schmidt, Werkst.Korros., 1978, 29, 11, 691-702.

3.   J.K. Tien and J.M. Davidson, Adv.Corr.Sci.Tech., 1979, 7, 1-51.

4.   J.H. Cleland and R. Rolls, Scripta Met., 1974, 8, 35-38.

5.   R. Rolls and J.H. Cleland, to be published.

6.   J.H. Cleland, R.R. Hough and R. Rolls, J.Phys.E: Sci.Instrum., 1976, 9, 628-630.

7.   R. Hales, P.S. Dobson and R.E. Smallman, Acta Met., 1969, 17, 1323-1326.

8.   P. Hancock, [Met.Soc.Conf.Proc.] Vacancies '76, 215-222, 1977, Bristol, The Metals Society.

9.   F.A. Mohamed, Scripta Met., 1979, 13, 1153-1155.

10. R.G. Stang, W.D. Nix and C.R. Barrett, Met.Trans.A, 1975, 6A, Nov. 2065-2071.

11. J.E. Harris, Acta Met., 1978, 26, 1033-1041.

12. S. Taniguchi and D.L. Carpenter, Corros.Sci., 1979, 19, 1, 15-26.

Fig. 1.　General view of creep testing facility.

R. *ROLLS* and M.H. *SHAHHOSSEINI*

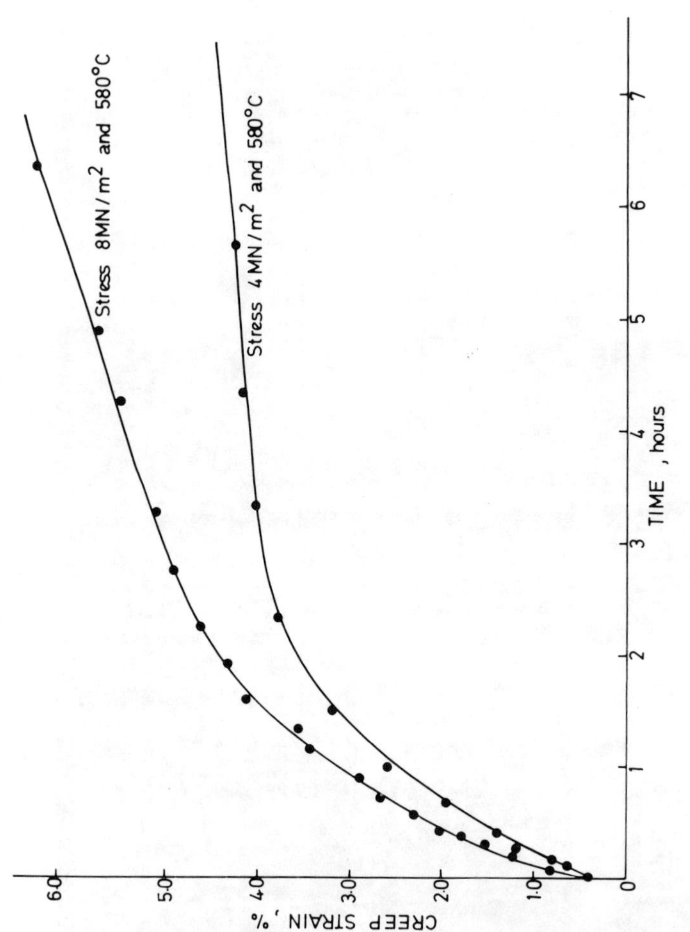

Fig. 2. The influence of constant stress on the creep of Fe-1Si at 853K with an oxygen pressure of 10^{-5} N m^{-2}.

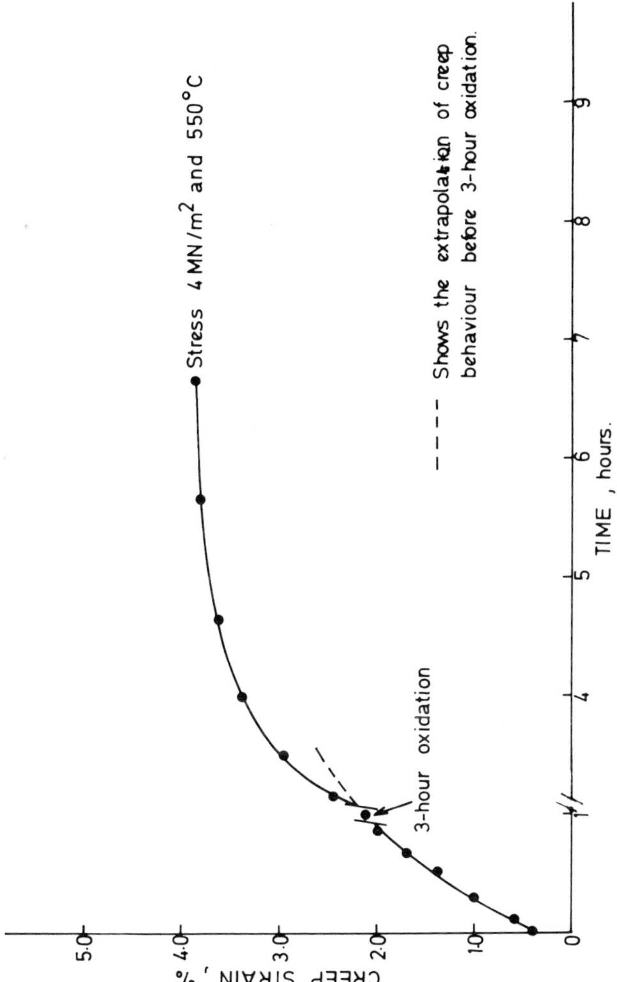

Fig. 3. The influence of additional oxidation on the transient creep of Fe-1Si at 823K with an oxygen pressure of 10^{-5} N m^{-2}.

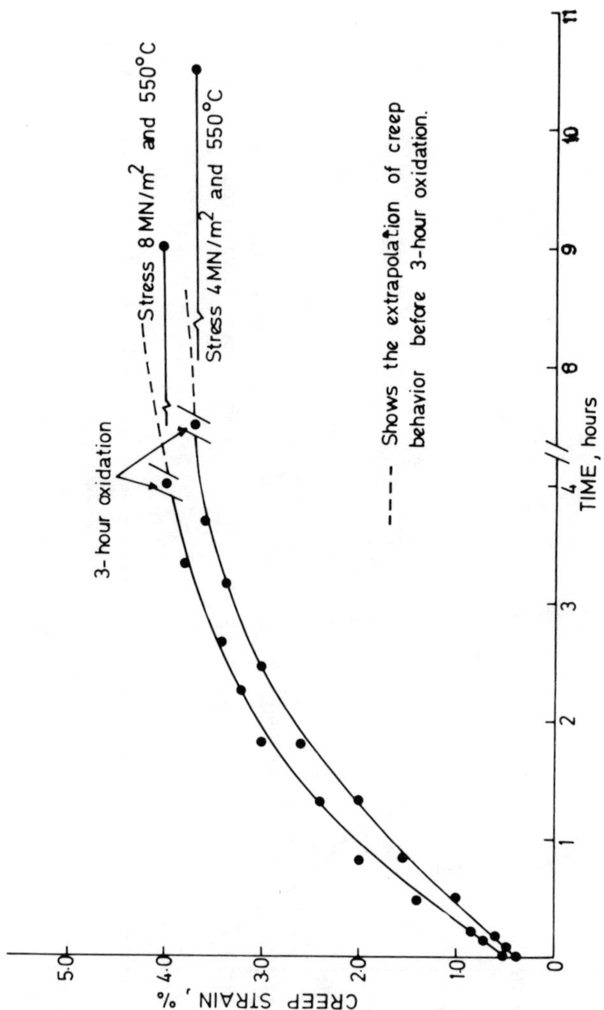

Fig. 4. The influence of additional oxidation on the steady-state creep of Fe-1Si at 823K with an oxygen pressure of 10^{-5} N m^{-2}.

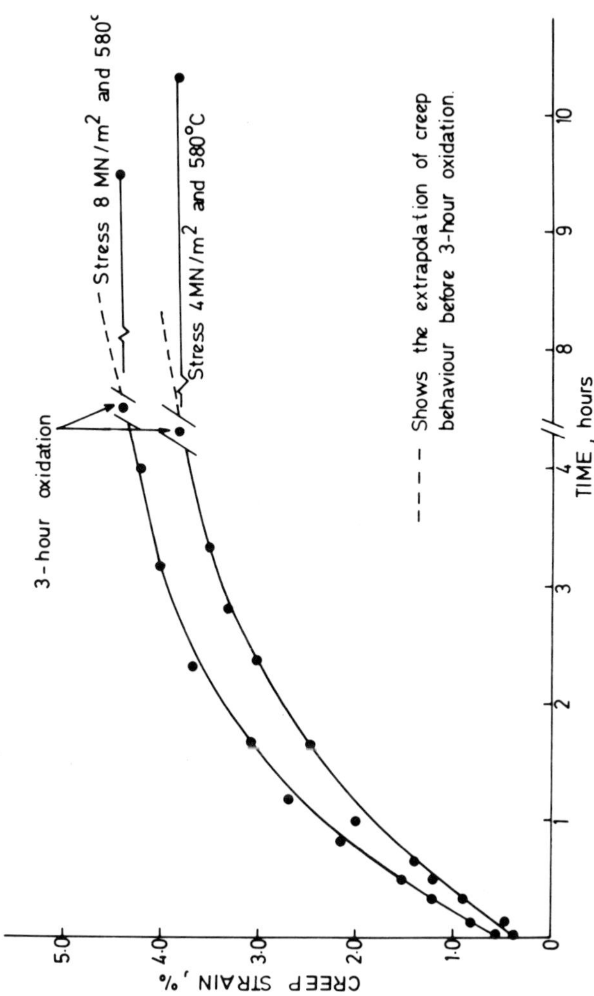

Fig. 5.   The influence of additional oxidation on the steady-state creep of Fe-1Si at 853K with an oxygen pressure of 10^{-5} N m^{-2}.

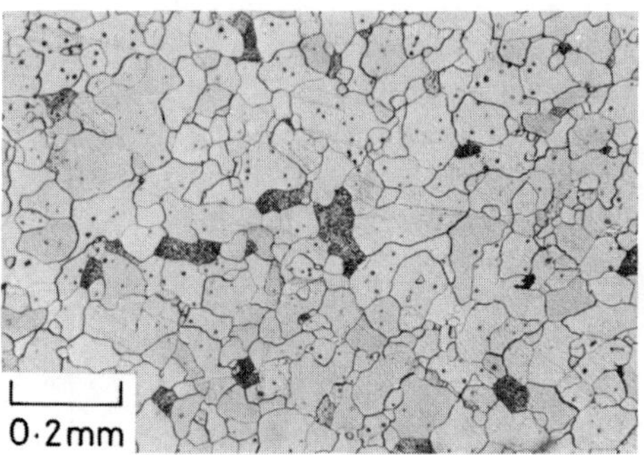

Fig. 6.   Fine recrystallized ferrite and internal oxidation in
           transverse section of Fe-1Si crept for 30 hrs. at 853K
           at stresses of 4 and 8 MN m^{-2} with an oxygen pressure
           of 10^{-5} N m^{-2}.

Fig. 7.   Elongated ferrite showing some recrystallization and
           internal oxidation in longitudinal section of Fe-1Si
           crept for 4 hrs. at 853K at 4 MN m^{-2} with an oxygen
           pressure of 10^{-5} N m^{-2}

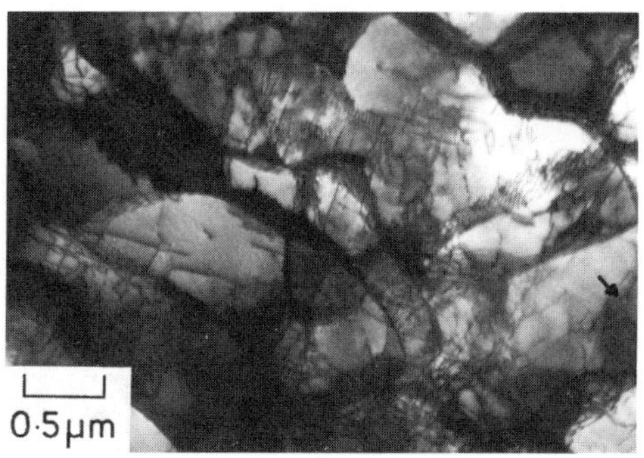

Fig. 8. Subgrains and dislocation networks in near-surface region
of Fe-1Si substrate, crept as per Fig. 3, showing dislocations pinned by internal oxide particle (arrowed).

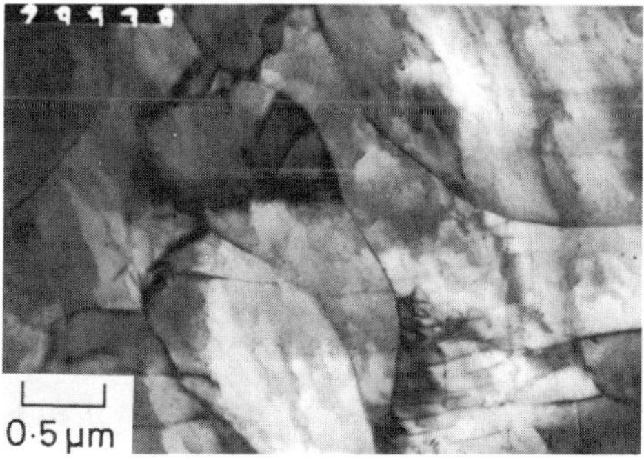

Fig. 9. Subgrains with fewer dislocation interactions (cf. Fig.
8) in centre of Fe-1Si specimen, crept as per Fig. 3.

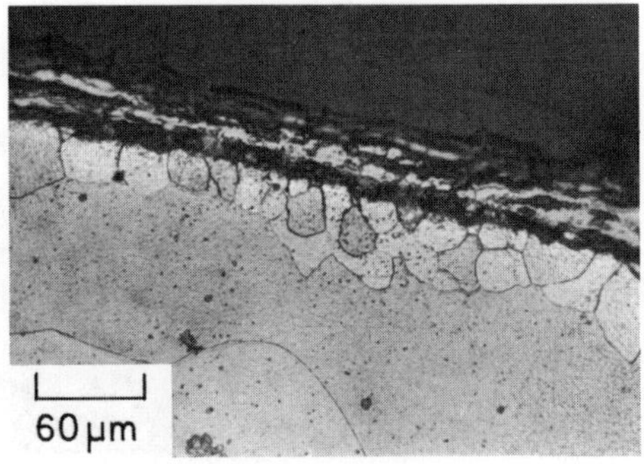

Fig. 10.   Fine grains of ferrite in near-surface of uncrept Fe-1Si oxidized for 5 hrs. at 998K with oxygen pressure of 6 kN m^{-2}

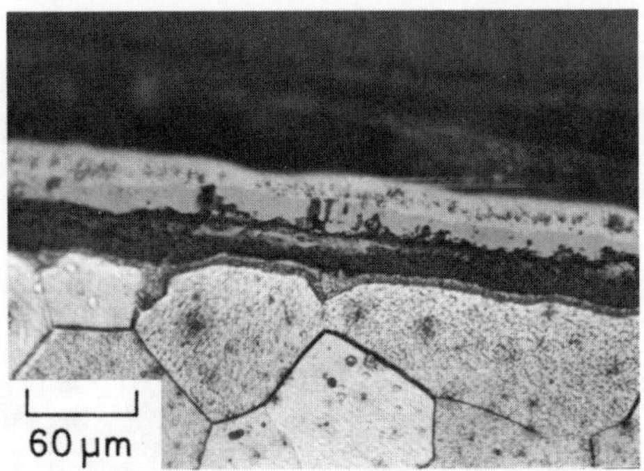

Fig. 11.   Adherent scale layers of outer magnetite, wustite and inner fayalite showing intergranular oxidation of Fe-1Si crept at 998K at stress of 16 MN m^{-2} with oxygen pressure of 10 kN m^{-2}

DISCUSSION

<u>M.I. Manning</u>: It should be possible, by studying creep behaviour of varying thicknesses of test pieces, to discover whether the re-initiation of primary creep following an interruption of oxidation was due to surface/oxidation effects or whether it was rather due to bulk effects such as static recovery. Have you investigated the influence of section thickness on the effects that you have observed?

<u>R.Rolls</u>: Unfortunately, we have not been able to investigate the influence of section thickness, but it is conceivable that had we used thin, rectangular coupon testpieces instead of cylindrical rod specimens our results may well have been different, if only because of the contrasting geometries of scale formation and the corresponding differences in scale adhesion. It is this feature, as well as testpiece thickness as you rightly suggest, that probably governs the extent to which bulk effects, like static recovery during the additional oxidation period, override surface effects like vacancy injection in controlling the renewed primary creep rate.

Investigations on the interactions between creep and corrosion of nickel base alloys in sulphate melts

U. Feld[+], A. Rahmel, M. Schmidt, Monika Schorr

Dechema-Institut, Theodor-Heuss-Alee 25

D-6000 Frankfurt (Main)

Abstract

Corrosion and creep tests have been performed with Nimonic 105 and IN 597 in $(Na_2, Ca, Mg) SO_4$ melt at $800\,^{\circ}C$ under potentiostatic conditions.

Both materials form protective scales at potentials $E_{Ag} < -100\text{mV}$. Above these potentials Nimonic 105 undergoes severe corrosion while IN 597 suffers from internal sulphidation. Strong external corrosion occurs with IN 597 at $E_{Ag} > + 200$ mV.

Creep tests performed with IN 597 revealed that at high stress levels the time to rupture is mainly determined by the mechanical stress and not by the corrosive attack. Corrosion affects, however, the creep rupture properties at lower stress levels. If a protective scale is formed ($E_{Ag} < -100$ mV) the specimen behaves similar as in air. Internal sulphidation (-100 mV $< E_{Ag} < + 200$ mV) reduces rupture time and elongation at rupture and changes the fracture mechanism from intergranular to transgranular.

+) Present address: Hoechst AG, Anlagenplanung, 6230 Frankfurt (Main) 80

1 Introduction

Interactions between mechanical stress and corrosion at high temperature have become increasing interest in the last years. A review
of possible interactions between creep deformation and hot gas
corrosion has been published recently[1,2].

In this paper the results of creep tests with the nickel base
alloys Nimonic 105 and IN 597 in sulphate melts at 800°C will be
reported. During the tests the electrode potential of the creep test
specimens was kept constant by means of a potentiostat and an auxi-
liary electrode system. By this kind of creep tests it was possible
to measure the effect of kind and extend of the corrosion on the
creep behaviour because the corrosion of superalloys and other
chromium containing high temperature materials in sulphate melts
depends on the electrode potential[3-8].

2. Experimental Procedure

Two kinds of tests have been carried out, mere corrosion tests
without mechanical stress and creep tests. All tests have been
performed in the eutetic melt of (in mole %) 53 Na_2SO_4, 7 $CaSO_4$
and 40 $MgSO_4$ (in the following marked $(Na_2, Ca, Mg)SO_4$).This melt
has been selected in agreement with other COST-50 partners. It
contains the main cations of sea salt as sulphates. Deposits containing these compounds can be found, e.g. on gas turbine blades
if the intake air is polluted with sea water droplets.

During the mere corrosion tests the melt was contained in an
Al_2O_3 crucible. The electrode potential of all test specimens was
kept constant during the test by means of a potentiostat and an
usual electrochemical cell arrangement in which the specimen was
the working electrode and a platinum sheet the counter electrode.
Reference electrode was was an Ag/Ag^+ electrode[9]. All potentials
were measured against this electrode and are marked E_{Ag} in the
following. Details of the mere corrosion tests will be described
elsewhere.[8]

A special cell arrangement has been developed for the creep tests
in the sulphate melt[10]. Fig. 1 shows a schematic cross section
of this cell. An Al_2O_3 crucible, 60 mm in diameter and 100 mm in
height with a 13 mm hole in the centre of the bottom was used to
contain the melt. One end of the creep test specimen was passed
through the hole and was tightened with a special nut. The crevice
between the specimen and the nut as well as between the nut and the
crucible was scaled with a gold washer. The specimen was surrounded
by a cylindrical platinum sheet which was connected as counter
electrode in the electrochemical cell. Between specimen (working
electrode) and counter electrode the reference electrode and a
thermocouple were placed. This cell was surrounded by a furnace

with three independently controlled zones and was part of the lever
arm creep test rig. The temperature fluctuations were $< \pm\ 3^{o}$C.

Because of the cell arrangement the creep strain of the specimen
could not be measured directly. Instead of this the increase of
length of the whole pull rod system was recorded outside the furnace. Is was assumed that the elongation of the pull rod is negligible compared with the creep strain of the specimen. The rupture of the specimen switched off the whole electrical power supply.

The dimensions of the creep test specimens are given in fig. 2 and
the chemical composition of the materials investigated in table 1.

After the corrosion and creep tests the specimens have been investigated metallographically, by scanning electron microscopy
(SEM) and by energy and wavelength dispersive microprobe analysis.

3 Experimental Results

3.1 Corrosion tests without mechanical stress

All details of the mere corrosion tests will be published elsewhere[8]. This paper contains only those results which are necessary to understand the influence of the electrode potentials
on the corrosion and the creep behaviour of the investigated
materials.

The influence of the electrode potential on the mass change of
the two materials tested after 100 h exposure in the $(Na_2, Ca, Mg)SO_4$
melt at 800°C is given in fig.3. Different potential regions can
be distinguished. At potentials $E_{Ag} < - 100$ mV a protective scale
rich in MgO and NiO is formed on the surface of both metals. The
mass change is very low and only little internal oxidation and
internal sulphidation is observed beneath the scale.

At $E_{Ag} > -100$ mV the two alloys show different corrosion behaviour.
Nimonic 105 with only 15% chromium undergoes localized breakdown
of the protective scale with the formation of excrescenses and
strong internal sulphidation beneath these excrescenses. This material has only two potential regions in this melt: protective scale formation at $E_{Ag} < -100$ mV and high, very often
localized corrosion with strong internal sulphidation at
$E_{Ag} > -100$ mV,

IN 597 with 25% chromium reveals three different potential regions. At potentials $E_{Ag} \approx -100$ mV the structure and composition of the scale undergo remarkable changes. Only little MgO
is deposited on the surfaces while NiO is extensively dissolved
in the melt so that the scale becomes rich in Cr-, Al-, and
Ti- oxides. These changes are summarized in fig. 4. The result is

a scale "permeable" for sulphur. Therefore, a zone of $(Cr, Ti)_3S_4$ precipitates formed by internal sulphidation can be observed beneath an interlocked metal/scale surface in the potential region -100 mV $< E_{Ag} < +200$ mV. However, the mass changes are small, fig. 3, and the external scales look even and unaffected. At potentials $E_{Ag} > +200$ mV IN 597 undergoes strong external corrosion and internal sulphidation by forming $(Cr,Ti)_3S_4$, $(Mo,Al,Cr)_3S_4$ and Ni_3S_2. Typical metallographic cross sections of specimens tested in the three potentials regions show figs. 5a - c.

To link the gap between potentiostatic test results in the laboratory and the corrosion behaviour in praxis, it is necessary to know the free corrosion potential of turbine components if they are covered with a thin layer of molten salt. From other measurements[6-8] it follows that the free corrosion potential is mainly affected by the SO_2-SO_3 content of the gas atmosphere and the vanadium (and molybdenum) content of the deposit. The free corrosion potentials can be expected approximately between $E_{Ag} \approx -100$ and $+200$ mV with the tendency that increasing SO_3 and V-contents shift the free corrosion potential to higher values (which in general leads to higher corrosion rates). From this follows that the free corrosion potential of Nimonic 105 is close or above the critical potential where breakdown of the protective scale occurs. Therefore, often localized corrosion can be observed after a critical time[7,10] which agrees with practical experiences in gas turbines[11]. The free corrosion potential of IN 597 is close or above the transition from protective scale formation to internal sulphidation.

3.2 Creep tests

Most of the creep tests have been done with IN 597. Some additional tests were performed with Nimonic 105.

3.2.1 IN 597

With this material creep tests were performed in all three potential regions with their different corrosion phenomenologies. By applying a constant potential during the creep tests it was possible to study the effect of different kinds of corrosion attack on the creep and the creep rupture behaviour.
All creep tests and their results are summarized in table 2. The relationship between stress σ, the corrosion potential E_{Ag} and the rupture time t_R, is also given in fig. 6. From table 2 and fig. 6, it follows:

a) The measured life time in air is in good agreement with the data given by the producer of this material.

b) At high stress levels ($\sigma = 348$ and 293 N.mm^{-2}) no significant influence of the corrosion potential on the rupture time t_R and on the elongation at rupture ϵ_R can be observed. Exceptions are the test runs at the potential of E_{Ag} = +250 mV with its extreme high corrosion rate.

c) At the lower stess of σ = 230 N.mm^{-2} the corrosion potential has a reproducible effect on rupture time and rupture elongation, figs. 6 and 7. The three potential regions with their different corrosion behaviour can also be distinguished in the creep rupture tests. In the potential region of E_{Ag} < -100 mV where a protective scale is formed on the surface rupture time and elongation at rupture are equal to those in air. In the region -100 mV < E_{Ag} < +200 mV where no external corrosion but internal sulphidation occurs both t_R and ϵ_R are distinctly reduced. The influence of the corrosion potential on t_R and ϵ_R at this stress level is illustrated in fig. 8.

d) At a potential of E_{Ag} = +250 mV with remarkable external corrosion the mechanical stress has little effect on the rupture time because the corrosion reduces the load carrying cross section rather fast.

SEM investigations of the fracture areas revealed the following results:
a) At the high stress levels of σ = 348 and 293 N·mm^{-2} no significant influence of the stress and the potential on the appearance of the fracture surfaces could be found. The mainly intergranular fracture looks similar to that in fig. 10a and can be compared to that in air. The crystal faces are covered with a honeycomblike structure of dimples as is demonstrated in figs. 9a and b. Such structures have also been reported after creep tests of IN 597 by Tipler et al.[12]

b) At the lower stress of σ = 230 N·mm^{-2} the kind of corrosion influences the appearance of the fracture area as shown in figs. 10a - c. At potentials E_{Ag} < -100 mV with protective scale formation the appearance is similar to that in air or at high stress levels, fig. 10a. An intergranular fracture predominates. However, if the potentials E_{Ag} > -100 mV and therefore internal sulphidation occurs the fractures look coarse grained and large parts are cracked transgranular, fig. 10b. In fig. 10c the load carrying cross section is evidently reduced by an excrescence because of the high corrosion potential of E_{Ag} = +200 mV. The remaining area is also cracked preferentially transgranular. A dimple-like structure has been found on inter- and transgranular faces. Figs. 11 and 12 demonstrate that particularly below the wall-like structures of the scale and subscale the fracture is transgranular.

Metallographic investigations of longitudinal cross sections through
and close to the fracture area revealed the following results:
a) At potentials $E_{Ag} < -100$ mV no internal sulphidation was observed beneath the scale of the undamaged part of the cylindrical
   surface though the creep strain was greater than 10%. Cracks
   in the base metal, however, were found to behave differently.
   The flanks of such cracks are internally sulphidized if they
   have been open to the melt, fig. 13. In general the depth of
   the internal sulphidation decreases in the direction to the
   crack tip. This indicates that the crack has grown and probably
   has started at the specimen's surface. There are however, some
   cracks with a rather constant zone of internal sulphidation
   along its flanks, fig. 14. Very probably these cracks have been
   initiated in the material and have been opened so that the melt
   could fill the whole crack spontaneously. In a few exceptional
   cases only one flank of the crack shows internal sulphidation,
   fig. 15. It seems that this kind of crack is filled mainly with
   oxidic corrosion products and is continuously reopened always
   at the same corrosion product/material interface. Because of
   the rather high creep strain all these specimens exhibit many
   intergranular secondary microcracks.

b) In the potential region -100 mV $< E_{Ag} < +200$ mV with internal
   sulphidation and reduced elongation at rupture only a few microcracks could be found close to the fracture. The internal sulphidation below scale on the cylindrical surface of the specimen
   was comparable with that of specimens tested without mechanical
   stress. Internal sulphidation could be found only in the outer
   zone of the fracture surface. No sulphide traces could be detected in the centre of the fracture area by metallographic methods.

c) The specimen creep tested at $E_{Ag} = +200$ mV and $\sigma = 230$ N·mm^{-2}
   revealed cracks in the internal sulphidation zone, fig. 16, with
   oxidized holes at the inward ends. In general this kind of cracks
   did not penetrate down to the unaffected base metal as well as
   to the surface. In addition to the internal sulphidation below
   the scale on the cylindrical surface area of the specimen also
   the outer part of the fracture area was affected by internal
   sulphidation, however, without recognizable external scale formation. This zone ends very suddenly, as is shown in fig. 17.
   From this it must be concluded that the outerpart of the fracture area was covered with the sulphate melt over a longer
   period while the remaining internal cross section cracked
   rather suddenly because of the high increase of stress.

d) Only a few secondary cracks were found in specimens with reduced
   elongation at fracture and a coarse grained fracture area.
   Therefore it is difficult to decide whether the crack starts
   and propagates inter- or transgranular. From the figs. 10a - c
   of the fracture areas it could be concluded that the grain size

of the test specimens is quite different. Metallographic examinations reveal, however, that the grain size of all specimens is rather similar, about ASTM 2 - 3. The different appearance is in fact a result of the different fracture mechanisms.

3.2.2 Nimonic 105

Only a few creep tests have been performed with Nimonic 105 with $\sigma = 348$ N·mm^{-2} which led in air to a rupture time of approx. 200 h and an elongation at rupture of approx. 8 - 10%.

As described in chapter 3.1 this material has only two potential regions, formation of a protective scale at $E_{Ag} < -100$ mV and remarkable external corrosion and internal sulphidation above this critical potential. If the specimens were creep tested at $E_{Ag} < -100$ mV the creep behaviour was comparable to that in air. The flanks of secondary cracks open to the melt have been internally sulphidized but not the uncracked cylindrical surface below the scale. The same behaviour was observed with NI 597.

At potentials $E_{Ag} > -100$ mV the rupture time is strongly reduced due to the decrease of the load carrying cross section. In addition the elongation at rupture is also diminished.

4 Discussion and Conclusions

Conclusions should be drawn with great care from the reported experiments and results and it is not yet clear whether they can be generalized and transferred to other nickel base alloys. These results have been obtained from wrought alloys having only a relative small amount of γ'-precipitates. In particular cast alloys could behave quite different because of their coarse graines and in general much higher amount of γ'-phase.

Nevertheless, the experiments demonstrated the strong effect of internal sulphidation on the creep properties of IN 597. The most important results seem to be:
a) It is not the sulphur and its compounds in general which damage the material. If a protective scale is formed on the material's surface liquid deposits rich in sulphates do not damage the material. However, it is unknown whether the scale in the protective potential region will remain unaffected for a period much longer than 1000 h or at higher temperatures.

b) It is suprising that even rather high creep strains do not affect the protectivity of the scale on the cylindrical surface of the specimen. Even clear grain boundary sliding in the metal does not result in preferential grain boundary attack, as long as a critical limit has not been exceeded. Figs. 18 and 19 show this effect below and above the critical limit. It is obvious that the scale must have been cracked more than one time above

the grain boundary of the metal. It must therefore be concluded
that the healing or "repassivation" of the scale is very fast
under potentiostatic condition in the potential range
$E_{Ag} < -100$ mV.

An exception are the crack surfaces along the grain boundaries of
the metal due to the creep strain, see figs. 13-15. They are
susceptible to internal sulphidation. The reason for this may be
the potential drop and changes of the melt composition in the
crack.

c) The results indicate that internal sulphidation worsens the creep
   properties of the material investigated only if the time to rupture is longer than an incubation period.

d) Internal sulphidation causes not only a reduction of the load
   carrying cross section but influences strongly the fracture
   mechanism from intergranular to transgranular after the incubation period.

It is clear that a reduction of the load carrying cross section,
e.g. by internal sulphidation, decreases the time to rupture because the stress in the unaffected zone of the specimen increases
with time. Grünling et al. [1,2] have proposed a simple equation
to estimate this effect if rate and rate law of the corrosion
reaction are known. In the case of IN 597 it was, however, difficult
to determine rate and rate law of the internal sulphidation. It
seems that it does not follow a parabolic rate law as would be
expected but a more complex one. Rough estimations reveal that the
reduction of life time can be explained in the first approximation
by the reduction of the load carrying cross section due to internal
sulphidation. Besides it must be taken into account that the formula
published by Grünling et al. [1,2] does not cover the tertiary creep
stage which is also affected by internal sulphidation resulting in
a lower strain rupture.

The most important observation, however, seems to be that internal
sulphidation reduces the rupture elongation and changes the fracture mechanism if a certain time has been exceeded. Both effects
occur simultaneously. The creep tests at different stress levels
and consequently different rupture times reveal that a critical
time exists above which the "sulphur effect" becomes evident. Comparable results are also obtained from tests with specimens of
four different diameters. Fig. 20 shows the creep curves of these
specimens which all were kept at the same initial stress of
$\sigma_o = 230$ N·mm^{-2} and the same potential $E_{Ag} = \pm 0$ mV. With decreasing
diameter the rupture time is shortened. The percentage of the load
carrying cross section damaged by internal sulphidation decreases
with increasing diameter after comparable times. The early rupture
of the $\emptyset$ 4,2 mm specimen, after about 480 h, leads to rather high
elongation at rupture, fig. 20, and an intergranular fracture,

fig. 21a. Though this specimen has the highest percentage of damage by internal sulphidation it reveals a normal creep rupture behaviour indicating that the time has been too short for sulphur to penetrate far enough into the metal. With increasing diameter and higher life times sulphur damages the material more and more by an unknown mechanism. The result is a decrease in the elongation at rupture and a transition from intergranular to transgranular fracture, figs. 21b - c. The time after which a critical sulphur affected depth has been reached is in the order of about 500 h at 800°C, and might decrease with increasing temperature which has not been determined yet. This indicates clearly that diffusion processes play a predominant role in the sulphur damage effect. The strain rate seems to be of secondary importance otherwise the sulphur effect should increase with rising stress.

Surprising is the change of the fracture mechanism from mainly intergranular to predominant transgranular. Normally intergranular cracking is typical for creep rupture fracture. Floreen and Kane[13] found in fatigue crack growth measurement with Inconel 718 at 650°C in different environments that sulphur enhanced the intergranular as well as the transgranular crack growth rate. In the here reported case the transgranular crack growth rate seems to be much more supported by the sulphur than the intergranular one. The mechanism is unknown. These observations are in contradiction to creep tests performed with IN 100 and René 41 in sulphur containing combustion gases[14,15]. The formation of dimples which have been reported in 14,15) to be typical for sulphur enhanced cracking have been observed in this investigation under all test conditions so that this statement cannot be generalized. Additional investigations will be necessary to understand the mechanism how sulphur affects the creep properties. It seems, however, to be clear from the presently known observations that sulphur can actually deteriorate the creep rupture properties of high temperature alloys if it penetrates through the oxide scale into base metal.

Acknowledgements

The authors would like to thank the Bundesminister für Forschung und Technologie, who sponsered these investigations in the frame of the COST-50 action "Materials for Gas Turbines" of the European Communities. The authors would like also to acknowledge the efforts of all those people who have contributed to the progress of this work.

References

1) H.W. Grünling, B. Ilschner, S. Leistikow, A. Rahmel and
   M. Schmidt: Werkstoffe und Korrosion <u>29</u> (1978) 691

2) H.W. Grünling, B. Ilschner, S. Leistikow, A. Rahmel and
   M. Schmidt in "Behaviour of High Temperature Alloys in Agres-
   sive Environments" (Eds.: Kirman, Marriott, Merz, Sahm and
   Whittle), The Metals Society, London 1980

3) A. Rahmel and W. Schwenk: "Korrosion und Korrosionsschutz von
   Stählen", Kapitel 6, Verlag Chemie, Weinheim 1977

4) U.Jäkel and W. Schwenk: Werkstoffe und Korrosion <u>26</u> (1975) 521

5) A. Rahmel and E. Tatar-Moisescu: Werkstoffe und Korrosion <u>26</u>
   (1975) 513

6) A. Rahmel: Werkstoffe und Korrosion <u>28</u> (1977) 299

7) A. Rahmel in "Ash Deposits and Corrosion due to Impurities in
   Combustion Gases" (Editor: R.W. Bryers), Hemisphere Publ. Co.,
   Washington, London (1978)

8) A. Rahmel, M. Schmidt and M. Schorr: Werkstoffe und Korrosion,
   to be published

9) A. Rahmel: Electrochem. Acta <u>15</u> (1970) 1267

10) U. Feld: "Einfluss des Elektrodenpotentials auf die Korrosion
    und das Zeitstandverhalten einer Nickel-Basis-Legierung in
    $(Na_2, Ca, Mg)SO_4$-Schmelze bei $800^{\circ}C$", Dissertation, TH Darmstadt
    1977

11) See e.g.: A.B. Hart and A.J.B. Cutler: "Deposition and Corrosion
    in Gas Turbines", Applied Science Publishers, London 1973;
    M. Semlitsch: Techn. Rundschau Sulzer, Forschungsheft 1970

12. H.R. Tipler, Y. Lindblom and J.H. Davidson in "High Temperature
    Alloys for Gas Turbines" (Eds.: D. Coutsouradis et al.), Applied
    Science Publishers, London 1978, p. 359

13) S. Floreen and R.H. Kane: Met. Trans. <u>10A</u> (1979) 1745

14) H.-J. Dorn: "Einfluss der Hochtemperaturkorrosion auf das Lang-
    zeitverhalten von Nickelbasislegierungen", Dissertation 1977,
    TU München

15) Kh.-G. Schmitt-Thomas, H. Meisel und H.-J. Dorn: Werkstoffe und
    Korrosion <u>29</u> (1978) 1

<u>Table 1</u>: Composition of IN 597 and Nimonic 105 (%)

	IN 597	Nimonic 105
C	0,049	0,125
Si	0, 23	0,22
Mn	0,09	0,04
P	<0,005	
S	0,006	0,003
Al	1,12	4,59
B	0,0063	65 ppm
Co	19,44	19,6
Cr	24,39	14,95
Cu	<0,01	0,01
Fe	0,64	0,42
Mo	1,6	5,13
N	0,006	
Ni	48,3	53,54
Nb/Ta	1,13	
Ti	2,63	1,28
V	<0,01	
W	<0,01	
Zr	0,025	0,095
Pb		0,9 ppm
Ag		<0,1 ppm
Bi		<0,1 ppm

U. FELD et al.

Table 2: Creep rupture data of IN 597 tested at 800°C in $(Na_2, Ca, Mg)SO_4$ -melt

	$\sigma = 348$ N/mm²		$\sigma = 293$ N/mm²		$\sigma = 230$ N/mm²	
	t_R [h]	Δl [mm]	t_R [h]	Δl [mm]	t_R [h]	Δl [mm]
in air	110,5	5,8	338,9	3,7		
no extern. current			281,0	2,24		
− 400 mV			245,9	2,02		
− 200 mV	144,6	6,0			1017,0	5,1
− 100 mV					932,3	4,37
± 0 mV			371,9	2,4	587,7	1,44
+ 100 mV					614,0	1,12
+ 200 mV	124,6	7,8	271,6	6,35	343,2	0,85
+ 250 mV	92,2	2,33	118,2	0,74	149,2	0,67
				d_o		
± 0 mV				4,2	479,0	2,88
± 0 mV				5,2	526,2	1,7
± 0 mV				6,0	587,7	1,44
± 0 mV				6,8	735,1	1,86

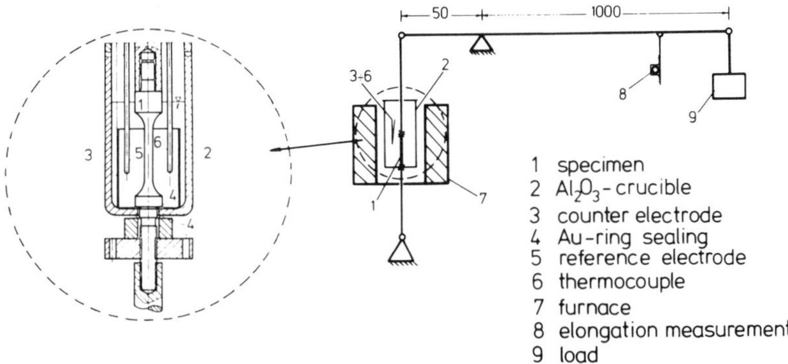

Fig. 1: Scheme of apparatus for creep tests under potentiostatic conditions in sulphate melt

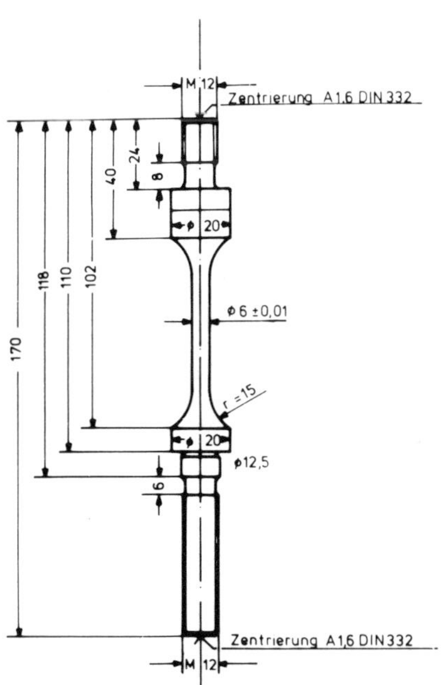

Fig. 2: Creep test specimen

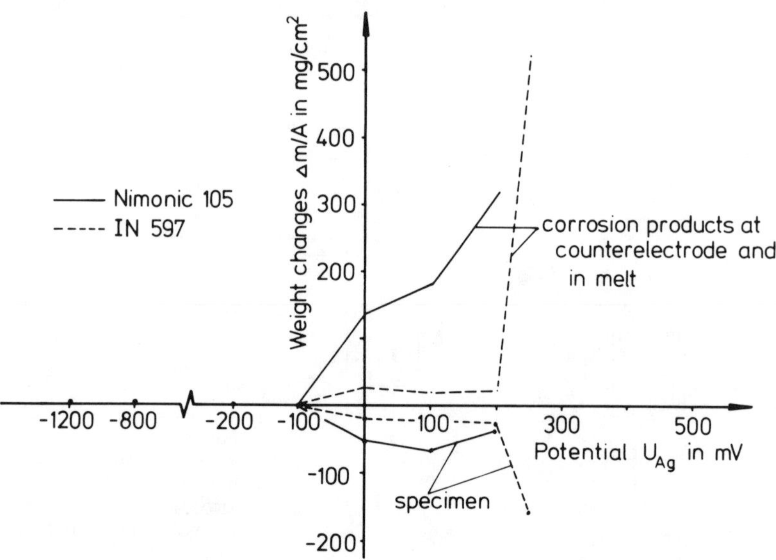

Fig 3: Influence of potential on the mass loss of Nimonic 105 and IN 597 and the amount of corrosion products found in the melt and at the counter electrode after 100 h potentiostatic measurements in $(Na_2,Ca,Mg)SO_4$-melt at 800°C

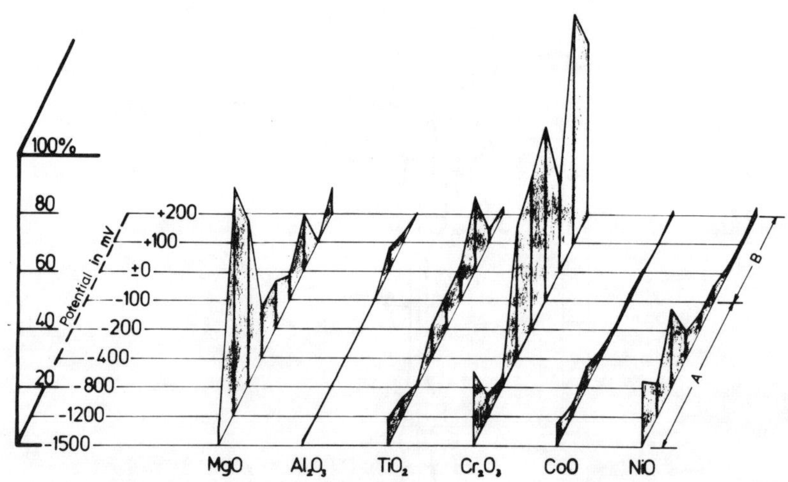

Fig.4a: Influence of potential on the composition of the oxide scale formed on IN 597 during 100 h potentiostatic conditions in $(Na_2,Ca,Mg)SO_4$-melt at 800°C

IN 597 in $(Na_2,Ca,Mg)SO_4$-melt at 800°C, 100 hrs

Maincomponents 46Ni, 24.5Cr, 20Co, 15Al, 3Ti, 15Mo, 10Nb

Potential U_{Ag} in mV	Color of Film	X-Ray Diffr. of Film	EDX of Film	
+ 200	green	Cr_2O_3	~90 Cr_2O_3 ~4 TiO_x ~6 (Ni,Co)O	+230 mV Break-through potential
+ 100	green	$(Mg,Ni,Co)Cr_2O_4$	~90 Cr_2O_3 ~6 TiO_x ~4 Al_2O_3 Tr.: (Ni,Co)O	(Ni,Co)O deposited at counterelectrode
± 0	green with black areas		~15 Cr_2O_3 ~25 MgO ~45 TiO_x ~13 Al_2O_3 ~2 (Ni,Co)O	
- 100	black a. green		~70 Cr_2O_3 ~10 MgO ~15 TiO_x ~5 (Ni,Co)O	
- 200	black with green areas	Cr_2O_3	~60 Cr_2O_3 ~15 MgO ~15 TiO_x ~10 (Ni,Co)O	
- 400	black		~50 Cr_2O_3 ~20 MgO ~15 TiO_x ~15 (Ni,Co)O	
- 800	black	Cr_2O_3 + $(Mg,Ni,Co)Cr_2O_4$	~4 Cr_2O_3 ~65 MgO ~1 TiO_x ~30 (Ni,Co)O	
-1200	black	Cr_2O_3 + $(Mg,Ni,Co)Cr_2O_4$ + (Mg,Ni,Co)O	~5 Cr_2O_3 ~80 MgO ~15 (Ni,Co)O	

Fig. 4b: Influence of potential on the colour, structure and composition of the oxide scale formed on IN 597 during 100 h potentiostatic conditions in $(Na_2,Ca,Mg)SO_4$-melt at 800°C

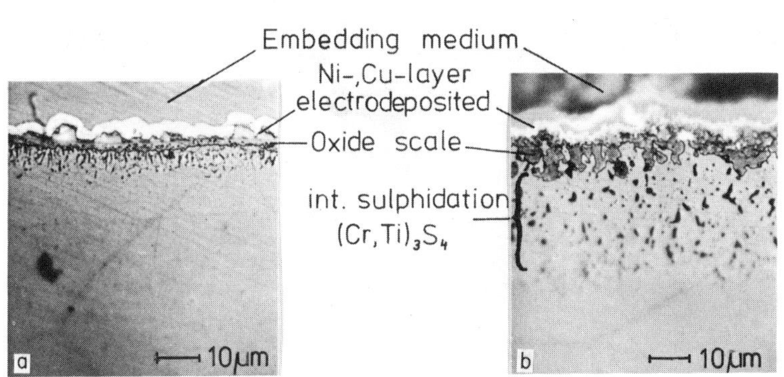

Fig. 5a: Cross section of specimen IN 597 kept for 100 h at E_{Ag} = -1200 mV in $(Na_2,Ca,Mg)SO_4$-melt at 800°C showing very little internal corrosion beneath a protective external scale

Fig. 5b: Cross section of specimen IN 597 kept for 100 h at E_{Ag} = ± 0 mV in $(Na_2,Ca,Mg)SO_4$-melt at 800°C showing remarkable internal sulphidation beneath a semi-protective scale, but only little external corrosion

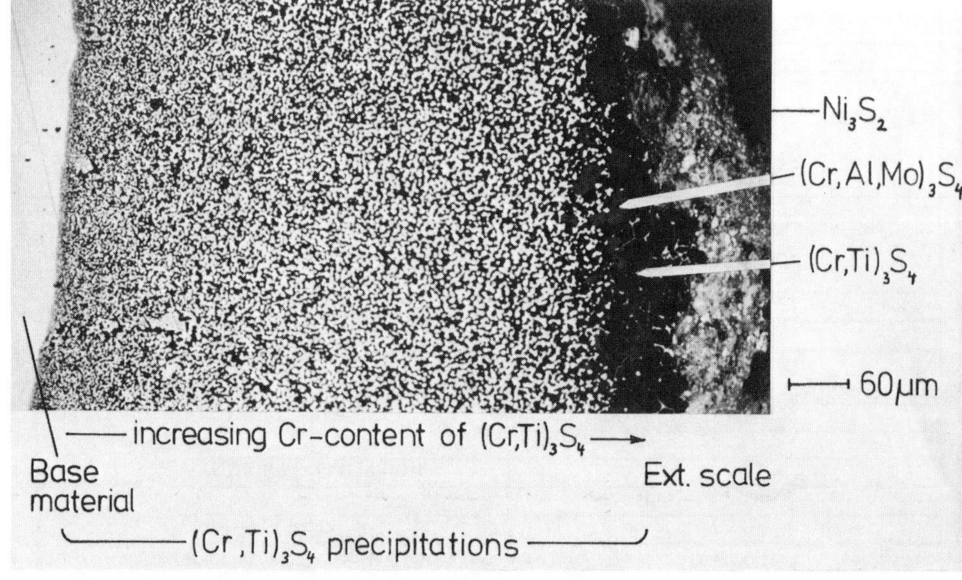

Fig. 5c: Cross section of specimen IN 597 kept for 100 h at E_{Ag} = +250 mV in $(Na_2,Ca,Mg)SO_4$ -melt at 800°C showing severe internal and external corrosion

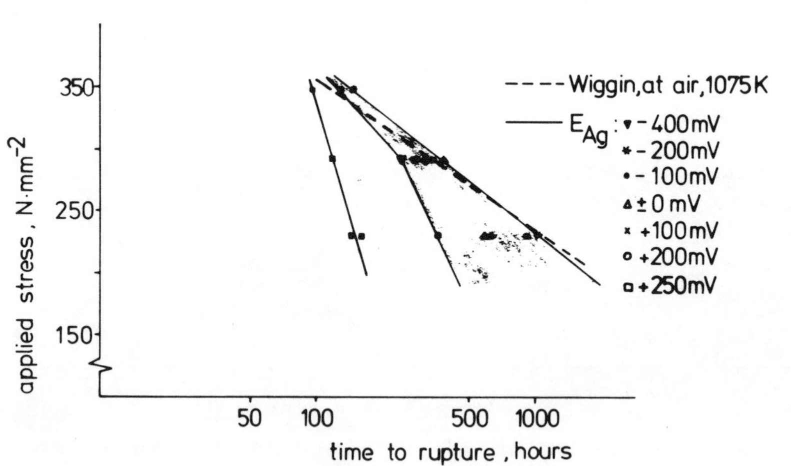

Fig. 6: Results of creep tests with IN 597 in air and at controlled potentials in $(Na_2,Ca,Mg)SO_4$ -melt at 800°C

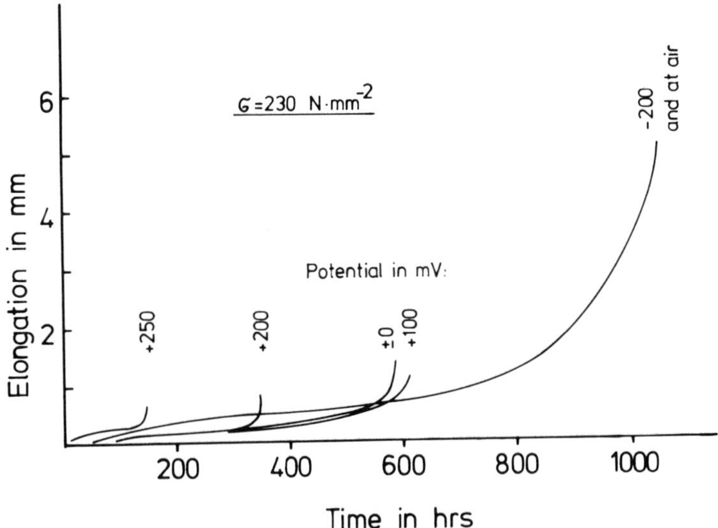

Fig. 7: Elongation-time relationship of IN 597 at different potentials (potentiostatic conditions) in $(Na_2,Ca,Mg)SO_4$-melt at 800°C, $G = 230$ N·mm⁻²

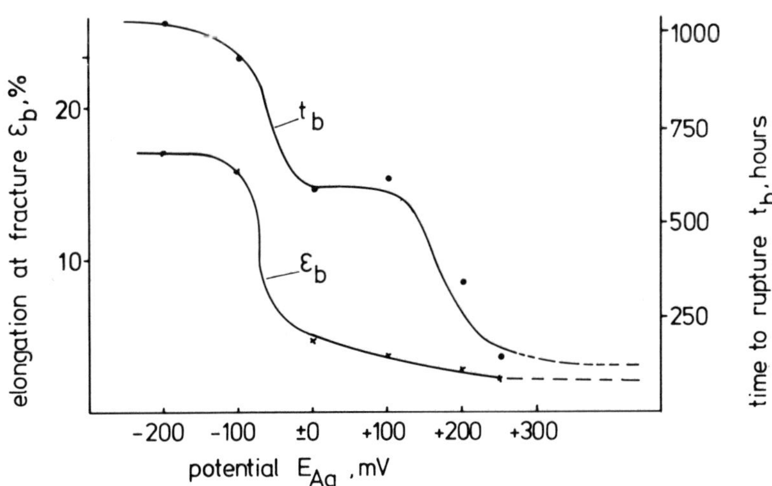

Fig. 8: Influence of potential on time to rupture and elongation at fracture of IN 597 in $(Na_2,Ca,Mg)SO_4$-melt

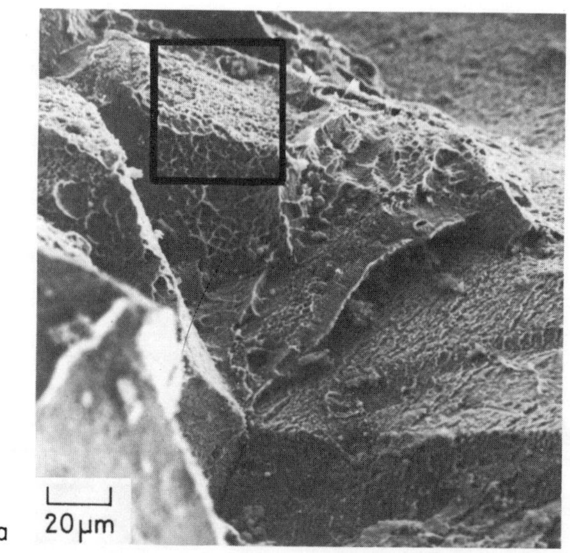

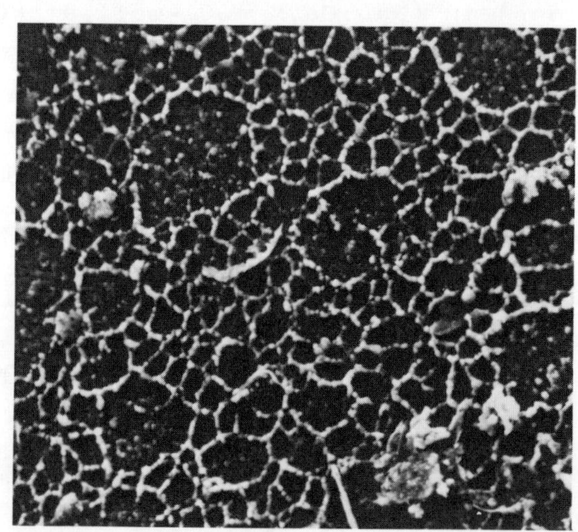

Fig. 9a, b: Dimple structure on fracture area of specimen IN 597 after creep test at 800°C in $(Na_2, Ca, Mg)SO_4$-melt at $E_{Ag} = -200$ mV, $t_R = 1017$ h, $\mathcal{G} = 230$ N·mm^{-2}

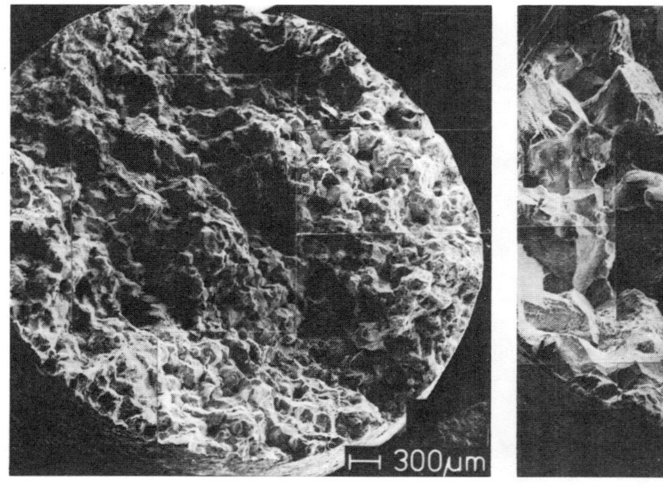
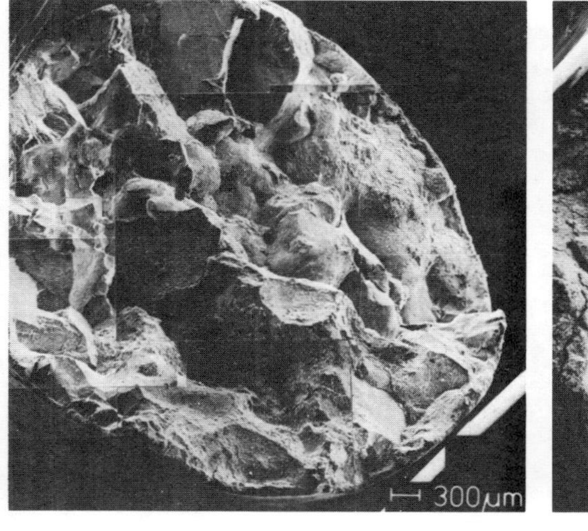
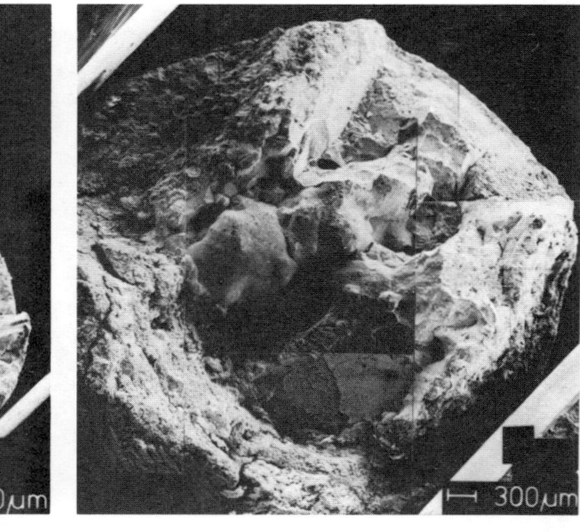

Fig. 10: Fracture area of specimen IN 597 tested at 800°C in $(Na_2,Ca,Mg)SO_4$-melt under

a) E_{Ag} = -200 mV, t_R = 1017 h, σ = 230 N·mm^{-2}

b) E_{Ag} = ± 0 mV, t_R = 588 h, σ = 230 N·mm^{-2}

c) E_{Ag} = +200 mV, t_R = 343 h, σ = 230 N·mm^{-2}

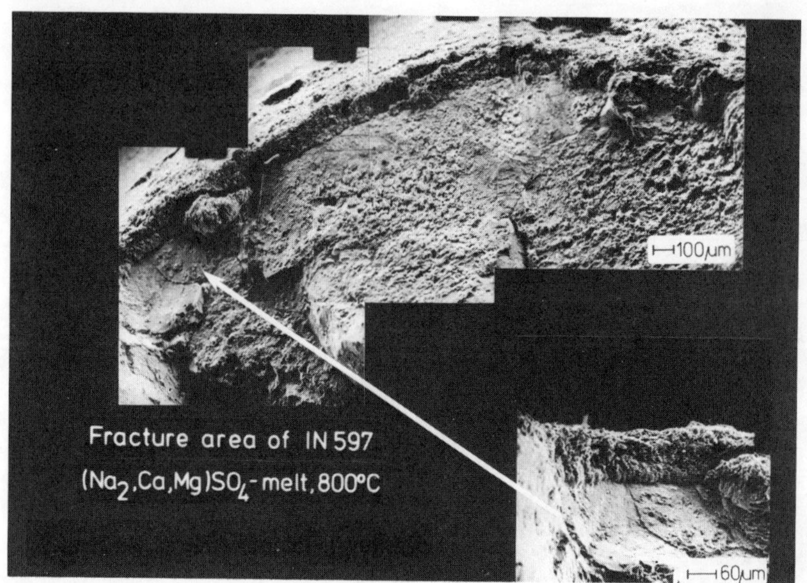

Figs.11,12: Outer zone of fracture area showing transgranular fracture regions below wall like
scale and subscale of specimen IN 597
tested at 800°C in (Na$_2$,Ca,Mg)SO$_4$-melt at
E_{Ag}= +200 mV, t_R= 343 h, G = 230 N·mm^{-2}

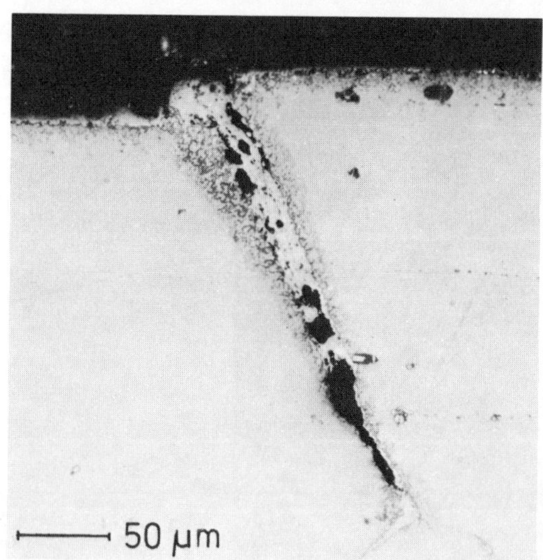

Fig.13: Crack open to the surface in a creep rupture
specimen of IN 597 with decreasing depth of
internal sulphidation towards crack tip. Testing conditions: (Na$_2$,Ca,Mg)SO$_4$-melt at 800°C;
E_{Ag}= -200 mV; G = 230 N·mm^{-2}

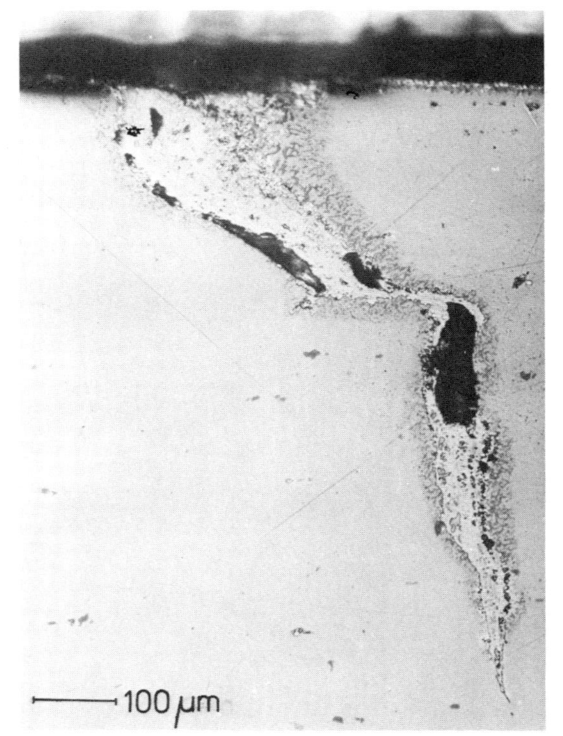

Fig.14:
Crack with constant depth of internal sulphidation along the crack flanks. Testing conditions: as fig. 13

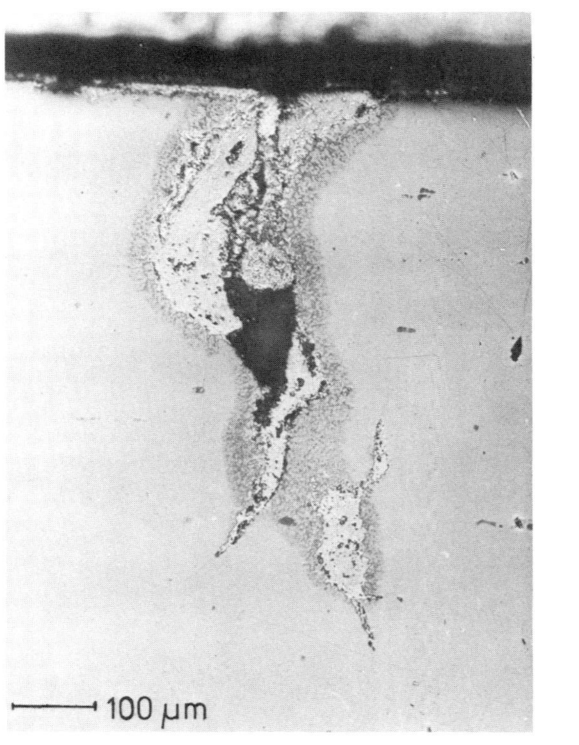

Fig.15:
Crack with internal sulphidation along only one flank of the crack. Testing conditions: as fig. 13

U. *FELD et al.*

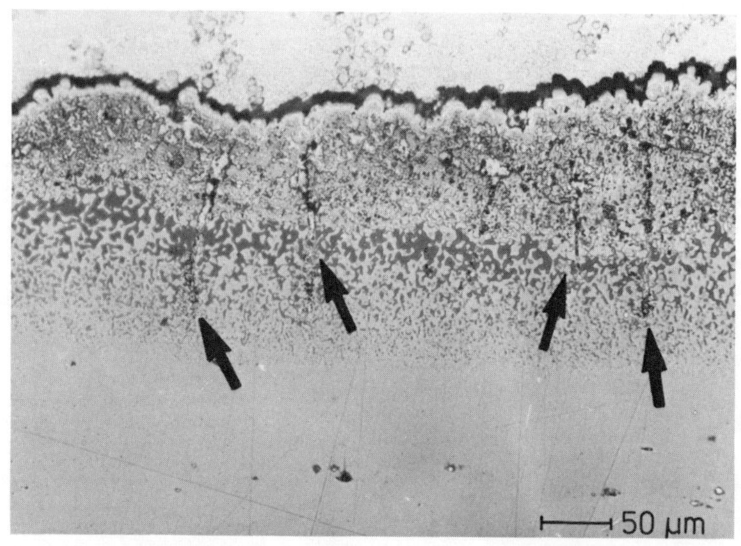

Fig. 16: Oxide filled cracks in corrosion affected zone in creep rupture specimen of IN 597 tested at 800°C in $(Na_2,Ca,Mg)SO_4$ -melt at E_{Ag}= +200 mV, σ = 230 N·mm^{-2}

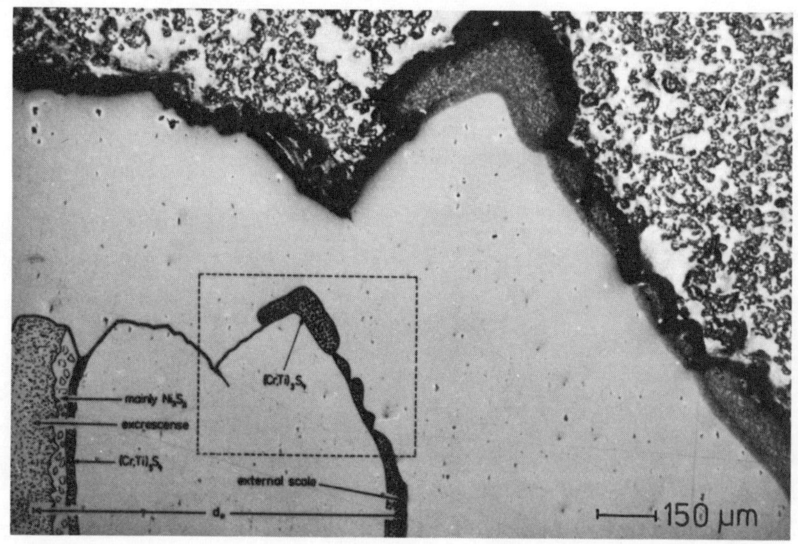

Fig.17: Cross section through fracture area of specimen IN 597 tested at 800°C in $(Na_2,Ca,Mg)SO_4$ melt at E_{Ag} = +200 mV, σ = 230 N·mm^{-2} showing internal sulphidized part of fracture

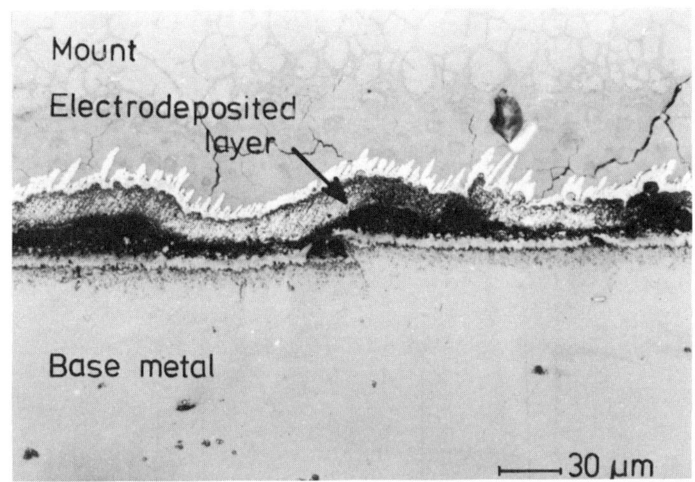

Fig.18: "Repassivation" of oxide scale above step
in surface of base material resulting from
grain boundary sliding. IN 597 specimen kept
at 800°C in $(Na_2,Ca,Mg)SO_4$ -melt at 800°C,
E_{Ag} = -200 mV, G = 230 N·mm^{-2}

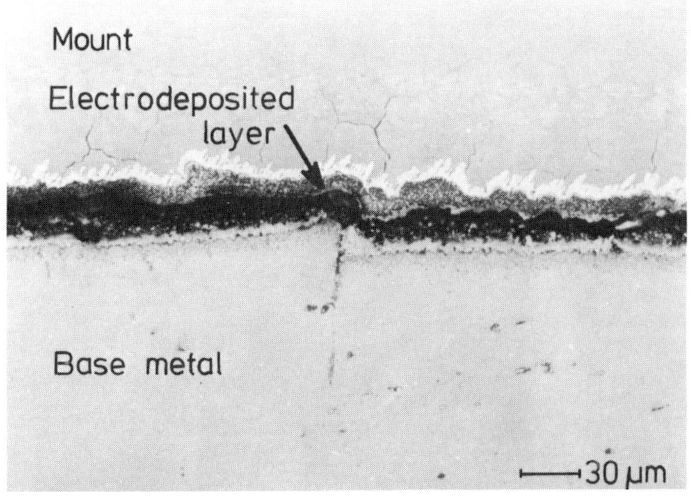

Fig. 19: Crack initiated by grain boundary sliding
with beginning internal corrosion along
grain boundary. The same specimen as
fig. 18

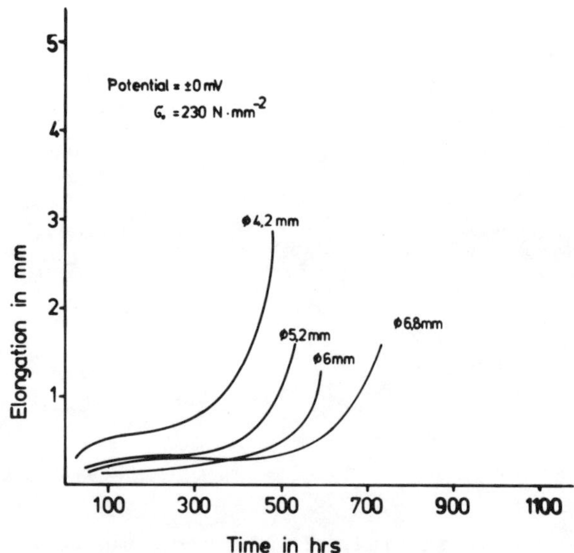

Fig.20: Elongation-time relationship of IN 597 specimens of different diameter tested at 800°C in $(Na_2,Ca,Mg)SO_4$-melt, $E_{Ag}= \pm$ 0 mV (potentiostatic conditions)

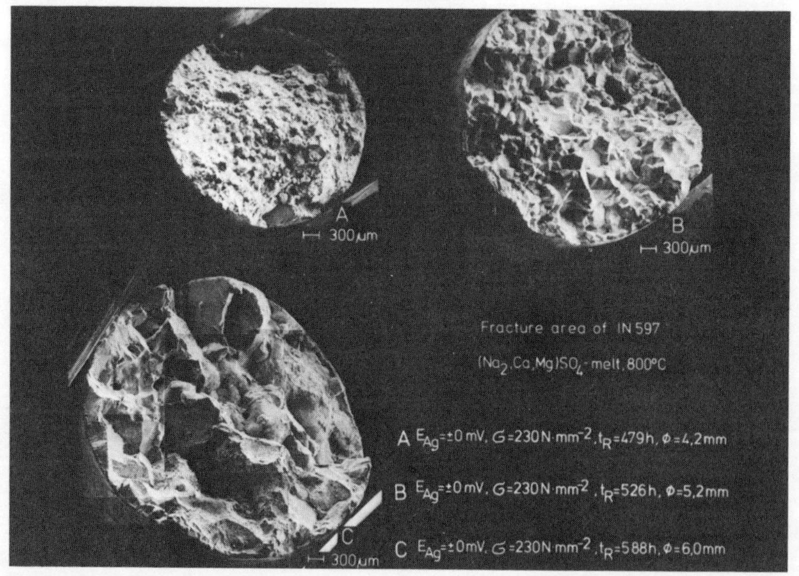

Fig. 21: A-C) Fracture areas of three specimens of fig. 20

DISCUSSION

P. Kofstad: Were the specimens completely submerged in the
sulphate melts? If so, would you expect different corrosion
behaviour if the alloys were covered by a thick film of the
sulphate melt and exposed to air or oxygen? Would the availability
of oxygen to a larger extent favour formation of a protective
film on the alloy surface?

Rahmel - Schmidt: The specimens were in fact completely submerged
in the sulphate melt. An answer to the second part of the question
is complex. If one assumes that the redox potential of the melt is
determined by the reaction

$$SO_4'' = SO_3 + 1/2O_2 + 2e' \qquad\qquad (1)$$

then it follows for the redox potential

$$E = E^O + \frac{RT}{2F} \ln P_{SO_3} \cdot P_{O_2}^{1/2} \qquad\qquad (2)$$

According to equ. (2) the redox potential depends on both, the O_2
and the SO_3 pressures, thermodynamic equilibrium assumed.

However, the free corrosion potential and therefore the kind and
the extend of the corrosion depends on the anodic and cathodic
partial current-potential curves (see e.g. A. Rahmel in "Ash
Deposits and Corrosion Due to Impurities in Combustion Gases"
(Ed. R.W. Bryers), Hemisphere Publ. Corp., Washington 1978, p. 185).
The cathodic partial current-potential curve is strongly influenced
by the SO_3 pressure but only very little by the O_2 pressure (see
e.g. A. Rahmel, Werkstoffe u. Korrosion 19 (1968) 750). From this
we conclude that oxygen does not significantly influence the corrosion potential and therefore the corrosion rate independent of
the thickness of the sulphate melt layer. The reason for this may
be the very small solubility of oxygen in the sulphate melt.

Relations between our potentiostatic measurements and the conditions in a real gas turbine have been briefly mentioned in the
paper.

H.J. Grabke: Would Schmidt or Rahmel please comment on the
occurance of transgranular fracture in the range of intermediate corrosion potential (range B)? It is somewhat surprising
that penetration of sulphur, which may be assumed under these
conditions, should cause transgranular and not inter-granular
fracture. Is it possible to explain this effect, maybe, by

precipitation of fine particles of hard sulfides, such as titanium
or zirconium sulfides - or eventually by the notch effect of the
cracks which you have shown?

Schmidt - Rahmel: We don't have any explanation for the occurrence
of transgranular fracture in the intermediate potential range.
Metallographic examinations didn't reveal sulfide precipitates
at the grain boundaries nor in the grains below the internal
sulphidation zone.

A. Palazzi: We have at CSM obtained nearly the same results.

Sometimes, vanadium is a tramp element for fuel oil; the question
is whether results for sulphates can be adapted or completed
with a survey on the behaviour of vanadates and/or of mixtures of
sulphates and vanadates.

Rahmel - Schmidt: We have not done creep rupture tests in
$(Na_2,Ca,Mg)SO_4$-melt containing any vanadates.
However, from earlier mere corrosion tests between $600^{\circ}C$ and $800^{\circ}C$
in $(Li,K,Na)_2SO_4$-melt we know that the addition of V_2O_5 shifts the
free corrosion potential of an alloy to more positive values
and therefore nearer to the critical break through potential
where corrosion starts. (A. Rahmel: Werkstoffe u. Korrosion
28 (1977) 299; A. Rahmel in "Ash Deposits and Corrosion Due to
Impurities in Combustion Gases" (Ed. R.W. Bryers), Hemisphere
Publ. Corp., Washington 1978, p. 185).
These results transferred to creep rupture tests could lead to
the conclusion that vanadates combined with sulphates stimulate
the corrosion attack even more than reported in this paper.

THE EFFECTS OF SEASALT ON THE HIGH TEMPERATURE CREEP PROPERTIES OF A NICKEL BASE GAS TURBINE BLADE ALLOY

J C Galsworthy

MOD(PE)

AMTE(HH), Poole, UK

SYNOPSIS

The high temperature blading components of gas turbines used in marine applications are subject to mechanical stress and may be susceptible to hot corrosion attack. The object of the work at AMTE is to establish the interactions between these processes on selected superalloys to assist in evaluation and selection. The aggressive engine environment caused by the ingestion of seasalt and sulphur from fuel results in hot corrosion attack of specific hot end components. The mechanisms of hot corrosion are complex and are still to be fully resolved. The environmental creep test methods must keep as close as possible to this undetermined reality as feasible. The various methods of dynamic test equipment, precorroded and salt coated tests are mentioned. Earlier results from salt coated experiments show the importance of seasalt on stress rupture properties and a more extensive programme has been formulated. The results presented in this paper are of salt coated experiments using various seasalt concentrations on the cast superalloy, IN738, over the temperature range 750-850°C. The results show a significant reduction in rupture ductility and life. Lives and creep rates are dependent on the concentration of seasalt present. The reality of the salt composition is discussed as are the corrosion effects on the mechanical properties of nickel base superalloys.

INTRODUCTION

Gas turbines have been used in marine applications since 1947
for propulsive machinery and auxiliaries. The high temperature
blading components of these engines are subject to high inertial
forces and may be susceptible to hot corrosion attack attributed
to the severe environment in which they operate.

The object of the work at AMTE is to establish the effects of
the marine gas turbine environment on the creep and stress
rupture properties of selected superalloys to provide more
realistic data to assist in evaluation and selection of marine
gas turbine blading and nozzle guide vane materials.

ENGINE ENVIRONMENT AND HOT CORROSION

The fuel used in naval gas turbines is normally diesel fuel or
gas oil with a maximum sulphur content of about 1.0% and likely
to have increasing sulphur contents as fossil fuels become
scarcer. The practice of ballasting fuel tanks with seawater
and in some instances the use of water displacement, gives rise
to seawater within the fuel. Fuel filtration systems are
required to reduce this salt content to acceptable limits.

The air contains a heavy sea spray burden and must be filtered
before entry to the engine. Filtration efficiencies depend on
factors such as sea state, wind and ship velocities,etc. The
three stage filtration system of spray eliminators, knitted
mesh filters and inertial vane separators is designed to maintain the salt level at below 0.01 ppm.

The constituents important in the hot corrosion attack observed
are seasalt entering by air and fuel and sulphur oxides produced
on combustion of the high sulphur fuel. Carbon, a product of
incomplete combustion may also be relevant. Many attempts to
establish by theoretical, experimental and engine sampling
programmes, the chemistry of the environment, the fluid dynamic
factors, temperature and pressure effects which result in the
formation of deposits on hot surfaces and the combustion
environment composition are still unable to resolve the complex
mechanisms although some progress has been made to a better
understanding.

The simplest theories of hot corrosion on the nickel base
superalloys involve the breakdown of the normally protective
oxide scale, usually chromia or alumina, attributed to the
presence of chloride, (be it NaCl (s), NaCl (v) or HCl), and
facilitated by a molten salt coating, eg fluxing. Sulphides
and sulphide - metal eutectics form and penetrate deep into the

metal alloy. The precise type of hot corrosion attack will
depend on the temperature of exposure and the precise environment present.

The various species existing at the hot components, specifically
the nozzle guide vanes and turbine blades is open to much
discussion. Despite the thermodynamic predictions that the
sodium chloride would be completely converted to sulphate in
the presence of SO_2/SO_3 at these temperatures research[1] has
shown that the short residence times characteristic of gas
turbine combustion systems (5 to 7 ms) are insufficient for
instantaneous conversion of chloride to sulphate.

ENVIRONMENTAL CREEP DATA

To obtain realistic environmental creep data the test conditions
must be as close to reality as feasible, under laboratory
conditions the engine environment at the specific engine location
of interest must be simulated. This environment is so often
unknown and can only be approximated until vital engine testing
is conducted.

Perhaps the most realistic environmental creep facility for gas
turbine materials is the use of dynamic testing rigs where the
combustion gases are produced in a combustion chamber, bled off
to various test ports and doped with seasalt. These facilities
involve considerable initial expenditure and are costly to
operate. An alternative method involves the mechanical testing
of specimens which have been exposed in corrosion combustion
rigs. The absence of stress during corrosion can however give
different corrosion behaviour and consequently misleading
results.

Attempts to simulate specific environments under controllable
laboratory conditions began at our laboratory with the use of a
hollow specimen through which the corroding constituents were
passed. In these tests seawater and SO_2 were used. Various
problems, not all experimental, necessitated the omission of the
SO_2 from the early experiments but results showed severe life
reduction in the presence of salt. Later solid sodium chloride
and seasalt were coated on the internal bore of the specimens,
again severe life reductions were observed. To help in
establishing the important chemical species influencing the
stress rupture life, a series of experiments on Nimonic alloy
90 salt coated round, cylindrical, specimens were conducted[2].
The stress rupture life data collected in this series of
experiments are shown in Figure 1. It can be seen from these
results that the presence of sodium sulphate had no effect on
the rupture life. However, the presence of sodium chloride

under these conditions caused a considerable reduction in life.
A further life reduction was observed when natural seasalt or a
sodium chloride/sodium sulphate mixture was present.

From these beginnings a more extensive programme has been
formulated using lower salt concentrations, a wider temperature
range and modern cast superalloy specimens. The work presented
here is part of a larger programme which will include the effects
of SO_2/SO_3 with and without seasalt and comparison with precorroded test specimens.

EXPERIMENTAL

The specimens used in these tests were of IN738, analysis as in
Table 1, cast to shape and surface grain refined. After being
given the recommended heat treatment of 2h/1120°C/AC and 24h/
845°C/AC, the parallel gauge length was machined to 5.045 mm $\pm$
0.005 mm giving a nominal cross sectional area of 20 mm^2. All
specimens were degreased prior to a Portland Harbour Seasalt
coating application, analysis as in Table 2. Thermocouples
were tied along the gauge length and an extensometer fixed onto
the specimen shoulders, spanning the machined gauge length.
Because the extensometer was not fixed directly to the gauge
length the creep strains obtained are quoted as 'Relative'.
The testing was conducted in conventional beam loading creep
machines. Creep testing was to BS3500 or better.

The temperature range covered by these studies was 750-850°C
representative of the operating temperatures of first generation
marine gas turbine engines. Each test temperature was paired
with an experimentally realistic stress to enable comparison of
results under the various concentrations of seasalt coating.
These concentrations were from a very thick layer, 1.0 mg mm^{-2},
to a surface glaze, 0.02 mg mm^{-2}; the later being closer to the
expected engine concentration with cumulative deposition on
blading.

The results, as shown in Figure 2 and Table 3, show a reduction
in rupture life at all temperatures in the presence of seasalt,
the severity of the reduction appearing to bear some relation on
the salt concentration and test temperature. The creep rates
determined increased with increasing salt concentration and the
rupture ductility data showed a considerable drop when seasalt
was present. In air values ranged from 3-4% but in the presence
of seasalt were reduced to around 2% at all three temperatures,
apparently independent of concentration.

DISCUSSION AND CONCLUSIONS

Since creep failures within engine components are rarely observed under normal marine operation it is evident that salt concentrations as high as 1.0 mg mm^{-2} are unlikely to exist. Smaller concentration of the order of 0.02 mg mm^{-2} could easily occur on blades particularly in localised regions. In all laboratory experiments the salt composition becomes of paramount importance and it could be argued that the chloride content of the salt used in these tests is too high. However, it has been observed[1] that chloride is not sulphated instantaneously by the presence of SO_2/SO_3 combustion products resulting in higher chloride concentrations on the blades from impacting particles. Also in the engine, fresh chloride rich seasalt deposits will continually arrive at the blade surface. Combustion rig sampling[3] and blade deposit analysis has indicated these high chloride levels.

It has been suggested that the salt concentration effects observed in this work should be attributed to the chloride content alone but the synergistic effects of sulphate and chloride as observed in earlier work[2] may contribute significantly to these effects. The observations of Hocking et al[4] have detected the evolution of sulphurous oxides from seasalt and salt mixtures containing chloride and sulphate, this enhances the argument and shows the importance of the next stage of this programme in which SO_2 and SO_3 alone or with seasalt are to be used.

The severity of life reduction was least at the 800°C test temperature. Thermodynamically, the rate of a reaction should increase with temperature and it would be expected that the life reduction showed a similar trend. However, the substantial reduction at 750°C may be attributed to evaporation rates of the chloride and the salt melting point. At this lower temperature a high chloride concentration can exist for longer and may effect the corrosion rate. As temperature increases the chloride concentration falls with time and the effects of sulphate increase. This observation may be related to the low and high temperature hot corrosion types being reported in the literature.

Metallographic examination being conducted on these specimens may identify the mechanisms of the interaction more precisely but it is considered unlikely that the life reductions result solely from the loss in load bearing cross sectional area. This trend was observed in the previous study.

Any scale formation will effect the mechanical properties of the material be it simple oxidation or complex accelerated hot corrosion. Surface reactions will alter the crack nucleation

and growth mechanisms, the outward diffusion of alloy elements
and the inward diffusion of the environment components into the
alloy. Rapid diffusion paths such as grain boundaries will
significantly enhance these processes. Superalloys are
particularly vulnerable to environmental weakening because any
changes in the matrix composition will alter solid solution
hardening effects and the stability of the precipitates, on which
the high temperature mechanical properties rely.

In conclusion, this work has shown our present lack of knowledge
of the precise environmental conditions in which the marine gas
turbine blades operate. With our limited knowledge care must be
taken in simulating these undetermined and inevitably varying
conditions but hopefully as more environmental creep data under
these various conditions is accumulated, our understanding of the
interactions of corrosion and creep will improve.

Copyright (C) Controller HMSO, London, 1980

REFERENCES

1.   HANBY VI, Paper No. 73-WA-CO-2, ASME Winter Annual Meeting
      Detroit, November 1973.

2.   GALSWORTHY JC, "Environmental Degradation of High Tempera-
      ture Materials, Vol 2", Institution of Metallurgists Spring
      Conference, Isle of Man, March 1980.

3.   CONDE JFG and McCREATH CG, Paper No. 80-GT-126, ASME Gas
      Turbine Conference, New Orleans, March 1980.

4.   HOCKING MG, VASANTASREE V, SWIDZINSKI M and CAREW JA, High
      Temperature Metal Halide Chemistry, ed Hildebrand DL and
      Cubicciotti DD, The Electrochem Soc Proc, Atlanta 1977.

TABLE 1 - Chemical Analysis of IN738 Test Material

ELEMENT	Ni	Cr	Co	Al	Ti	C	Mo	Nb	Ta	W
per cent	bal	16.3	8.35	3.34	3.38	0.17	1.76	0.87	1.78	2.62

TABLE 2 - Chemical Analysis of Natural Portland Harbour Seasalt

CONSTITUENT	SO_4^{--}	Cl^-	Na^+	K^+	Mg^{++}	Ca^{++}
per cent	6.5	56.1	32.5	0.9	2.7	1.2

TABLE 3 - Experimental Results for IN738

		AIR	Seasalt Concentration mg mm^{-2}		
			0.02	0.10	1.0
Rupture Life (% life in air)					
750°C	400 MPa	100	19	24	0.8
800°C	300 MPa	100	58	20	0.3
850°C	200 MPa	100	7	14.5	0.5
Creep rate (h^{-1} x 10^{-5})					
750°C	400 MPa	1.21	1.62	1.63	9.49
800°C	300 MPa	1.56	1.32	4.19	51.5
850°C	200 MPa	0.87	1.72	5.71	28.5

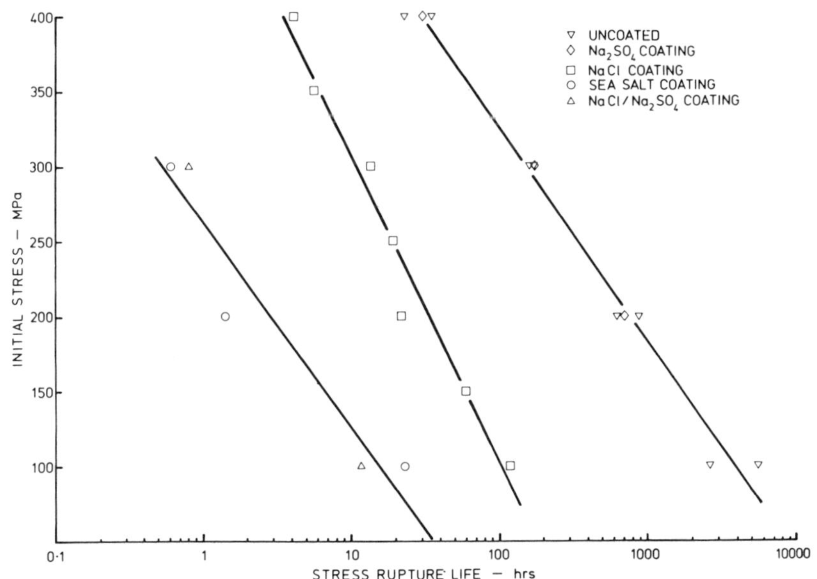

FIGURE 1 - Stress rupture data on salt coated
Nimonic alloy 90 at 750°C

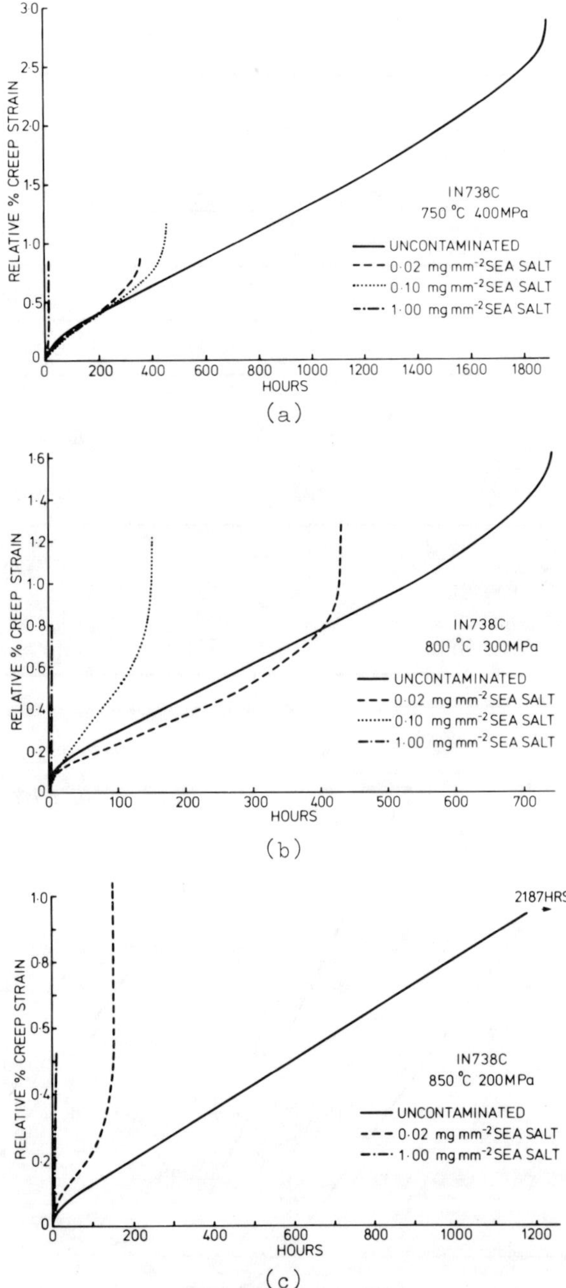

FIGURE 2 - Creep strain against time for seasalt coated IN738

DISCUSSION

P. Kofstad: What is the effect of the chloride?   Where is the
chloride located (or concentrated) in the corroded specimens?   How
does chloride degrade the protective properties of compact scales?

J C Galsworthy: The immediate and significant effect of chloride
on the integrity of the protective scale layers has been reported
[1,2]. Suggested mechanisms for inducing cracking and spalling
involve either contaminant interactions with the scale or substrate alloy, or the action of stresses generated during growth
of the scale or superimposed by operating conditions.   It has
been proposed by a number of workers in the field that vapour
phase species may be involved in the degradation of protective
scales.[1, 2, 3]. Studies by Wareham[4] provide evidence that vapour
phase species might be involved in scale degradation by contaminants.   This was based on the observation of titanium rich crystals
which appeared to have formed by a pyrolitic or vapour deposition route when preoxidised superalloy samples were exposed to
chloride containing salt media in the range 750 to 950°C.   Wareham's
work has been further substantiated by other unpublished joint
studies by MOD and Imperial College, London workers who have
shown TiO_2 and Cr_2O_3 crystals generated by exposure of superalloy
materials to SO_2/seasalt environments at similar temperatures.
The mechanism was associated with the stability of chlorides
and sulphides at low oxygen pressure and decomposition to form
the oxides at positions of higher oxygen pressure.   Mentioned
in the text is Hocking's work showing the evolution of sulphurous oxides from seasalt and salt mixtures containing sulphate
and chloride.   The location of the chloride in the corroded
specimens has not been determined since the presence of chloride
is difficult to confirm due presumably to the high volatility
of chlorides which results in very low residual levels.

1.   Condé J F G, AGARD Conference on High Temperature Corrosion
      of Aerospace Alloys, Denmark, April 1972 AGARD-CP-120.

2.   Hurst R C et al, Proc of CEGB Conference on Corrosion and
      Deposition in Gas Turbine, London, December 1972 Ed A B Hart
      and J B Cutler, Appl Sci Publishers Ltd, London, 1973.

3.   Gray P S, Proc 1st Int Conf on Metallic Corrosion,
      Butterworths, London 1962.

4.   Condé J F G and Wareham B A, Proc of the 1974 Gas Turbine
      Materials in the Marine Environment Conference, Castine,
      Marine, USA.   MCIC Rept July 1974, MCIC-75-27.

H.W. Grünling: My question is addressed to both Galsworthy and Götzmann, since I had the impression that they did not tell us enough about the influence of simple cross section reduction, due to corrosion attack on the reported results. I would like to know how this cross section reduction is influencing the reported stress rupture or creep results.

J C Galsworthy: The effect of loss in cross section due to corrosion will cause an increase in the applied stress under the constant load test conditions. The precise magnitude of this increase has not been determined in the work presented but work previously reported on Nimonic 90 has shown the life reduction cannot alone be attributed to this stress increase. The suggestion that the environment has an important effect on crack initiation could be substantiated by the results presented. The secondary creep rate, when low salt concentrations were present, is initially similar to the static air rate, the strain then increases rapidly and failure occurs. This change could be environment enhanced crack initiation resulting in premature failure.

O.Götzmann: The cross section reduction was an integral part of the test results. There was an reduction in cross section observed but not measured on all individual test specimens. The reduction in rupture time is due to reduction in cross section and possible other corrosion effects (e.g. stress concentration due to grain boundary attack). The influence of simple cross section reduction was not specifically investigated.

B. Ilschner: This is a suggestion rather than a question. It might be of interest for the interpretation of the phenomena observed with corrosive atmospheres if some of the experiments reported here could be repeated with pre-cracked specimens according to fracture-mechanical procedures. This would allow to isolate the effect of corrosive reaction on crack propagation rate from the effect on crack initiation probability. Data like rupture life or (average) creep rates are, of course, integrating over a variety of contributing sub-mechanisms.

Corrosion Enhanced Stress Rupture of Cr-Ni Stainless Steels at High Temperature

V.Gibs, O.Götzmann

Kernforschungszentrum Karlsruhe
Institut für Material- und Festkörperforschung
Postfach 3640, D 7500 Karlsruhe 1
Federal Republic of Germany

SYNOPSIS

Burst rupture and creep rupture tests were carried out in atmospheres contanimated by halogen and chalcogen vapour species at temperatures of 500 and 700°C. The corrosive species were available at various chemical potential levels. The test results clearly indicate that corrosion effects on the specimens increase with increasing chemical potential of the reactive species in the atmosphere. There are also some indications that the corrosiveness of the elements in the halogen and chalcogen groups decrease with increasing atomic number (decreasing electronegativity). In the presence of oxygen, corrosion by halogens becomes stronger. This, however, seems to depend both on the chemical potential of the halogens and that of oxygen. Corrosion shortens the creep rupture life of stainless steel specimens by either increasing the creep rate during the secondary creep stage, or by accelerating crack growth, or both. Secondary creep rate is enhanced by a corrosive action on the specimens' surface increasing incipient crack density, and crack growth perhaps by a corrosive action at the tip of the crack.

INTRODUCTION

In our investigations on the corrosion behaviour of stainless steels and other candidate materials for claddings of nuclear fuel pins we have learned that the most sensitive method to indicate wether chemical interactions with the corrosive environment have taken place is to determine the change in mechanical properties, especially the ductility of the metallic materials. In order to

demonstrate how sensitively the load bearing capability of the
cladding is affected by small quantities of fission products or
impurities in the fuel we have utilized testing methods by which
the test parameter was not the concentration of aggressive products but the chemical potentials of such products or elements in
the environment. We assume that it is not primarily the amount of
reactive products available but their chemical state in which they
are present that accounts for the corrosion effects.

The basic conception in this investigation was the idea that for
a low reactivity in the corrosive environment (i.e., small driving
force for reaction) the combination with stress would generate
greater effects, a behaviour seen in stress corrosion cracking,
for instance. The elements most reactive against the metallic
materials are the elements of the seventh and sixth group of the
periodic table, i.e., the halogens and chalcogens. The elements
of the first group in the periodic table, the alcaline metals, also regarded as aggressive - cesium, for instance, plays an important role in cladding attack of oxide pins - react with the metallic materials only with the help of oxygen.

The capability of the halogens and the chalcogens to cause a critical stress rupture situation with stainless steels was assessed
as being related to their electronegativity with respect to
chromium. In this sense the reactivity in the two element groups
should decrease with increasing atomic number, and the halogens
should be more reactive with the stainless steels than the
chalcogens. Above all, however, it was assumed that the level of
the chemical potential at which the corrosive species are available has the most significant effect.

EXPERIMENTAL

To demonstrate such a behaviour different kinds of tests were performed. One test series employed tubular specimens that were
filled with reactive chemicals, pressurized and annealed at 500°C
(see table 1 and fig. 1). The inner gas pressure at temperature
was in the range of some 35 MPa. The resulting stress in the
tubing wall exceeded the yield strength of the material. As reactive chemicals, powder mixtures of chromium, iron, nickel and
copper with their respective halides were used (see table 2). The
metals and metal halides were selected in such a way that a
differentiation in reaction potential was possible.

To study the influence of oxygen on the halogen corrosion, an
oxide of the same metal was added to the metal/halide mixture
(see table 2).

In a test series to study the creep rupture behaviour in more
detail, flat sheet specimens were heated in a corrosive environment up to 700°C, and loaded to above their yield strength. The
corrosive environment was supplied by similar powder mixtures
as in the above described test series. These mixtures were kept
at a temperature somewhat below the test temperature. The vapour
species evaporating from these mixtures were carried to the test
specimen by a gas stream of purified argon.

Finally, in a third test series, strip metal specimens of various
materials (see table 3) were bent and fixed in the bent position
by point welding (fig. 1), then annealed in a closed capsule
filled with one of the corrosive powder mixtures (see table 2).
These mixtures were positioned in the capsule such that contact
with the test specimen occurred only by vapour phase. This last
test series was meant to show how stress relaxation is affected
by corrosion, and also to show how corrosion is influenced by
stress.

RESULTS AND DISCUSSION

Burst tests

The presence of the reactive chemicals shortened the rupture time
of the stainless steel tubes. As can be seen from fig. 2, and
more so in the following figures 3 to 6, the effect of the reaction potential is abvious. Of the halides used the chromium
compounds were the ones that yielded the lowest chemical potential in each halide group. That means, a halogen reaction with
the stainless steel tubes was thermodynamically the least probable when a chromium halide was present as reactive agent. The
iron, nickel, and copper halides, generally in this order, are
less stable than the chromium halides and thus support a higher
chemical potential (activity) of the reactive elements. The fact
that the higher reaction potential yields a shorter rupture time
is clearly indicated by the results. The chromium halide always
has the least effect on the mechanical behaviour and the copper
halide the most, except in the case of the iodides. CuI supports
a lower iodine pressure than iron iodide (see table 4) which is
in agreement with the outcome of the rupture tests. The diagram
in fig. 2 was conceived as an arrangement of assumed behaviour.
Under the concept of electronegativity, fluorine was suposed to
be the most reactive species among the halogens, thus it was
assumed that the fluorides yield shorter rupture times than the
corresponding chlorides, and these again shorter ones than the
borides and iodides. Such an order of behaviour, however, is not
fully supported by the results of the burst tests.

The influence of oxygen on the halogen corrosion is well demonstrated by the test results as shown in fig. 3 to 6. In systems

with chromium halides, there was only a slight effect of oxygen
on the corrosiveness of the halogen environment. In systems with
somewhat higher halogen potentials, as in those with iron and
nickel halides, the effect of oxygen was significant. In systems
with the highest halogen potentials, which was normally the case
in systems with a copper halide, the already high corrosion effect
was not enhanced much more by the presence of oxygen. However,
one has to consider that by adding oxygen to the halide environments, oxides of the same metals were used (see table 2). Thus,
the oxygen potential in the different system varied according to
the halogen partner employed. It was lowest in systems with
chromium halides and highest in systems with copper halides. Moreover, Cr_2O_3 buffered with metallic chromium is thermodynamically
not capable of oxidising stainless steel, whereas all other metal
oxides employed are. Thus, it is well conceivable that in systems
with chromium as halogen partner the effect of oxygen was only
weak.

Similar results were obtained by tests with prestressed (bent)
specimens for which the weight gain was measured after anneals in
various corrosive atmospheres. As fig. 7 shows, the influence of
oxygen is most significant for an intermediate chemical potential
level as achieved in an atmosphere contaminated by iron fluoride.
If added to nickel sulphide the presence of oxygen even at a
high oxygen potential does not change the low reactivity situation. Also if added to the highly corrosive atmosphere produced
by $CuCl_2$ the oxygen is not capable to cause much more corrosion.
However, one can see from fig. 7 that different corrosive environments lead to different corrosion behaviour of the stainless
steels and similar alloys. One corrodent may attack one alloy
more than an other one. In another corrosive atmosphere it may
just be the other way around. That is to say, if one material
proves itself superior to another one in one corrosive atmosphere
it may be inferior to the same material in a different corrosive
environment.

An illustrative example of how oxygen influences halogen corrosion (or vice versa) can be seen from pictures of corrosion
zones in fig. 8.

Creep rupture tests with flat specimens

The results of this test series are given in figs. 9 to 12. In
fig. 9 the creep elongation over time is shown for specimens
tested in corrosive environments produced by copper halides. Here,
the results seem to give come credit to the concept of an increase in corrosion effect with increasing electronegativity of
the corrosive species. The rupture time is being decreased in

going from the reference specimen (exposed to an inert atmosphere) to specimens exposed to an iodide, to a bromide, to a chloride, and to a fluoride atmosphere (in this order). However, this seemingly convincing picture is somewhat dimmed by the fact that CuI yields a rather low reaction potential compared with the other iodides used. The same is the case for CuBr; whereas CuF_2 and $CuCl_2$ yielded by far the highest reaction potential in their respective groups.

The mechanism that led to the premature failure of the specimens exposed to the corrosive atmospheres was causing an increased strain rate during the stage of secondary creep. This effect is becoming stronger in going from the iodide atmosphere to the fluoride atmosphere (see fig. 10).

A similar set of results is shown in fig. 11 for tests with a somewhat higher load on the specimens. Here, we also see the impact of chalcogens as corrosive agents. Their potential was offered at a higher level since the pure elements and not compounds were used to contaminate the atmosphere. For selenium and tellurium the partial pressure in the atmosphere corresponded to the vapour pressure at 520^OC, and for sulfur to the vapour pressure at 130^OC. Hence, tellurium and selenium were available at a higher potential than sulfur. For comparison (and to show the effect of a higher load if compared with fig. 9) the creep rupture curves for specimens tested in $CuCl_2$ and CuI atmospheres are also shown in this diagram. Again, these results seem to confirm the concept of an increasing corrosion effect with increasing electronegativity of the reactive species, especially by comparing the results of the chalcogen atmospheres. Selenium was more effective than tellurium and sulfur would have certainly been more effective if it had been offered at a comparable chemical potential level.

What makes the results in this diagram interesting is the different rates of secondary creep for the specimens exposed to different atmospheres (see fig. 12). Some of the values for the secondary creep rates fall on a straight line (drawn over rupture time) which suggests that the same creep rupture mechanism was at work for these specimens. For specimens with shorter rupture times a different mechanism was effective. If one takes a look at scanning electron microprobe pictures of surfaces of ruptured specimens the different mechanisms become visible (see fig. 13). Those specimens that experienced an increased secondary creep rate but still fall on the straight line with the reference specimen exhibit a higher density of incipient cracks on the surface (example given in fig. 13 for Te-atmosphere). The specimen exposed to sulfur, which did not experience a higher secondary creep rate, however the same reduced rupture time, shows much fewer cracks on the surface. In this case, it is believed that the corrosive environment affected not so much the specimen's surface but the tip of the crack and thus furthered crack growth.

CONCLUDING REMARKS

The tests have shown that the halogens and chalcogens are reactive agents that shorten the lifetime of stainless steels under load even in the absence of oxygen. Variation of their chemical potential in the atmosphere has a significant effect on the load bearing capability of stainless steels. Their impact becomes more critical in the presence of oxygen. There are some indications by the results of these tests that the corrosiveness of the elements in the halogen and chalcogen groups decrease with increasing atomic number (decreasing electronegativity). The corrosive action under load is brought about by different mechanisms. One affects the surface of the specimen and leads to an increase in incipient crack density which results in a higher strain rate during the secondary creep stage. Another one effects an acceleration of the crack growth perhaps by an action on the tip of the crack. The first mechanism is predominant for lower loads and weaker reactivity of the corrodent. The second one dominates at higher loads and higher reactivity in the corrosive atmosphere. The test results presented here are still preliminary. By further investigations we hope to corroborate these findings.

Table 1: Test Conditions

Burst test:	1.4970 type stainless steel tubular specimens, 15% c.w.; test temperature 500°C, pressure 32 to 37 MPa; hoop stress 490 to 570 MN/m^2; metal halides support corrosive atmosphere
Tensile creep tests:	1.4970 type stainless steel flat specimens s.a.; test temperature 700°C, stressed to 210 and 250 MN/m^2 (yield strength of material at test temperature 135 MN/m^2); metal halides and chalcogens heated to 520°C (130°C) support corrosive atmosphere
Bent specimen annealing tests:	Bent flat tensile specimen of 1.4970, 15% c.w.; 1.4988, 50% c.w., and Inconel 718; test temperatures: 400 and 500°C for 1000 h; metal halides and chalcogens support corrosive atmosphere

Table 2: Powder mixtures used in tests to produce corrosive atmospheres

CrF_2/Cr	$CrCl_3/Cr$	–	CrI_2/Cr
$CrF_2/Cr_2O_3/Cr$	$CrCl_3/Cr_2O_3/Cr$		$CrI_2/Cr_2O_3/Cr$
FeF_3/Fe	$FeCl_2/Fe$	$FeBr_2/Fe$	FeI_2/Fe
$FeF_3/FeO/Fe$	$FeCl_2/FeO/Fe$	$FeBr_2/FeO/Fe$	$FeI_2/FeO/Fe$
NiF_2/Ni	$NiCl_2/Ni$	$NiBr_2/Ni$	–
$NiF_2/NiO/Ni$	$NiCl_2/NiO/Ni$	$NiBr_2/NiO/Ni$	
CuF_2/Cu	$CuCl_2/Cu$	$CuBr/Cu$	CuI/Cu
$CuF_2/CuO/Cu$	$CuCl_2/CuO/Cu$	$CuBr/CuO/Cu$	$CuI/CuO/Cu$

Table 3: Composition of stainless steels and Inconel 718 tested [w %]

Material	Fe	Cr	Ni	Mo	V	Ti	Nb	Si	Mn	C
1.4970	rest	14.9	15	1.19	<0.05	0.43	–	0.4	1.61	0.092
1.4988	rest	16.02	13.62	1.34	0.8	–	0.9	0.29	1.1	0.075
Inc.718	19.2	18.63	51.92	3.14	–	0.86	5.05	0.3	0.19	0.04

Table 4: Chemical potentials of halides at 500°C [kJ/Mol]

Product	$\Delta \bar{G}^{\circ}_{f}$	Product	$\Delta \bar{G}^{\circ}_{f}$	Product	$\Delta \bar{G}^{\circ}_{f}$	Product	$\Delta \bar{G}^{\circ}_{f}$
CrF_2	−673	$CrCl_3$ $CrCl_2$	−136 −308	$CrBr_2$	−233	CrI_2	−136
FeF_3 FeF_2	−537 −599	$FeCl_2$	−246	$FeBr_2$	−183	FeI_2	− 80
NiF_2	−533	$NiCl_2$	−189	$NiBr_2$	−139	NiI_2	− 39
CuF_2	−424	$CuCl_2$ $CuCl$	2 −195	$CuBr_2$ $CuBr$	15 −165	CuI	−101

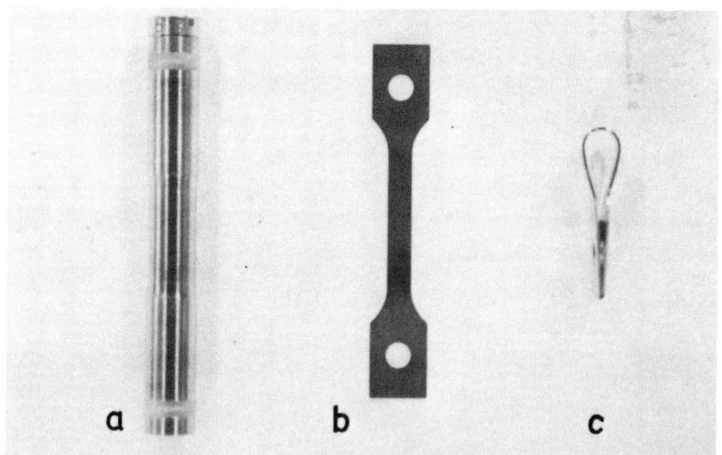

Fig. 1: Form of test specimens: a) tubular specimen with thinned wall in the middle part used in burst tests; b) flat sheet specimen used in creep rupture tests; c) bent flat sheet specimen used in annealing tests

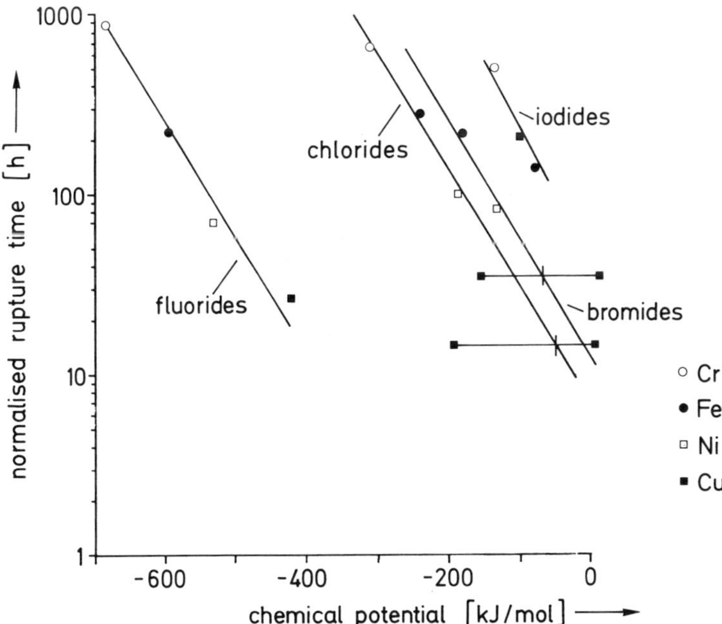

Fig. 2: Rupture times for tubes of type 1.4970 stainless steel, 15% c.w., pressurized to about 35 MPa and corroded at 500°C by halogens of different chemical potential

                    V. GIBS and O. GÖTZMANN

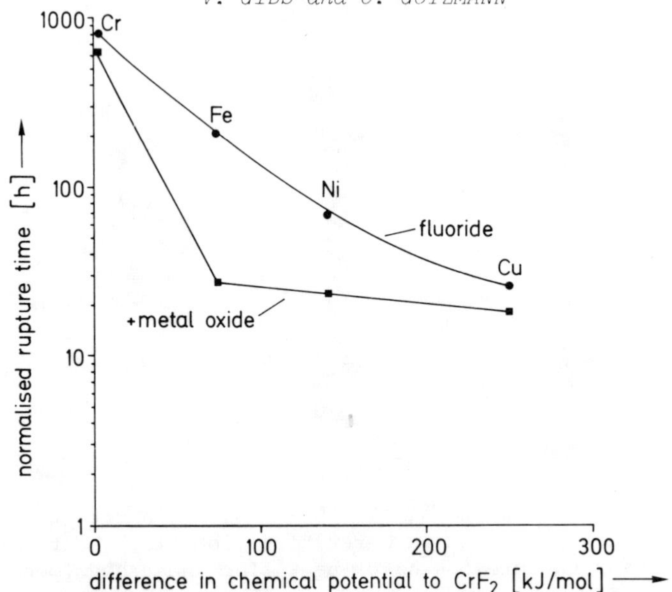

Fig. 3: Rupture times for tubes of type 1.4970 stainless steel, 15% c.w., pressurized to about 35 MPa and corroded at 500°C by fluorine (upper curve) and by fluorine + oxygen (lower curve) at various potential levels

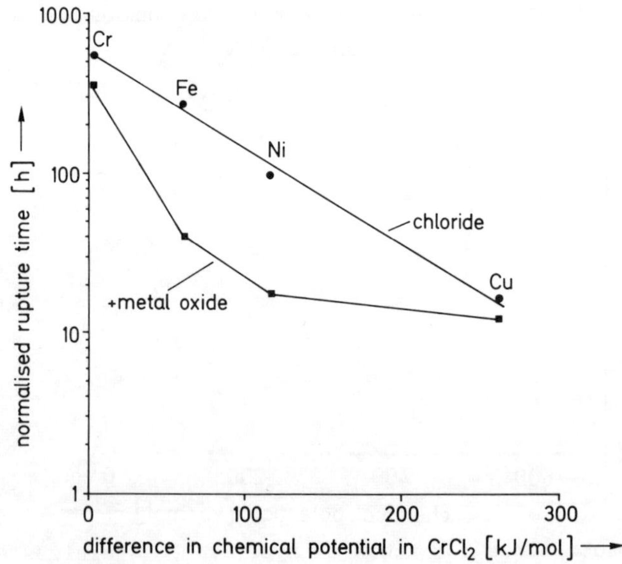

Fig. 4: Rupture times for tubes of type 1.4970 stainless steel, 15% c.w., pressurized to about 35 MPa and corroded at 500°C by chlorine (upper curve) and by chlorine + oxygen (lower curve) at various potential levels

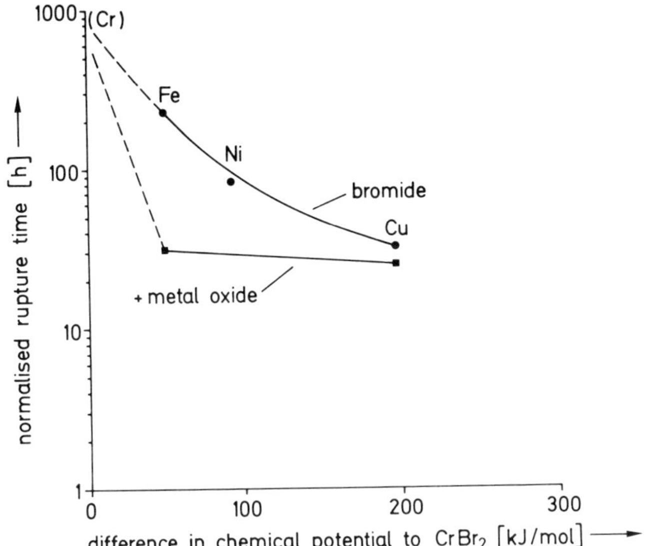

Fig. 5: Rupture times for tubes of type 1.4970 stainless steel, 15% c.w., pressurized to about 35 MPa and corroded at 500°C by bromine (upper curve) and by bromine + oxygen (lower curve) at various potential levels

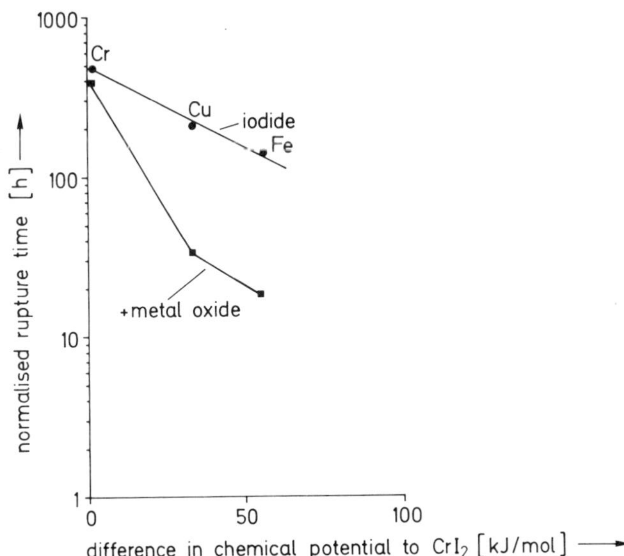

Fig. 6: Rupture times for tubes of type 1.4970 stainless steel, 15% c.w., pressurized to about 35 MPa and corroded at 500°C by iodine (upper curve) and by iodine + oxygen (lower curve) at various potential levels

V. GIBS and O. GÖTZMANN

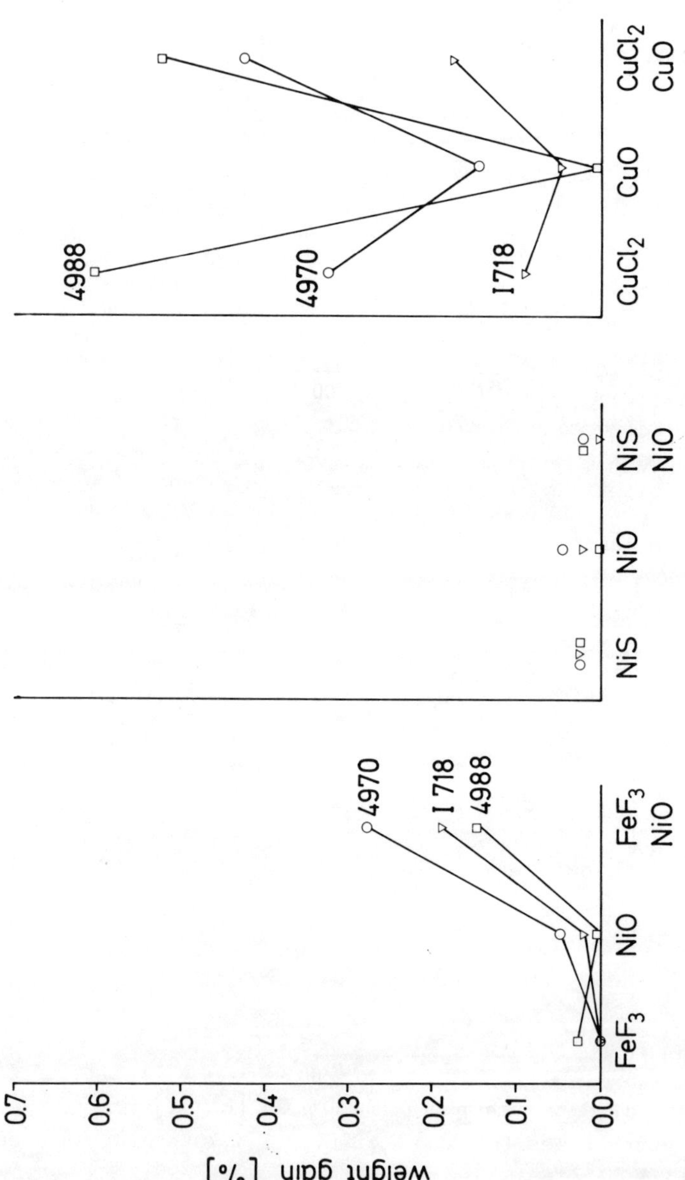

Fig. 7: Weight gain of stainless steels and Inconel 718 exposed to corrosive atmospheres with and without oxygen at 500°C for 1000 h

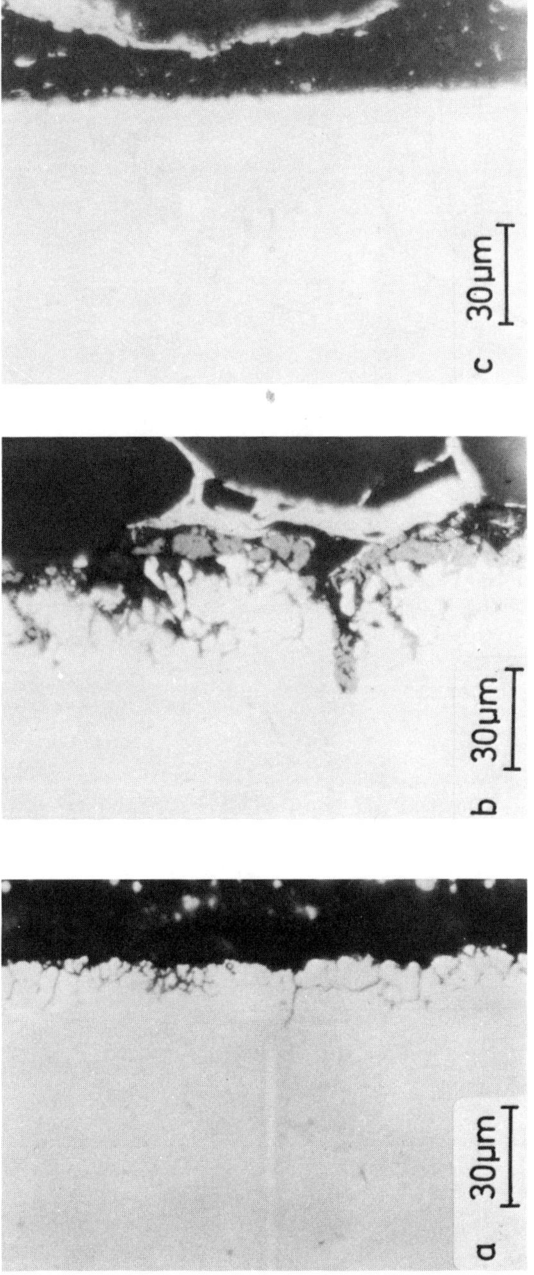

Fig. 8: Corrosion zones on 1.4970 type stainless steel produced by various atmospheres at 500°C/50 h; a) fluorine, with potential of NiF_2/Ni pairing; b) fluorine and oxygen, with potential of $NiF_2/NiO/Ni$ powder mixture; c) oxygen with potential of NiO formation

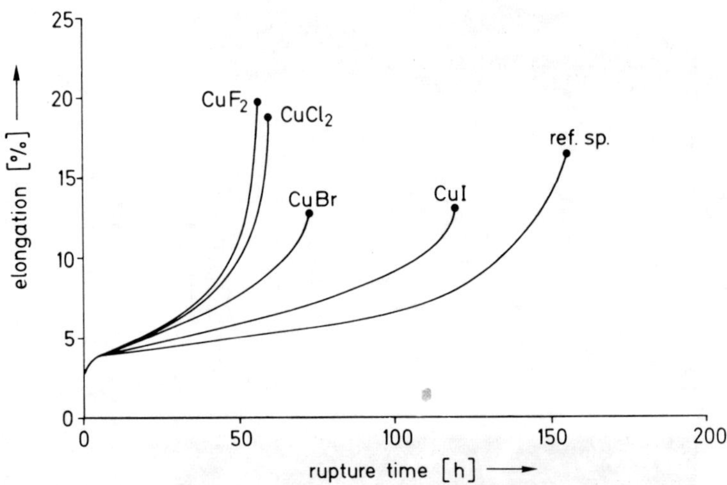

Fig. 9: Creep rupture behaviour of flat sheet specimens of type 1.4970 stainless steel, s.a., stressed to 210 MN/m² and exposed to corrosive atmospheres of various halogen activities at 700°C (ref. sp. = reference specimen exposed to inert atmopshere)

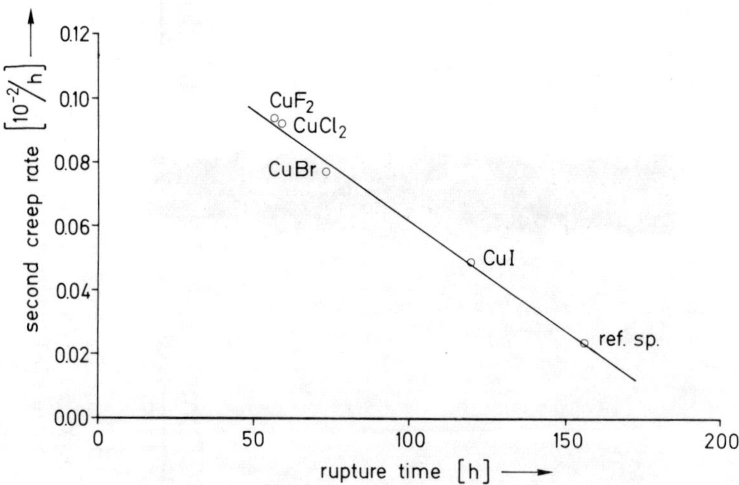

Fig. 10: Secondary creep rate of specimens exposed to corrosive atmospheres of various halogen activities (data from fig.9)

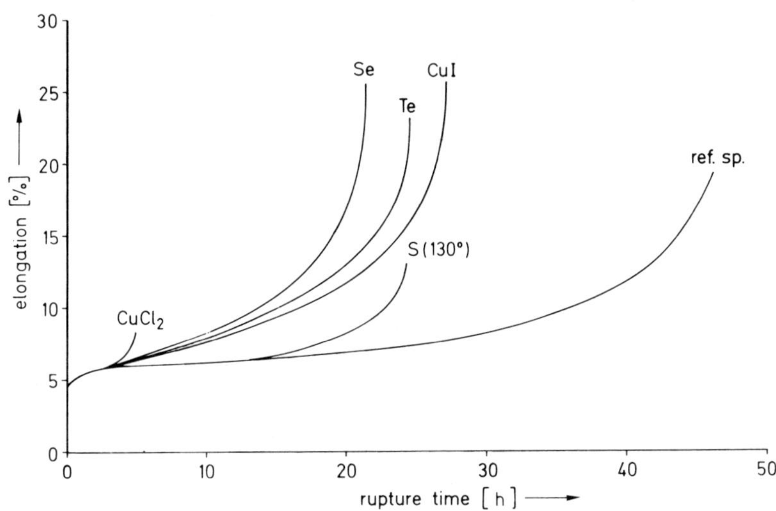

Fig. 11: Creep rupture behaviour of flat sheet specimens of type 1.4970 stainless steel, s.a., stressed to 250 MN/m^2, and exposed to various corrosive atmospheres at 700°C (ref. sp. = reference specimen exposed to inert atmosphere)

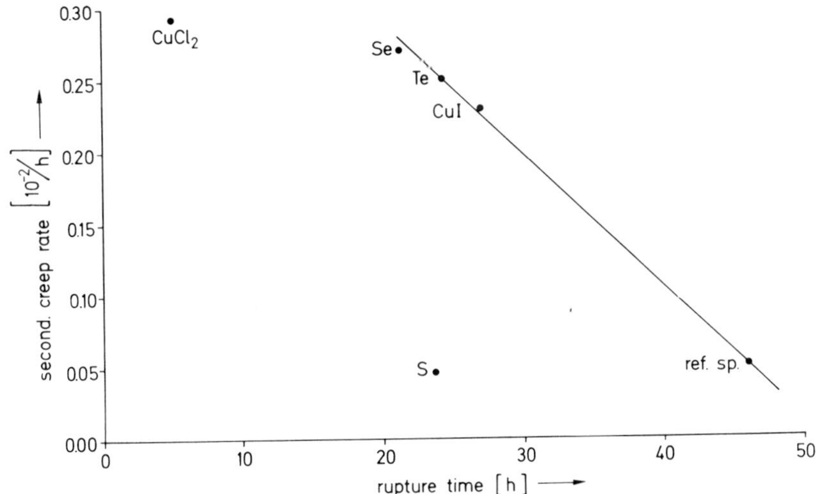

Fig. 12: Secondary creep rate of specimens exposed to various corrosive atmospheres (data from fig. 11)

V. GIBS and O. GÖTZMANN

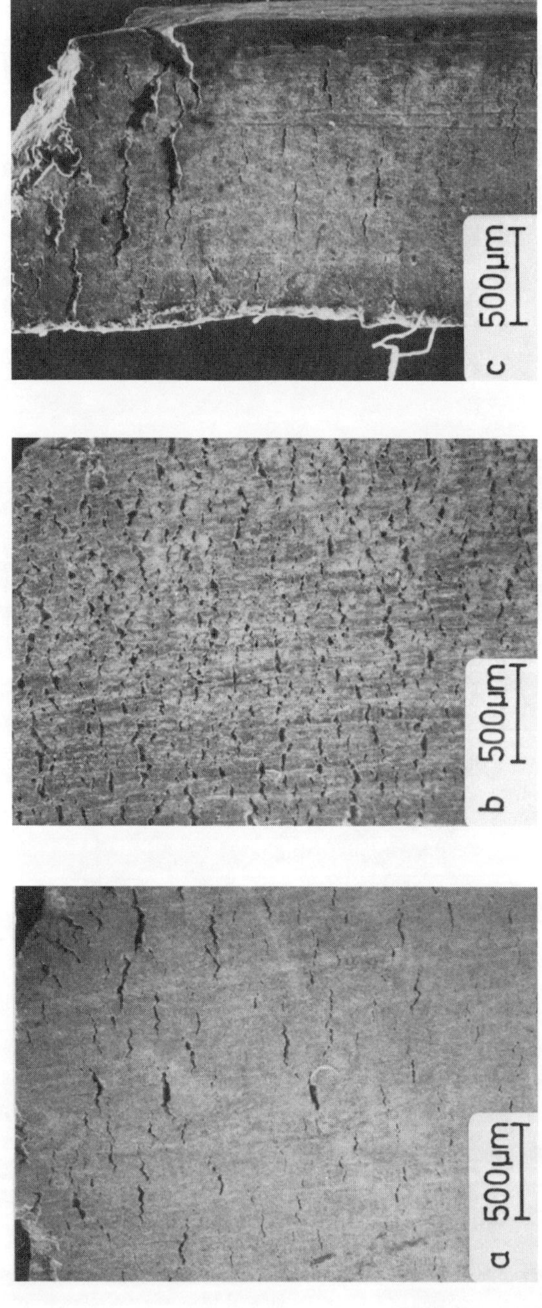

Fig. 13: Scanning electron microprobe pictures of ruptured 1.4970 type steel specimens: a) reference specimen; b) ruptured in tellurium atmosphere; c) ruptured in sulfur atmosphere

Longterm High Temperature Tests with Simultaneous Mechanical Stress on Hot Corrosion Resistant Materials for Land-based Gas Turbines.

Friedhelm Schmitz, Karl-Heinz Keienburg

Technik-Werkstoffe

Kraftwerk Union AG, Mülheim/Ruhr

Since the beginning of the seventies we have been carrying out longterm high temperature tests in respect of the interaction of corrosion and mechanical stresses on high temperature materials for land-based gas turbines. Results obtained in laboratory tests are continuously being compared with experience gained from turbines in operation and test bed experiments.

These investigations show that there is no great reduction in the creep rupture strength due to alkaline sulphate corrosion that can be explained from the loss in cross-section due to corrosion. Under the influence of the hot gas with increased chloride content, damage occurred in the areas near the grain boundary which led to a great reduction in the creep rupture properties. Investigations on land-based gas turbine blades stressed in operation have not yet shown any signs of a possible worsening in the creep rupture behaviour.

Land-based single shaft gas turbines for electricity supply are currently being built for outputs up to 130 MW and far larger units are being developed [1-6]. The blades of the turbines belong to the highly stressed components which are subjected not only to thermo-mechanical stresses but also erosive and corrosive ambient influences.

The essential properties of the blades are fulfilled today by the complex Ni- and Co-based alloys [7].

With gas turbine blades, the high temperature corrosion caused by the alkali sulphates and heavy metal compounds is often the influence which limits the length of life. The mechanism of high temperature corrosion is not described further here but a reference is made to the literature [8-16]. Fig. 1 shows an example of such an attack of high temperature corrosion on a blade made from the alloy Udimet 520 that was subjected to sulphurous natural gas for approx. 27.000 h at $750°C$. The individual sectional figures represent results of the metallographic and microprobe investigations.

The choice and development of suitable materials with sufficiently good resistance to corrosion is an important part of the work of a material engineer in gas turbine field. To guarantee the correct choice, corresponding investigations are made, ranging from laboratory experiments to rig tests up to plant tests, and the results of these correlated with each other.

Fig. 2 and 3 reproduce an example of such tests. Fig. 2 shows the known influence of the Cr-content on the corrosion rate of Ni- and Co-based alloys [17-19]. The photographs show the macroscopic findings for a Ni-based alloy IN 939 with high Cr-content and good resistance to corrosion and a Co-based alloy S 816 with low Cr-content for Co-based alloys and correspondingly lower resistance to corrosion. As a comparison to these laboratory investigations, we have inserted various alloys and coatings in a blade set of the machine previously described and examined them after different periods of service time.

Fig. 3 shows the results from the two above-mentioned alloys IN 939 and S 816 after approx. 25.000 h operating time. Good agreement was obtained in the evaluation scale of the high temperature corrosion behaviour between experiments in the laboratory and in the plant.

Alloys with good corrosion resistance were selected for further tests with combined corrosion and simultaneous mechanical stress.

The reason for performing tests of this sort were the results obtained by Huff and colleagues from the creep rupture tests in hot gas [20]. With these, considerable reductions in creep rupture strength were found on alloys of aircraft engines under hot gas conditions.

As the range of application for land-based gas turbine blades deviates greatly from that of aircraft turbines [21] we have been working within the framework of the COST 50 programme with creep rupture tests under the influence of corrosion on alloys for land-based gas turbines since 1973. During this time various working groups have been employed with this important topic, the results or partial results of which have already been reported [21-37].

Fig. 4 gives a survey of the investigation programme performed by us. The chemical composition of the most important elements of the materials examined by us are reproduced in Table 1. Table 2 shows the corrosion conditions under which the investigations were performed.

Only excerpts from the investigation results obtained can be given here. Individual results can be seen in the COST 50 reports [38-43].

As an example of one of the alloys investigated, the results with the Ni-based alloy IN 738 LC at 850°C are to be submitted. Fig. 5: With all alloys examined, as is shown with the alloy IN 738 LC, the running time of the test until rupture under a cover of ashes (sulfidation) was uniformly reduced in comparison to air. However, all rupture points are still in the scatter band. To clarify whether the reduction in the creep rupture strength was caused by the reduction in the cross-section due to high temperature corrosion or by damage to the alloy beyond this, metallographic investigations followed.

Fig. 6 shows the metallographic section through a broken sample which was stressed with 200 N/mm^2 over a period of 1.270 h. The known type of high temperature corrosion is present in Ni-based alloys. The estimated amount of increase in stress due to the reduction in cross section led to the result that the creep rupture values corrected in this way do not point to any recognisable reduction as opposed to air. This result does not agree with the results from creep rupture tests on blades of aircraft engine under hot gas corrosion [20,23] that have already been mentioned. Based on Auger-Microprobe investigations it was concluded then that sulphur from the hot gas had caused the reduction observed in the creep rupture strength due to its interaction with the creep mechanism. This investigation result has repeatedly given rise to discussions. Therefore analogous AES-tests - on the samples tested in the high temperature corrosion creep rupture test - were made by the MPI-Düsseldorf.

The notched samples were broken at $\sim -100^{\circ}$C in the Auger instrumentation under ultra high vacuum and subsequently analysed. The rupture surface was situated in the test area of the creep rupture samples in both cases.

Both rupture surfaces showed in the REM an almost exclusively transcrystalline rupture structure with intercrystalline secondary cracks.
Fig. 7 shows a typical section of the rupture surfaces. Chrome-, tungsten- and titanium-carbides as well as zirconium sulphide were found on the intercrystalline cracked grain boundaries of both samples. The AES spectrum of the samples tested under hot temperature corrosion and air were identical. As an example, Fig. 7 shows the Auger spectrum and the Auger map for nickel and sulphur for a grain surface with zirconium sulphide deposits. The occurrance of zirconium sulphide in IN 738 LC is known 7,45; a metallographic examination of this melt confirms the AES findings.
Furthermore it must be stated that with these AES investigations no indication whatsoever of a deep penetration into the interior of the sample of sulphur from the testing medium was found. It is also difficult to imagine such a process thermodynamically. These findings correspond with the results of our HTK creep rupture tests and the experience meanwhile gained with GT blades with operation times of up to over 100.000 h. To date, no creep rupture damage has occurred to KWU blades.
We assume that an interaction between the creep fracture and corrosive environment will take place only when a macroscopic creep cracking occurs. In this final phase of the creep life the influence of the test duration can be neglected.
Fig. 8 shows a comparison of the influence of the high chloride content of the testing gas on the same alloy IN 738 LC. With this, it is clear that a great reduction in the creep rupture strength occurred with individual samples. This could not be caused to the reduction in cross section. The same also occurred with other alloys tested.
Fig. 9 shows the metallographic investigation of a creep rupture sample. It can clearly be shown that, with samples having a reduced running time, intercrystalline damage over the whole length in the area near the grain boundary is present. The reasons for this behaviour have not been clarified at the present time.

From this example it is especially clear how important it is to correlate findings from laboratory tests with investigations on blades after different service times.
Here are some examples of this:
Fig. 10 presents the results of non-finalized creep rupture tests on samples from the operationally stressed blades shown in Fig. 1. This concerns comparative examinations of the state after operation and after two additional different "regenerating" heat treatments.

The creep rupture strength of the operationally stressed
state is on the upper limit of the scatter band. This result
corresponds to the low theoretical life consumption which
results from the operation life of about 27.000 h up to now,
which is low in comparison with the designed life > 200.000
hours. In this case, then, it is not expected that the
"regenerating" heat treatment will yield any considerable
improvement in the creep rupture strength. Apart from this,
the creep rupture tests show that the blade in the area of
the sample location, i.e. in the interior of the blade, has
not suffered from any creep rupture damage due to the
strongly corrosive sulphurous medium.
A further example is to show how metallographic investigations of creep stressed Ni-based alloys can provide
qualitative indications of the life consumption if necessary.
An example of a creep test specimen of the alloy Udimet 520
is shown in Fig. 11, where γ'-free light grain boundary
zones with chromium carbide deposits are formed under creep
stress [44,45] . Depending on the formation of these zones,
the stability of the grain boundary, which mainly determines
the creep rupture strength, is more or less greatly reduced.
In the longitudinal microsection of the creep rupture specimen, it was found that the width of the light zones decreased
on the one side from rupture to the clamped end and on the
other side from the specimen surface to the specimen interior.
The formation of the γ'-free zones is obviously an interaction with the various local creep processes.
In this connection, specific examinations were performed on
the transverse microsection of the blades of the afore-
mentioned operationally stressed blades made of the same
alloy. The result is given in Fig. 12. It was found that
indications of commencing zones formation were present up to
a depth of approx. 0,4 mm. To provide a comparison, the
metallographic examination of a broken creep rupture test
specimen is shown. The reason for the commencing damage in
the boundary area is considered to be mainly the alternating
thermal stress on start-up and shut-down and on load changes
during gas turbine service. In this case the surface area
shows only very slight damage in comparison with the creep
rupture test specimen.
To summarize the results it can be stated that no decrease
in the creep rupture strength occurs beyond the average
reduction due to sulfidation. The, in part, great decrease
in the creep rupture strength under the influence of chloride
must continue to be thoroughly investigated. Subsequent work
in this field should take dynamic stress, higher testing
temperatures and longer test durations into consideration.

Such investigations can no longer be financed by industrial
firms alone but, as already planned, must be performed as
cooperative experiments with public financing (e.g. IfW
Darmstadt).

Acknowledgement:

The authors would like to thank Prof. Grabke and his co-
workers for carrying out the AES-examinations and for helpful
discussions.
This work was supported in part by the German Ministry of
Research and Technology (BMFT).

References:

1   H. Maghon,
VGB-Kraftwerkstechnik 53, 1973, Heft 10, Oktober,
S. 651-656

2   O. Schmoch,
Gasturbinen, Brennstoff-Wärme-Kraft 29, 1977, Nr. 4,
April, S. 161-166

3   O. Schmoch,
Gasturbinen, Brennstoff-Wärme-Kraft 30, 1978, Nr. 4,
April, S. 184-188

4   J. Leopold,
Maschinenschaden 49, 1976, Heft 5, S. 3-15

5   B. Becker,
Motortechnische Zeitschrift 40, 1979, Heft 5, Mai,
S. 247-249

6   Gas Turbine World, 1980, März, S. 32

7   V. Thien, U. Schieferstein, W. Voss,
Mikrochimica Acta, 1976, 7, S. 389-403

8   F. Umland, H.P. Voigt, H. Fechner,
Forschungsbericht der Forschungsvereinigung Verbren-
nungskraftmaschinen E.V., 1968, Heft 90, Januar,
S. 2-238

9   H. Fechner,
Forschungsbericht der Forschungsvereinigung Verbren-
nungskraftmaschinen E.V., 1972, Heft 128

10   H. Böttger,
Forschungsbericht der Forschungsvereinigung Verbren-
nungskraftmaschinen E.V., 1972, Heft 129

11   F. Umland, W. Möller, H.P. Voigt,
MTZ-Motortechnische Zeitschrift, 1964, September,
Heft 9, S. 347-354

12   F. Umland, H.P. Voigt,
Bericht vom Institut für Strömungsmaschinen,
TH Hannover, 1964, Bericht Nr. 73

13   F. Umland, H.P. Voigt,
Werkstoffe u. Korrosion, 1965, Oktober, Heft 10,
S. 871-875

14   F. Umland, H.P. Voigt,
Werkstoffe u. Korrosion, 1970, Heft 21, S. 254-263

15   K. Wickert,
VGB-Mitteilungen, 1968, Heft 3, Juni, S. 149-154

16   J.A. Goebel, F.S. Pettit, G.W. Goward,
Deposition and Corrosion in Gas Turbines, 1, S. 96-114,
1973, London, A.B. Hart, H.J.B. Cutler

17   P. Felix,
Deposition and Corrosion in Gas Turbines, 1,
S. 331-349, 1973, London, A.B. Hart, H.J.B. Cutler

18   P.A. Bergmann,
MEL-sponsored Report

19   A.M. Beltran,
Kobalt, 1970, Nr. 46, März, S. 3-13

20   H. Huff, F. Schreiber,
Werkstoffe und Korrosion, 1972, Heft 5, S. 370-372

21   W. Betz,
High Temperature Alloys for Gas Turbines, 1,
S. 409-422, 1978, London, D. Coutsouradis, P. Felix,
H. Fischmeister, L. Habraken, Y. Linelblom, M.O. Speidel

22   A. Rahmel, M. Schmidt,
Schlußbericht Cost 50 Programm, 1976, Projekt D1/9

23   K.H.G. Schmitt-Thomas, H. Meisel, H.J. Dorn,
Werkstoffe und Korrosion, 1978, 29, S. 1-9

24   W. Track, W. Betz, K. Schweitzer,
Schlußbericht Cost 50 Programm, 1977, Projekt D1/3

25   K.H.G. Schmitt-Thomas, H. Meisel, H.J. Dorn,
Praktische Metallografie, 1976, 13, S. 427-435

26   W. Betz, H. Huff, W. Track,
Werkstoffe und Korrosion, 1976, Heft 5, Mai, S. 161-196

27   D. Sunamoto, T. Nishida,
MTB, 1970, 65, April, S. 1-8

28   K.H. Kloos,
Symposium über erste Ergebnisse des Forschungs- und
Entwicklungsprogrammes "Korrosion und Korrosionsschutz",
1977, April, S. 141-142, Braunlage

29   K.H.G. Schmitt-Thomas,
Symposium über erste Ergebnisse des Forschungs- und
Entwicklungsprogrammes"Korrosion und Korrosionsschutz",
1977, April, S. 143-144, Braunlage

30 H.W. Grünling,
    Symposium über erste Ergebnisse des Forschungs- und
    Entwicklungsprogrammes "Korrosion und Korrosionsschutz",
    1977, April, S. 145-146, Braunlage

31 K.H. Kloos, J. Granacher, H. Demus,
    Forschungsbericht der Forschungsvereinigung Ver-
    brennungskraftmaschinen E.V., 1978, Heft 329

32 H.W. Grünling, B. Ilschner, S. Leistikow, A. Rahmel,
    M. Schmidt,
    Werkstoffe und Korrosion, 1978, 29, S. 691-703

33 K.H.G. Schmitt-Thomas, H. Meisel, H.J. Dorn,
    Materials and Coatings to Resist High Temperature
    Corrosion, 1, S. 87-102, 1978, London, D.R. Holmes,
    R. Rahmel

34 F. Schmitz,
    Materials and Coatings to Resist High Temperature
    Corrosion, 1, S. 102, 1978, London, D.R. Holmes and
    R. Rahmel

35 W. Betz,
    Materials and Coatings to Resist High Temperature
    Corrosion, 1, S. 185-197, 1978, London, D.R. Holmes
    and R. Rahmel

36 K.H. Kloos,
    2. Symposium über Ergebnisse des Forschungs- und Ent-
    wicklungsprogramms "Korrosion und Korrosionsschutz"
    des Bundesministers für Forschung und Technologie,
    1980, April, Willingen

37 K.H. Schmitt-Thomas,
    2. Symposium über Ergebnisse des Forschungs- und Ent-
    wicklungsprogrammes "Korrosion und Korrosionsschutz"
    des Bundesminister für Forschung und Technologie,
    1980, April, Willingen

38 F. Schmitz,
    Schlußbericht Cost 50 Programm, 1976, Projekt D1/6

39 F. Schmitz,
    VDI-Veröffentlichung, Forschungsberichtsreihe
    "Technologische Forschung und Entwicklung", 1980,
    BMFT-FBT 79-106

40 F. Schmitz, U. Heinrich, S. Kurth,
    Cost 50 Programm, 1. Zwischenbericht der 2. Runde,
    1977, TWA 66 700

41   F. Schmitz,
Cost 50 Programm, 1978, November, Seminar CEGB London

42   F. Schmitz, U. Heinrich, S. Kurth,
Cost 50 Programm, 2. Zwischenbericht der 2. Runde, 1978,
TWA 68 618

43   F. Schmitz, U. Heinrich, W. Slotty,
Cost 50 Programm, Schlußbericht der 2. Runde, 1980,
April, TWC 70 962

44   C.T. Sims, W.C. Hagel,
The Superalloys, 1, 1972, New York, J. Wiley & Sons

45   P. Brezina et al.,
Praktische Metallographie 10, 1973, S. 343 und 377

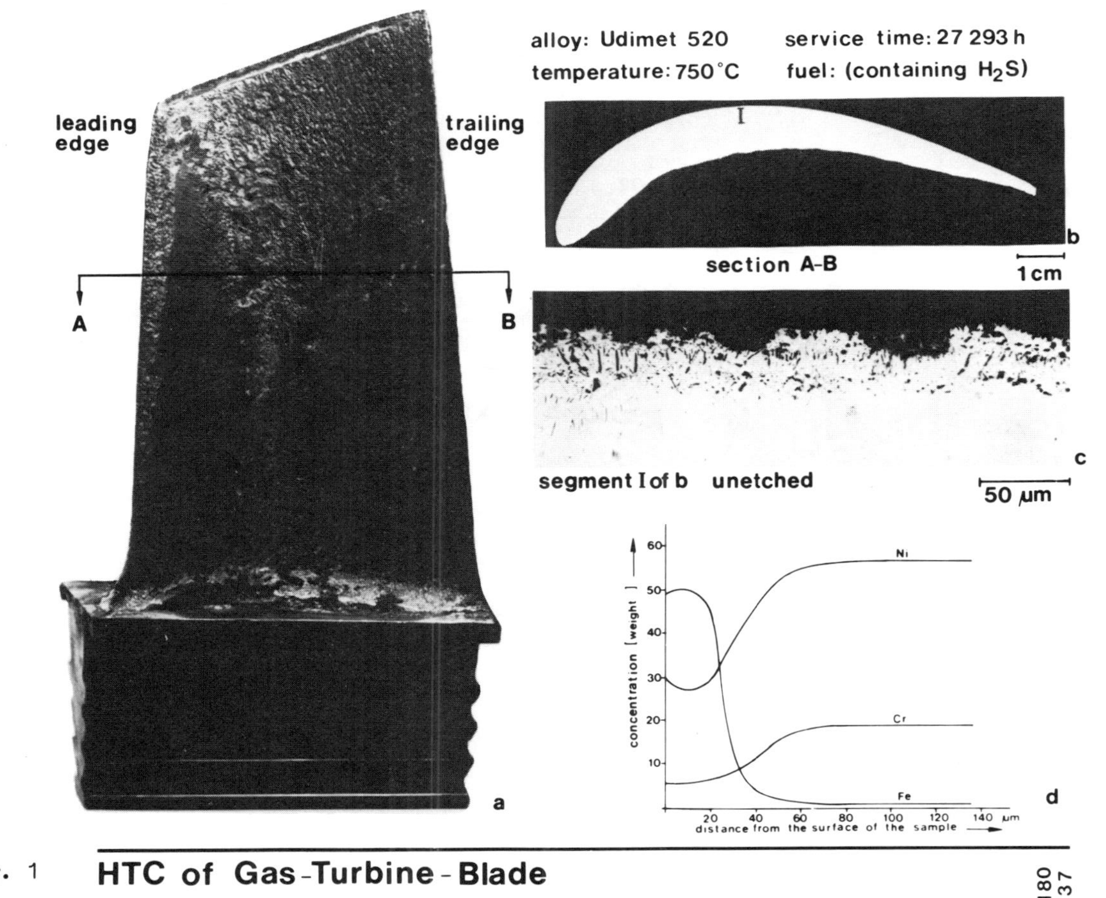

Fig. 1　HTC of Gas-Turbine-Blade

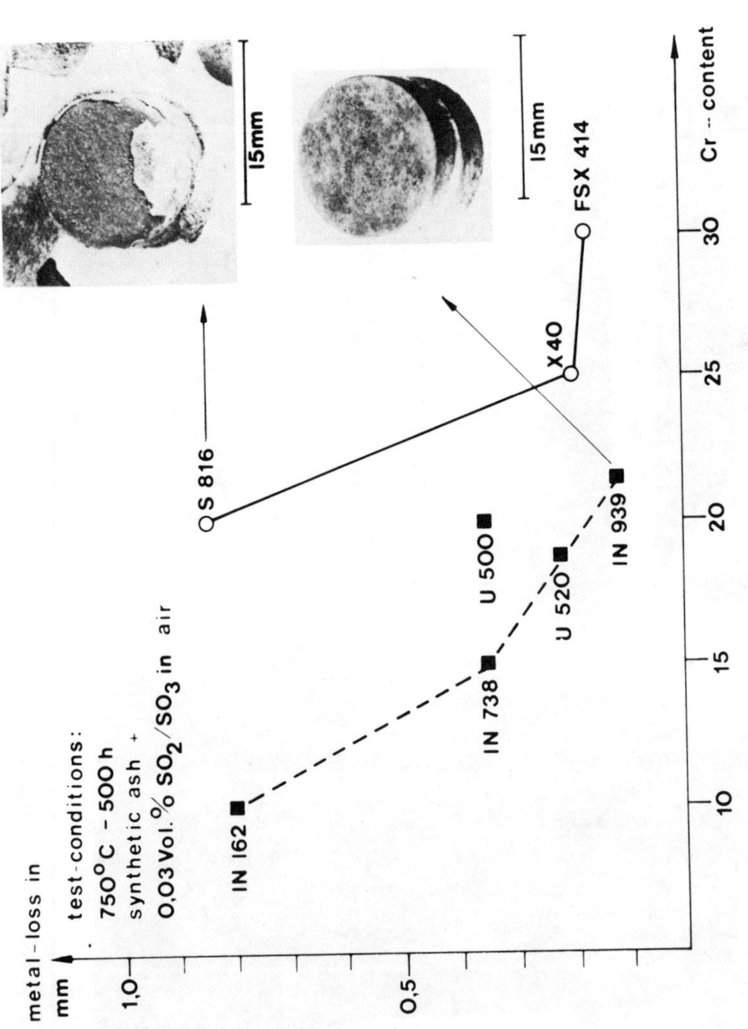

Fig. 2   HTC-laboratory-tests of Ni- and Co-based-alloys for gas-turbine-blade-materials

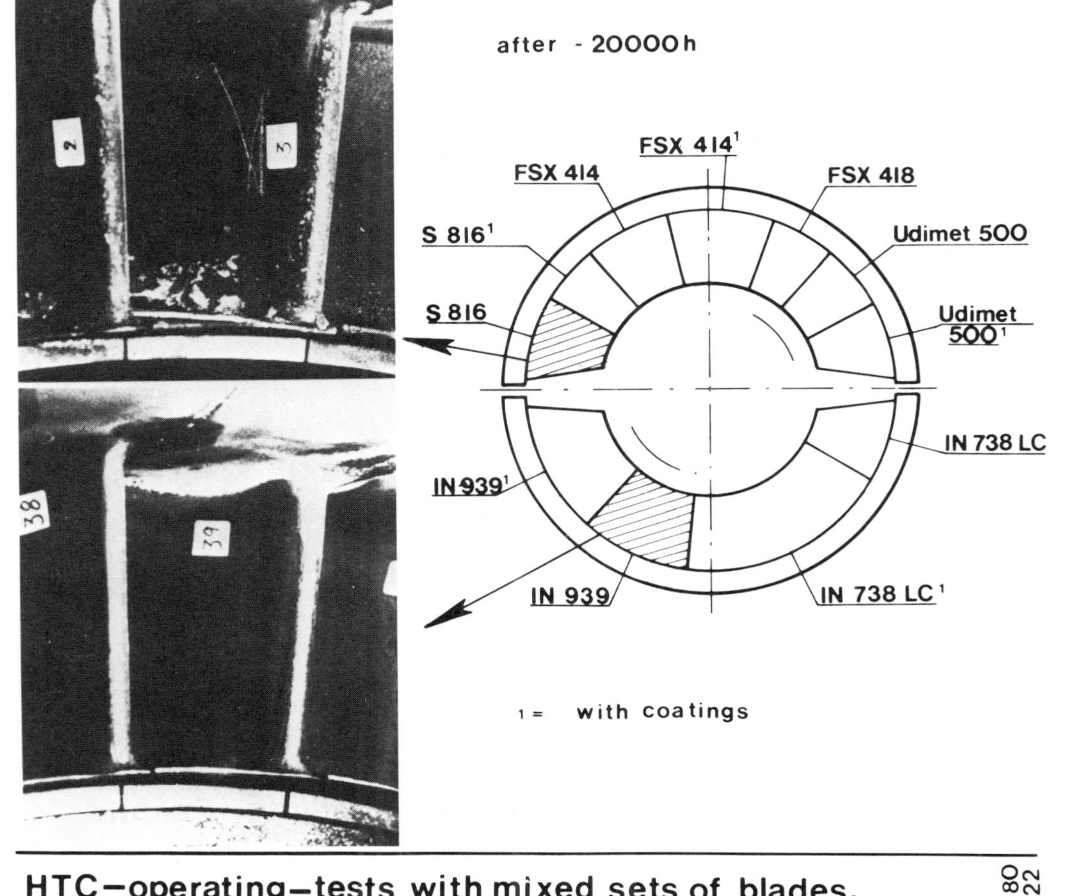

Fig. 3 HTC—operating—tests with mixed sets of blades. fuel: S-containing natural-gas, temperature ~750°C

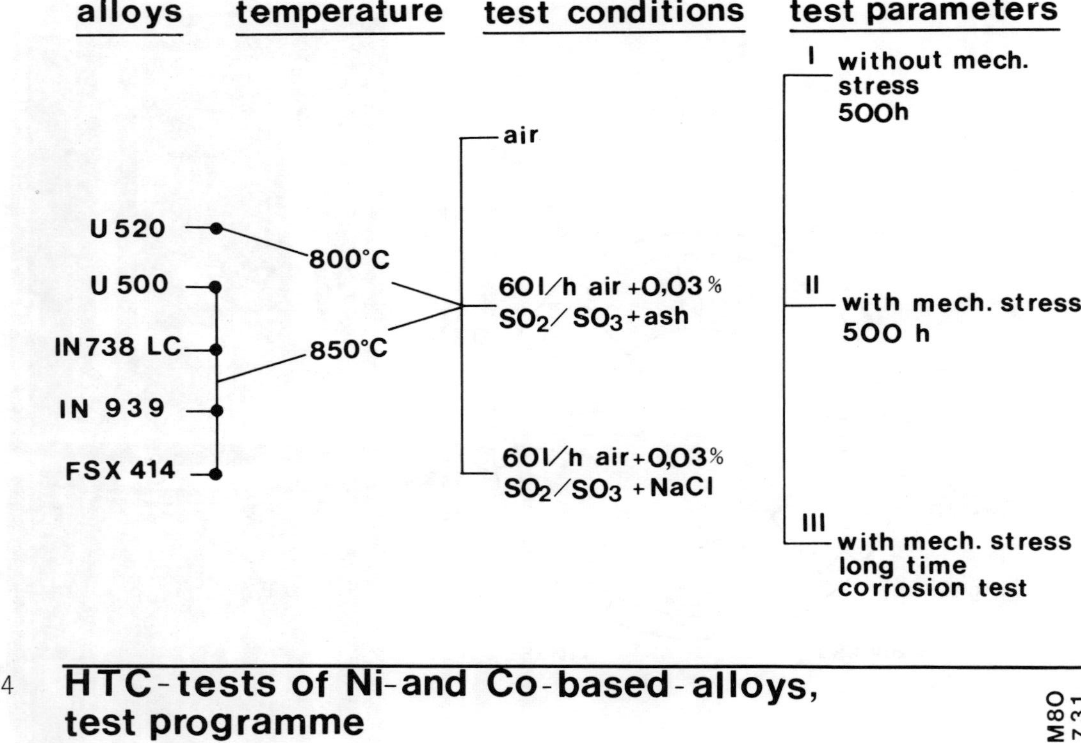

Fig. 4   **HTC-tests of Ni-and Co-based-alloys, test programme**

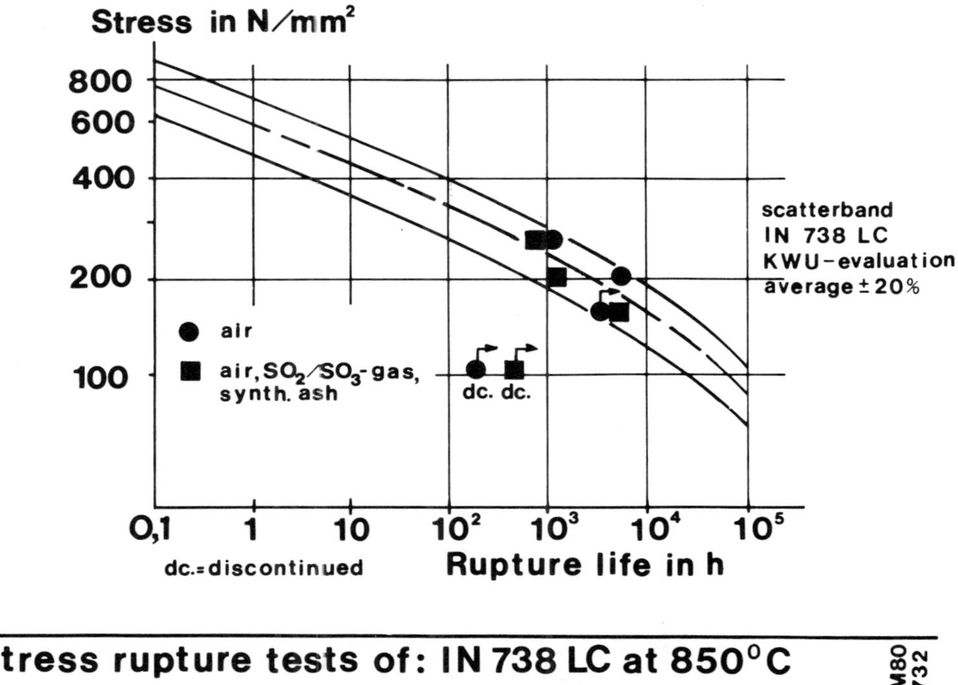

Fig. 5 **Stress rupture tests of: IN 738 LC at 850°C**

alloy   : IN 738 LC
sample-No.   : 5200
temperature   : 850°C
stress   : 200 N/mm^2
time to rupture : 1270 h
test-condition : ash.air + 0.03% SO_2/SO_3

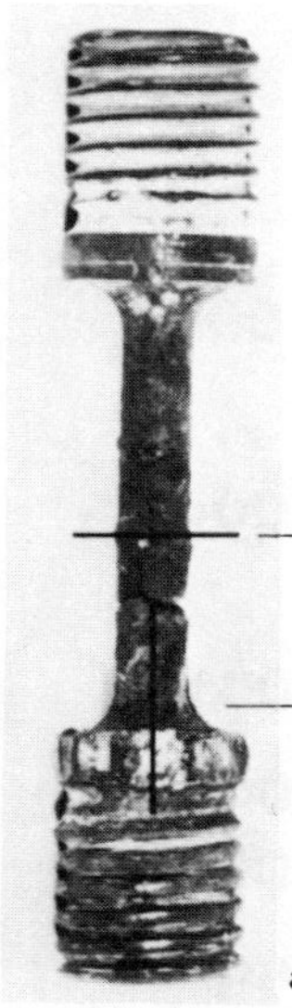

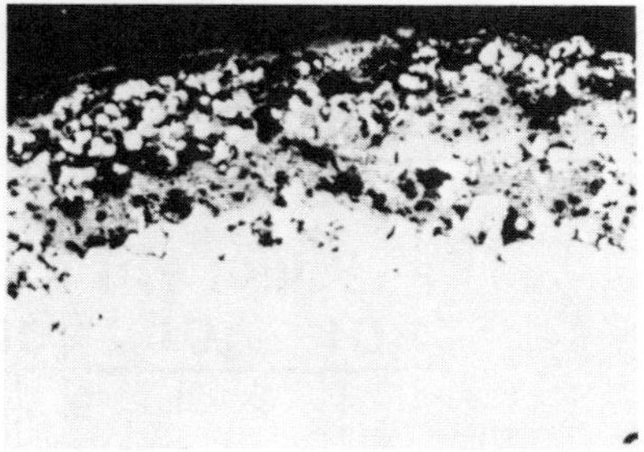

Fig. 6   Metallographic examination of high-temperature corrosion-tests with mechanical stress

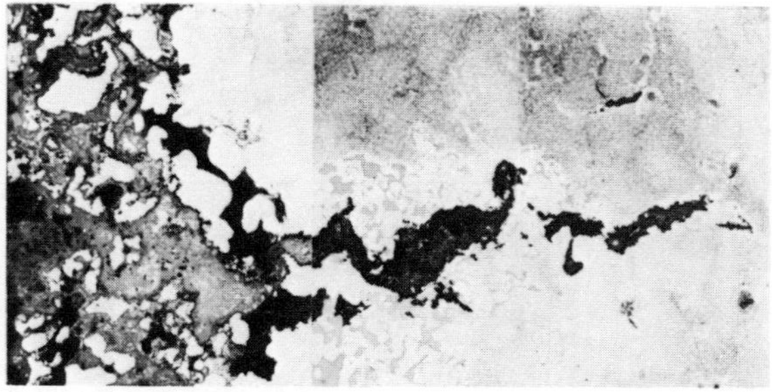

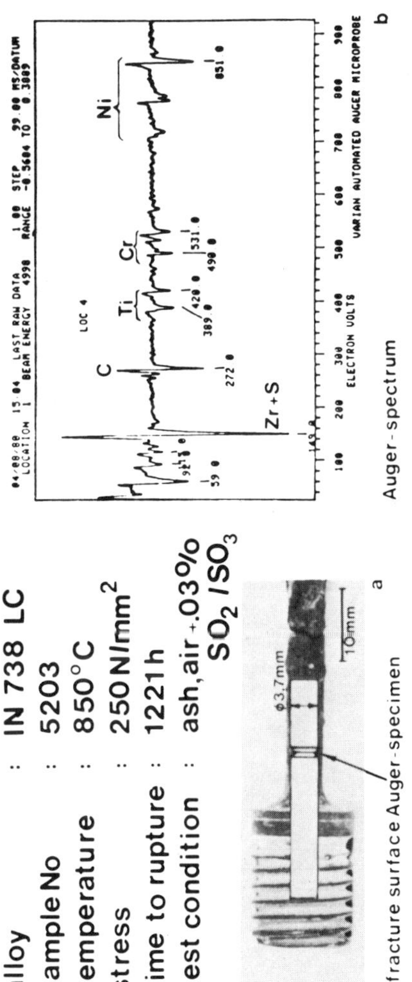
alloy : IN 738 LC
sample No : 5203
temperature : 850°C
stress : 250 N/mm²
time to rupture : 1221h
test condition : ash, air +.03% SO₂/SO₃
10mm
ø3.7mm
a
fracture surface Auger-specimen
LOC 4
Ni
Cr
Ti
C
Zr + S
Auger-spectrum
ELECTRON VOLTS
VARIAN AUTOMATED AUGER MICROPROBE
b

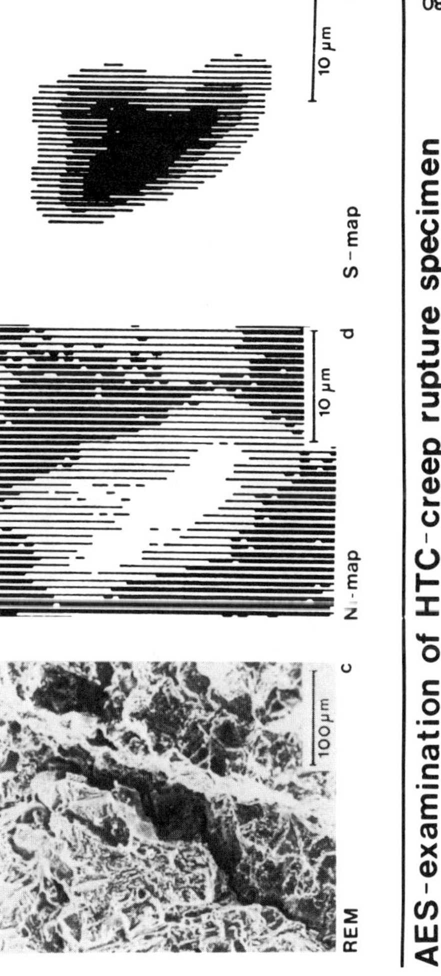
10 µm
e
S-map
10 µm
d
Ni-map
100 µm
c
REM
M80
730
Fig. 7 AES-examination of HTC-creep rupture specimen

F. SCHMITZ and K.H. KEIENBURG

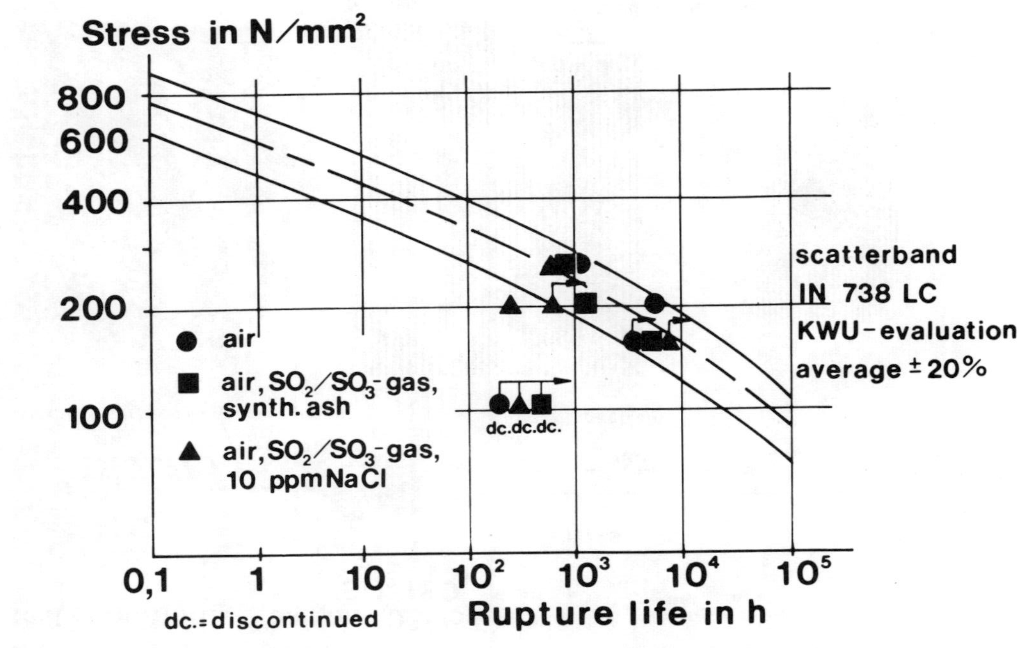

Fig. 8   Stress rupture tests of: IN 738 LC at 850°C

M 80
703

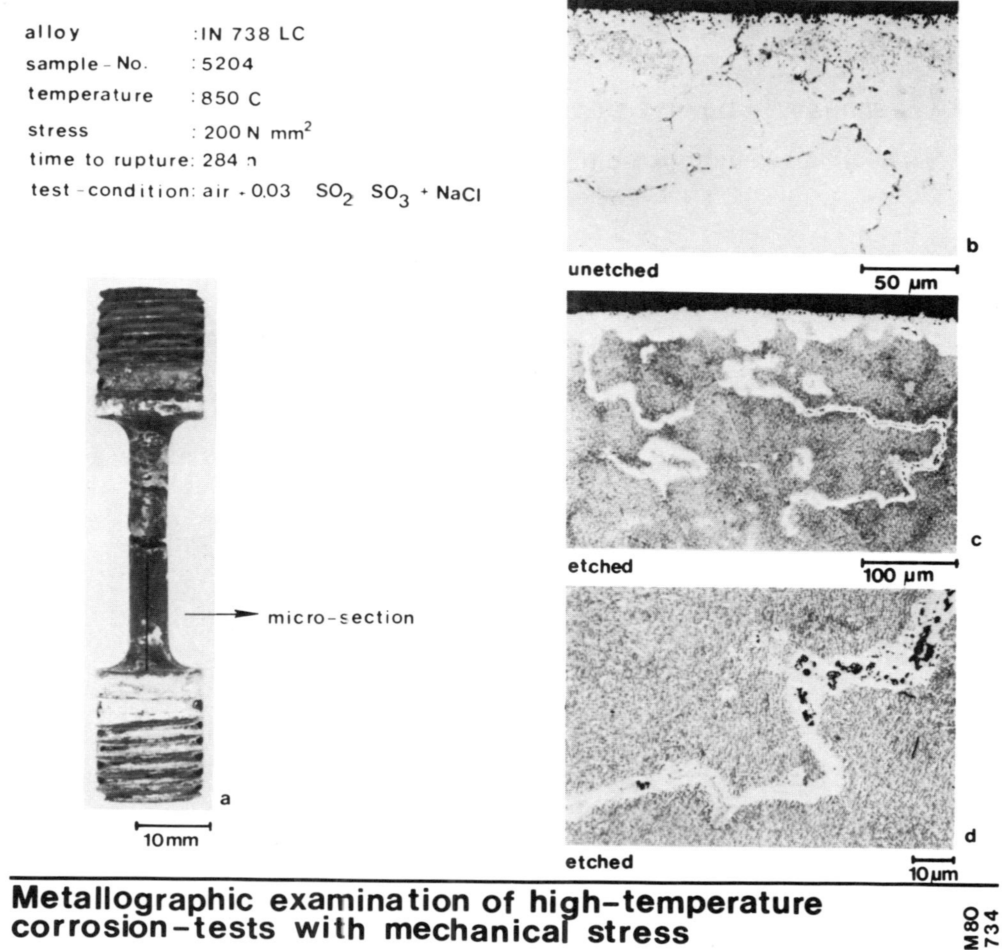

Fig. 9 Metallographic examination of high-temperature corrosion-tests with mechanical stress

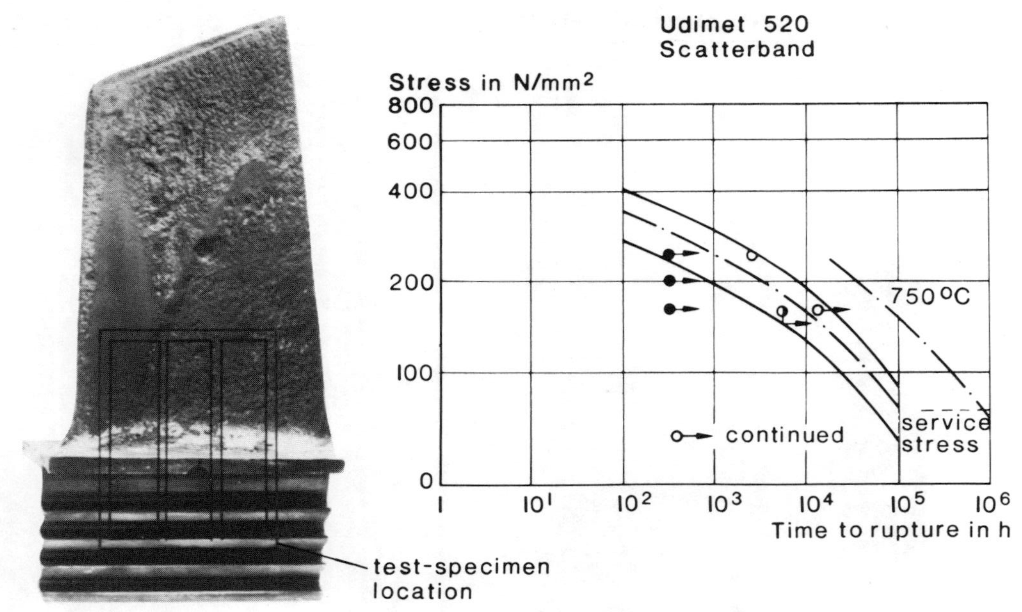

Fig. 10 Stress rupture test of : Udimet 520 at 800 °C

M80
723

Creep rupture strength of cast alloys for gas turbine blades under hot gas corrosion

K.H. Kloos, J. Granacher and H. Demus

Institut für Werkstoffkunde der Techn. Hochschule Darmstadt

Grafenstraße 2, Postfach 11 09 49, 6100 Darmstadt

Synopsis

An important question concerning the use of alloys for gas turbine blades is, whether the creep rupture strength can be lowered by high temperature corrosion due to streaming hot gas. For experimental examination two hot gas test rigs were constructed, each containing 8 test units. Creep rupture tests under streaming hot gas produced by the combustion of fuel JP 4 and fuel oil containing corrosive additives were conducted on several high temperature alloys for aircraft and land based gas turbines. The results of these tests reaching several thousand hours and conducted on round and flat specimens of investment cast materials show no corrosion induced decrease of the creep rupture strength under hot gas as compared with the creep rupture strength normally measured under calm air. Metallographical examination of specimens after rupture indicates, that severe corrosion acting for longer times may lower the creep rupture strength.

1. Introduction

Gas turbine blades are simultaneously exposed to a high static stress due to centrifugal load and to a high temperature corrosion attack due to the streaming hot gas. For the alloys used the question arises whether this combination of mechanical stress and corrosion attack leads to the same creep rupture strength as normally determined in creep rupture tests conducted under calm air. Earlier investigations had shown that

the creep rupture strength under high temperature corrosion
can be lowered to a greater degree than is explainable by the
loss of cross-section due to corrosion [1,2].
Other investigations did not show this effect in all cases [3].

From the metallurgical viewpoint, deterioration of microstructure is possible due to the action of corrosive elements
as sulfur, sodium, chlorine and vanadium, which enter the
turbine via fuel or combustion air [4,5]. When the oxide layer
or an artificial protective coating of the turbine blade is
deleted, grain boundary attack, depletion of γ'-precipitate,
internal oxidation and sulfidation can severely reduce the
strength of the alloy [6,7]. So several corrosion induced
mechanisms are able to rise the creep rate, to accelerate
the crack propagation and to lower the time to rupture [8].
On the other hand creep can accelerate the deletion of oxide
layers and coatings. Because the problem of interaction
between corrosion attack and creep stress is important for
both land based and aircraft gas turbines, a special research
project has been started under the DECHEMA research and development program "corrosion and corrosion protection".

The objectives of this project are the evaluation of the creep
rupture strength of typical gas turbine alloys under streaming
hot gas and the comparison of the measured "hot gas rupture
strength" with the creep rupture strength normally measured
under calm air. The test program comprises creep rupture tests
on uncoated specimens and on specimens covered with protective
coatings. These coatings can raise the rupture stress under
hot gas, because they improve the corrosion resistance [9].
On the other hand the heat treatment during coating can lower
the rupture stress [10]. Different corrosive conditions for aircraft and land based gas turbines are simulated by combustion
of different fuels and by adding corrosive substances into
the hot gas or into the fuel. Several investment cast alloys
used for turbine blades have been chosen for the investigations. Chemical and metallographical examination of the
specimens tested under hot gas are planned in order to study
the interaction between high temperature corrosion and creep
rupture behaviour.

2. The hot gas test rigs

Starting from a hot gas test rig developed by MTU München [1]
two improved test rigs were constructed. With these rigs it
was possible to conduct long time tests with different fuels,
corrosive additives, temperatures and loads. In each test rig
(fig. 1) the hot gas is produced by a combustion chamber of a

small gas turbine. After temperature equalization in a vessel,
the hot gas stream is distributed to 8 test chambers. In each
test chamber the hot gas stream is directed against the specimen with a velocity of about 130 $^m/s$. The specimen temperature
is controlled by injection of small amounts (<5 % of hot gas
quantity) of cooling air. Each test chamber in connection
with a loading frame builds a test unit. In each test rig
there are 7 units provided for static loading by means of
a spring and 1 test unit for dynamic loading by means of a
servohydraulic system. One test rig is equipped for the combustion of aviation fuel JP 4, the other for fuel oil. The
development and construction of both test rigs took about
2.5 years. The JP 4 burning rig has reached about 18 000 hours
of service, the rig burning fuel oil was finished later and
has reached about 8000 hours of service. The shorter service-
time of this rig is caused – among others – in difficulties
with the combustion of fuel oil with additives.
More detailed informations of the hot gas test rigs have been
published earlier [11].

3. <u>Test program</u>

In the test rig A a "weak corrosion for aircraft and land
based turbine alloys" is realized (<u>fig. 2</u>). "Severe corrosion
for aircraft turbine alloys" can be simulated by the direct
addition of artificial seasalt dissolved in water. In the
test rig B "Severe corrosion for land based turbine alloys"
is realized by the combustion of fuel oil containing additions
of sulfur, sodium, vanadium and chlorine bound to substances
soluble in fuel. The content of tho corrosive elements in the
fuel oil corresponds with the upper limits allowed for usual
applications and amounts to 0.5 masspercent sulfur, 15 PPM
sodium, 10 PPM chlorine and 5 PPM vanadium. In comparison
with the commercial fuel oils this content of corrosive
elements is elevated. So the lower gas pressure in the test
rig as compared with a gas turbine may be partially compensated
by the higher contents of corrosive substances in the hot gas.

The alloys for land based turbines selected in the test program are the nickel-base alloys IN-738 LC (60Ni-16Cr-8.5Co-1.
5Mo-2.5W-2.5Al-3.5Ti) and IN-939 (48Ni-22Cr-19Co-2W-2Al-3.7Ti)
and the cobalt-base alloy FSX 414 (51Co-10Ni-29Cr-7W) only
used for low stressed turbine parts such as vanes. The alloys
selected for aircraft turbines are the nickel-base alloys
IN-713 (73Ni-12Cr-1Co-4Mo-6Al-0.7Ti) and IN-100 (60Ni-10Cr-
15Co-3Mo-5.5Al-5Ti).
The land based turbine alloys are tested with round specimens
(<u>fig. 3</u>), while a flat specimen of same section is provided

for testing the aircraft turbine alloys in this way taking into account the smaller wall thickness of aircraft turbine blades.

The test plan provides creep rupture test under air, hot gas of JP 4 and hot gas of fuel oil at 700 and 800 $^\circ$C on the alloys for land based turbines and at 800 and 900 $^\circ$C on the alloys for aircraft turbines. The longest test times provided in the program are 3000 hours at 700 and 900 $^\circ$C and 8000 hours at 800 $^\circ$C.

4. Test results

Normal creep rupture tests under calm air have been carried out on the alloys IN-738, IN-939, FSX 414 and IN-713. The tests have reached times to rupture of about 20 000 hours. The results of these tests gave the reference values for comparison with the results of the tests under hot gas.

The creep rupture tests under hot gas have reached about 2000 hours at 700 $^\circ$C and about 5000 hours at 800 $^\circ$C. In these tests the test temperature was shown to be the most problematic parameter. This can be explained by the test arrangement chosen (fig. 4) nozzle, specimen and working thermocouple being in the same order of magnitude. Calibration measurements with coaxial thermocouples incorporated in the specimen by brazing gave a difference between the hot gas temperature and the maximum specimen temperature of about 20 K. The difference between the maximum specimen temperature T_s and the temperature T_w of the working thermocouple used for temperature control and positioned behind the specimen amounts to about 25 K. Due to a temperature gradient within the hottest cross sectional area of the specimen the "true specimen temperature" T_i characterizing the creep rupture behaviour is about 4 to 7 K lower than the maximum temperature T_s. Combining the temperature differences T_w-T_s (-25 K) and $\Delta T_i = T_s$-T_i (4 to 7 K) the so called temperature deviation $\Delta T_w = T_w$-T_i is a characterizing testing quantity. Additionally to this temperature deviation the individual error of the working thermocouple used is taken into account.

Systematical mearsurements indicated that the above defined temperature deviation sensitively depends on the positions of nozzle, specimen and working thermocouple within the test chamber. At a nominal temperature of 800 $^\circ$C the temperature deviation was $\Delta T_w = -20 \pm 5$ K (fig. 4) for round specimens as well as for flat specimens with an angle of 20° of inclination against the direction of the hot gas flow (normal test condition up to this time). The tolerances indicated consider

the usual variations of the specimen position in normal test
service. At 700 °C similar deviations have been observed.
According to first calibrating measurements the deviation at
900 °C doesn't seem to be essentially smaller.

Calibrating temperature measurements according to fig. 4 and
carried out after test made it possible to determine the true
test temperature of the hot gas creep rupture tests performed
till now to a little higher accuracy than indicated by the
limits above. In this way the true test temperatures sometimes
show a correction with regard to the nominal test temperature.
In order to compare the results of the hot gas tests with the
results of the tests under calm air at 700 and 800 °C, the
temperature correction when existing has been converted into
an equivalent stress correction with the time to rupture being
kept constant.

Considering this stress correction the test results gained until now at 700 and 800 °C under calm air and under hot gas of
JP 4 and of fuel oil are plotted in fig. 5 (IN-738 LC), fig. 6
(IN-939) and fig. 7 (FSX 414). Considering the larger scatter
of the hot gas tests compared with air tests neither the combustion of aviation fuel JP 4 nor the combustion of fuel oil
containing corrosive additives reduced the hot gas creep
rupture strength at 700 and 800 °C in comparison with the calm
air creep rupture strength. The current 800 °C-tests are to be
continued till a maximum test time of about 8000 hours. The
tests planned at 900 °C for test times of about 3000 hours and
also the tests with salt addition to the hot gas of JP 4 have
not yet been performed.

All specimens ruptured in the time-temperature range indicated
above show only a small reduction of cross sectional area due
to creep. The specimens tested under air or under hot gas of
JP 4 do not show a considerable additional loss of cross
sectional area due to corrosion. On the other hand specimens
of the alloys IN-738 and IN-939 and ruptured at 800 °C under
hot gas of fuel oil after times longer than 1000 hours show
an additional loss p_c of cross sectional area due to corrosion
and increasing with time (fig. 8). When the loss of cross
sectional area exceeded 2 % the rupture stress in fig. 5 and 6
was raised by the half percentage of this loss of area $(p_c/2)$.
In this way the corrosion induced loss of area was supposed
to increase linearily with time and the resulting stress increase was assumed to act linearily. Both assumptions are
rough approximations but seem to be justified regarding the
temperature scatter of the hot gas tests, the relatively small
losses of area observed and the fact that only small reductions
of area due to creep were observed.

Results of tests carried out on flat specimens are available
for 800 °C nominal test temperature, hot gas of JP 4 and the
alloys IN-738 and IN-713 LC (<u>fig. 9</u>). These test results also
show no corrosion induced decrease of the rupture stress.
Tests on coated specimens, on the aircraft alloy IN-100,
tests at 900 °C and tests with hot gas containing salt
additions have not yet been performed.

5. <u>Results of metallographical examination</u>

Metallographical examinations are helpful to analyse wether a
decrease of creep rupture strength is due to a loss of cross
sectional area or to a deterioration of microstructure. First
micrographs were prepared from the cross and longitudinal
sections of specimens tested at 800 °C. Specimens tested under
air and under hot gas of JP 4 did not show differences in
structure or in corrosion layers. This agrees well with the
results of the creep rupture tests.

Regarding specimens tested under hot gas of fuel oil containing corrosive additives the alloys IN-738 (<u>fig. 10</u>) and
IN-939 (<u>fig. 11</u>) showed the above mentioned corrosion attack
which was concentrated at the side exposed to the hot gas
flow and produced a loss of cross sectional area increasing
with time. A comparison of fig. 10 and fig. 11 in connection
with the rupture times also shows that corrosion attack is
increasing more rapidly on alloy IN-738 than on alloy IN-939.
Beyond the zone depleted of γ'-precipitate a corrosion attack
partially leading to grain boundary damage can be seen. Such
a grain boundary attack acting for longer times is one of the
possible mechanisms which can decrease the creep rupture
strength to a greater extent than is explainable by the
corrosion induced loss of area.

Micrographs from the longitudinal section of broken specimens
were made for electron microprobe analysis. The longitudinal
sections of the specimens tested unter hot gas were cutted
through the zone of maximum corrosion attack (in front of the
hot gas nozzle) beginning at the rupture surface.
The longitudinal sections of the specimens tested under calm
air were cutted near the outer diameter and also beginning at
the rupture surface. First results can be seen on BSE- and
x-ray-images in <u>fig. 12</u> from specimens tested under different
corrosion conditions and broken with comparable times to
rupture. With increasing severity of the corrosion conditions
there is to be see a slightly increasing zone of γ'- and
Cr-depletion. The oxidation- and corrosion layers are enriched
with Alumina and Cr-oxide. S was only found in the corrosion

layer of the specimen broken under hot gas of fuel oil.
Further investigations will be made by electron microprobe
analysis using quantitative methods. Additional examinations
in the scanning electron microscope are planned with the aim
to asess the microcrack configuration.

6. Conclusion

The results of the creep rupture tests carried out under two
different conditions of corrosion and reaching 2000 hours at
700 $^{\circ}$C and 5000 hours at 800 $^{\circ}$C up to now show no corrosion
induced decrease of creep rupture strength under hot gas compared with creep rupture strength under calm air. A conti-
nuation of the 800 $^{\circ}$C-tests and tests at the higher temperature of 900 $^{\circ}$C in connection with further metallographic
examination are expected to give a more detailed understanding of the interaction of creep and high temperature
corrosion.

Acknowledgements

Thanks are due to the Bundesminister für Forschung und Technologie for his financial support of the work within the
research and development program "corrosion and corrosion
protection".

K.H. KLOOS et al.

<u>References</u>

1   Track, W., Betz, W. und Schweitzer, K.:
MTU-Schlußbericht 1977 zu COST 50, Projekt D1/3, BCT 31.

2   Schmitt-Thomas, K.H., Meisel, H. und Dorn, H.J.:
Heißgaskorrosion und Zeitstandfestigkeit an einer Nickel-
basislegierung unter betriebsnahen Bedingungen bei
750 °C bis 950 °C,
Werkstoffe und Korrosion 29 (1978) p. 1/9.

3   Schmitz, F.:
KWU-Schlußbericht 1976 zu COST 50, Projekt D1/6, BCT 32.

4   Boettger, H. und Umland, F.:
Untersuchungen zur Hochtemperaturkorrosion durch Chloride,
Werkstoffe und Korrosion 25 (1974) p. 805/16.

5   Just, C.:
Comparison of Corrosion Test Methods to Determine the
Temperature Dependance of High Temperature Corrosion,
in: High-Temperature Alloys for Gas Turbines, Proceedings
of a Conference (COST 50) Liège (1978) p. 147/77.

6   Whittle, D.P.:
High Temperature Oxidation of Superalloys,
see ref. 5, p. 109/25.

7   Morbioli, A. und Glider, H.:
Sulfidation Behaviour of Ni- and Co-Base-Superalloys,
see ref. 5, p. 125/47.

8   Grünling, H.W., Ilschner, B., Leistikow, S., Rahmel, A.
und Schmidt, M.:
Wechselwirkung zwischen Kriechverformung und Heißgaskorrosion,
Werkstoffe und Korrosion 29 (1978) p. 601/703.

9   Betz, W., Huff, H. und Track, W.:
Zur Bewertung von Schutzschichten gegen Heißgaskorrosion
an Gasturbinenschaufeln,
Z. f. Werkstofftechnik 7 (1976) p. 161/96.

10   Schneider, K.:
Einfluß von Beschichtungsmaßnahmen auf das Zeitstandver-
halten der Werkstoffe IN-100, Udimet 520 und IN-738.
Bericht auf der Sitzung des Lenkungsausschusses der
Arbeitsgemeinschaft für Hochtemperaturwerkstoffe am
02.09.1976 in Düsseldorf.

11   Kloos, K.H., Granacher, J. und Demus, H.:
Rupture Stress of Alloys for Gas Turbine Blades under
Hot Gas Corrosion,
International Conference Petten (October 1979).

Fig. 1. Partial view of both hot gas test rigs

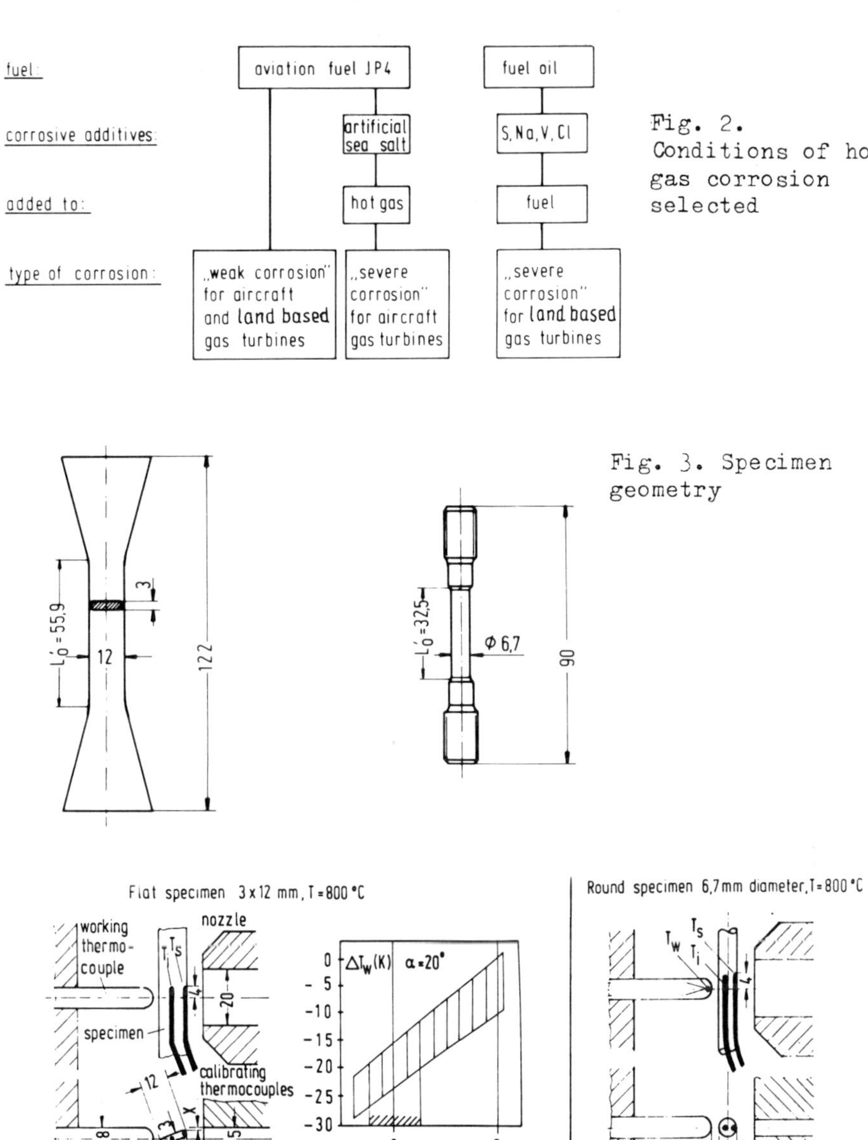

Fig. 2. Conditions of hot gas corrosion selected

Fig. 3. Specimen geometry

$$\Delta T_W = T_W - T_S + \Delta T_i \; ; \; \Delta T_i = T_S - T_i$$

T_W : Temperature measured by the working thermocouple

T_S : Maximum Temperature at the hot gas specimen

T_i : Temperature responsible for the rupture-time and measured in the cross sectional area with maximum temperature

Fig. 4. Test arrangement in the hot gas test chamber and definition of the significant temperature deviation ΔT_W of the working thermocouple

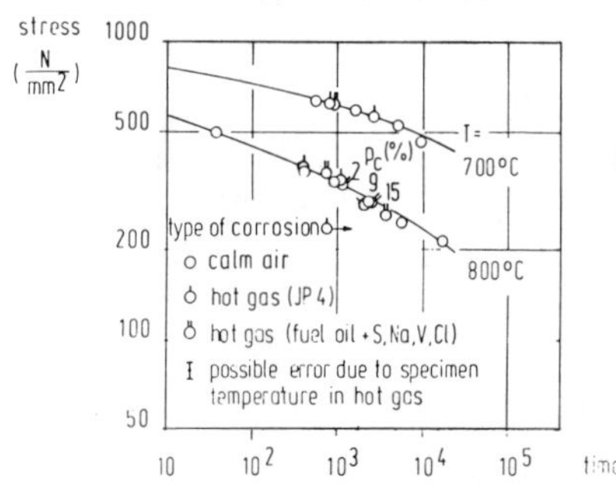

p_c: Loss of cross sectional area due to corrosion measured after rupture

Fig. 5. Creep rupture strength of alloy IN-738 LC under different conditions of corrosion

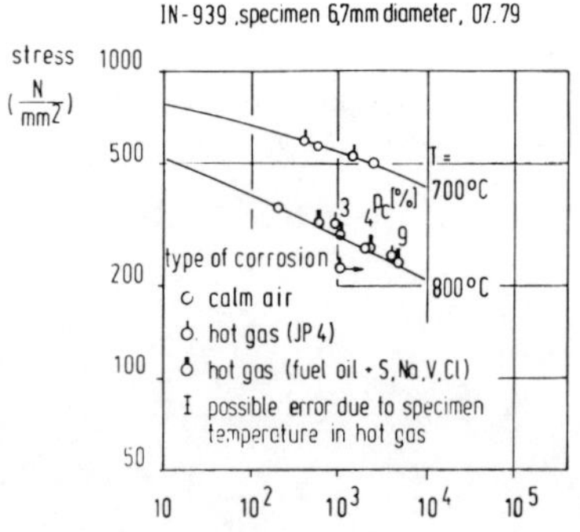

p_c: Loss of cross sectional area due to corrosion measured after rupture

Fig. 6. Creep rupture strength of alloy IN-939 under different conditions of corrosion

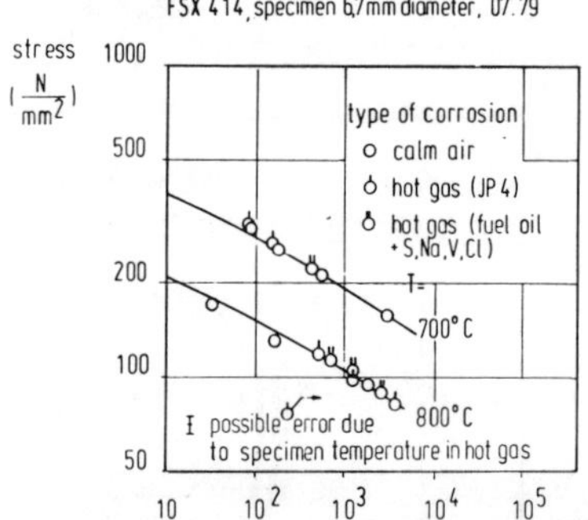

Fig. 7. Creep rupture strength of alloy FSX 414 under different conditions of corrosion

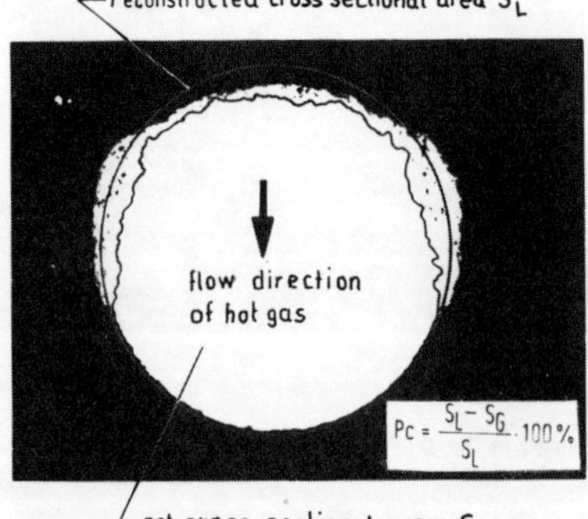

Fig. 8. Evaluation of the loss of cross sectional area p_c

alloy   : IN-738 LC
typ of corrosion: fuel oil + S, Na, V, Cl
test temperature: 800°C
stress   : 258 N/mm^2
time to rupture : 3680 h

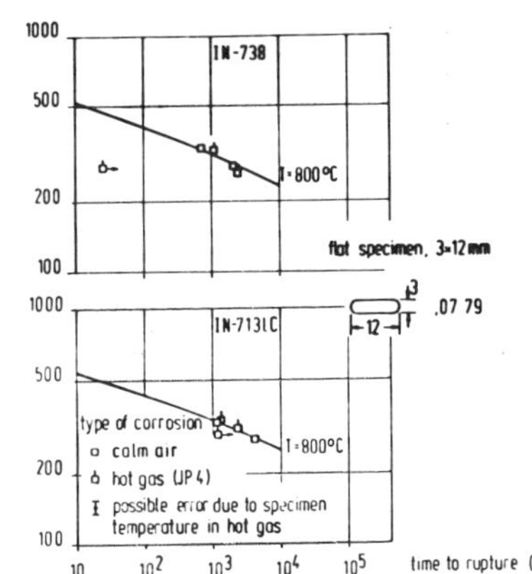

Fig. 9. Creep rupture strength of alloy IN-738 and IN-713 LC under different conditions of corrosion

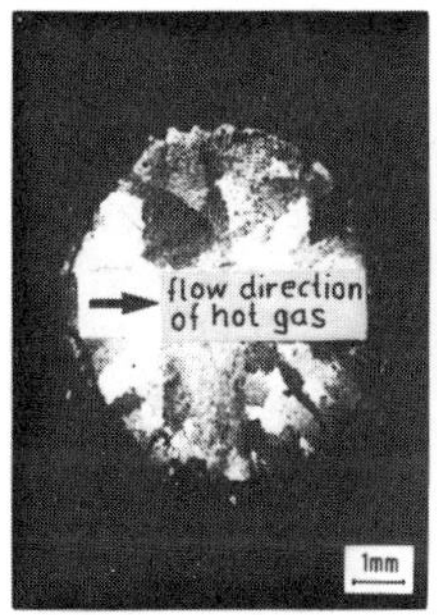

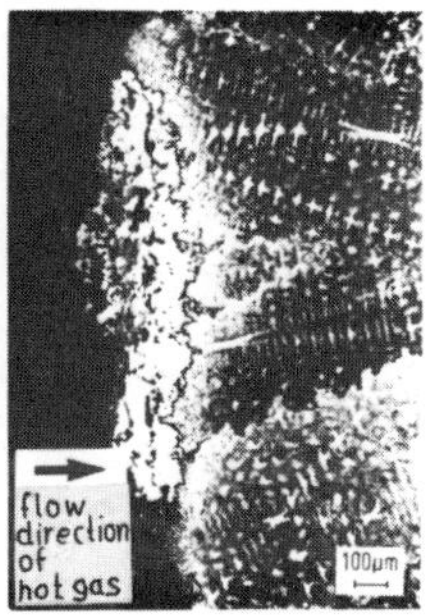

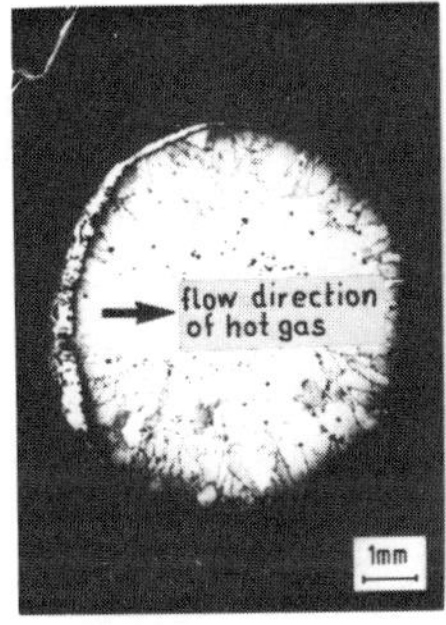

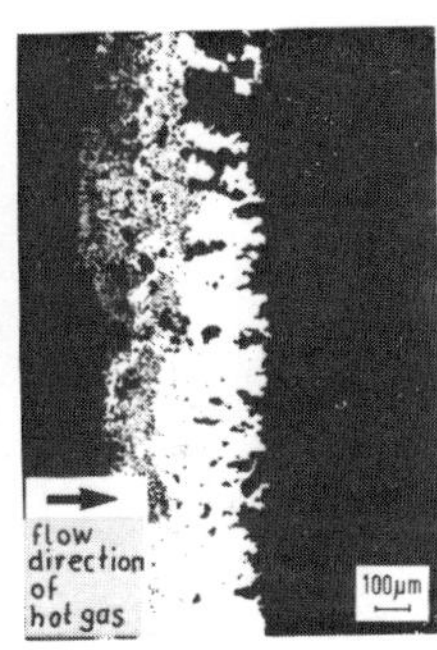

cross section in 5mm distance from the rupture surface

microstructure in front of the hot gas nozzle

cross section in 5mm distance from the rupture surface

microstructure in front of the hot gas nozzle

Fig. 10. Micrographs of alloy IN-738 LC

type of corrosion: fuel oil + S, Na, V, Cl

test temperature : 800 °C

stress : 293 N/mm²

time to rupture : 2450 h

Fig. 11. Micrographs of alloy IN-939

type of corrosion: fuel oil + S, Na, V, Cl

test temperature : 800 °C

stress : 240 N/mm²

time to rupture : 4885 h

calm air

outer shape
t_m = 2421 h

hot gas burning
JP 4
area in front of the hot gas nozzle
t_m = 2038 h

fuel oil + S,Na,V,Cl
t_m = 2454 h

Back scatter electron images

Al

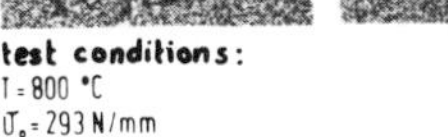
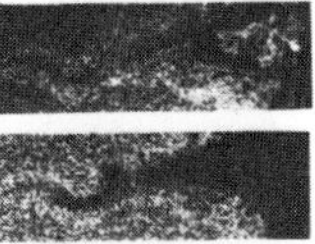

Cr

S

Fig. 12. BSE- and x-ray-images by electron microprobe analysis on specimens of alloy IN-738 LC broken under different conditions of corrosion

test conditions:
T = 800 °C
σ_o = 293 N/mm²

specimen diameter 6,7mm

Constant Strain Rate Creep Tests with Gas Turbine
Blade Materials under Hot Corrosion Environmental
Conditions

W. Hartnagel, R. Bauer and H.W.Grünling
Brown, Boveri & Cie (F. R. G.)

0.Synopsis

Constant strain rate methods, well approved in aqueous SCC-investigations have been used to study the
interaction between mechanical stresses and high
temperature corrosion. Subjected to strain rates
from 10^{-9} s^{-1} to 10^{-7} s^{-1} the gas turbine blade materials IN 738 LC and IN 939 exhibited a loss of
strength and time to rupture under hot corrosion
conditions, compared to air test results. A critical
strain rate cannot be established due to the severe
conditions, leading to sulfidation within very
short exposure times. On the basis of the tests conducted so far in this investigation it can generally be concluded that the average cross section reduction due to high temperature corrosion attack
cannot account solely for the observed degradation
in creep properties. This phenomenon is attributed
to enhanced creep crack initiation and propagation.
Metallographic examination of stressed and un-

stressed specimens showed similar corrosion morphology revealing a porous oxide scale and γ'-de-
pleted subsurface regions with internal sulfide precipitations preferably at grain boundaries. From
metallography no significant acceleration of the
corrosion rate due to mechanical loading could be
deduced so far.

1. Introduction

Common design properties of gas turbine blade materials are creep and creep rupture properties derived
from air test results. Under hot corrosion environments these data may be condiderably altered due
to the accelerated loss in material cross section.
Different mechanisms of possible synergistic interactions between corrosion and creep deformation have
recently been summarized by Grünling et al[1].

Most of the experimental work was therefore concentrated on creep rupture tests under simulated corrosive conditions. Nicholls et al[2] found a marked reduction of creep rupture life and loss of ductility
with salt coated tests on 316 stainless steel and
Nimonic 108. From metallografic examinations it is
concluded that loss in the load bearing section
cannot account solely for the observed degradation
in creep properties.

Similar observations have been made by Rahmel and
co-workers [3] who carried out electrochemical tests
in molten alkaline sulfate mixtures with defined
hot corrosion inducing potentials. More conflicting

results are reported from burner rig tests. Huff and Schreiber [4] found a significant premature failure at both nickel base alloys IN 100 and René 41, particularly when sea salt was injected into the gas stream. The observed reduction in creep strength was not related to the loss in cross section but to the weakening of grain boundaries by penetration of sulfur which will increase the likelihood of grain boundary sliding [5]. Otherwise no corrosion induced decrease of creep rupture strength could be detected in comparable burner rig tests simulating severe industrial hot corrosion environments [6].

The basic idea of the present work was to determine whether a critical strain rate does exist above which detrimental creep/hot corrosion interactions will occur. This may be due to either cracking and spalling of protective oxide layers followed by grain boundary attack, enhanced creep crack propagation or other unknown mechanisms.

For this purpose constant strain rate creep tests well approved in aqueous SCC-investigations have been performed. Compared to creep rupture tests with continuously changing strain/stress parameters this method provides a more useful tool for the sophisticated study of these complex interaction mechanisms.

2. Test procedure

Fig. 1 illustrates one of the six creep test machines used in this work. Strain rates from 10^{-6} s^{-1} (o. 36 %h) down to 10^{-9} s^{-1} (0.00036 %h) can be realized with complete units of synchronous electric motors coupled with special mechanical gears which are easily exchangeable within a few minutes. The constant strain rate was continuously controlled by extensiometric measurements. The ground specimens had a gauge length of 40.0 mm between the extensiometer flanges and 8.0 mm diameter. Fig. 2 shows the mounting of the specimen inside the furnace. By means of an alumina crucible the gauge length of the specimen is embedded in a synthetic slag simulating hot corrosion environmental conditions.

The chemical composition of this slag (w/o: Na_2SO_4-4.3; $CaSO_4.2H_2O$-22.7; Fe_2O_3-22.3; $ZnSO_4.H_2O$-20.6; K_2SO_4-10.4; MgO-2.8; Al_2O_3-6.5; SiO_2-10.4) is based upon an average analysis of actual deposits on first stage buckets. The slag does not melt up to temperatures of about $1000^{\circ}C$. Due to the decomposition of zinc sulfate beginning at a temperature of approximately $765^{\circ}C$ the ash is acting as a SO_2/SO_3 donator system.

The SO_3 partial pressure at $850^{\circ}C$ is approximated to 6 x 10^{-3} atmospheres [7] which is comparable to gas turbine service environments. Small cylindrical samples were also placed within the crucible during the tests to serve as comparison for strained and unstrained conditions.

The tests were carried out with the nickel base

superalloys IN 738 LC and IN 939. They are both used
as blade and vane materials in the hot stages of
stationary gas turbines.

3. Results

From stress vs. strain curves measured in air and
slag environment the influence of hot corrosion on
creep properties can be derived. Figures 3-5 show
examples of these measurements at different strain
rates. A marked reduction of flow stress could be
found with both alloys accompanied by a decrease
of fracture elongation and hence a decrease of fracture lives.
Compared to experiments with high or medium strain
rates (fig. 3-4) different curve characteristics
were obtained with the lower strain rate range
(fig. 5). Whilst in air a slight increase of flow
stress with increasing strain was observed, the flow
stress values are rather strongly decreasing with
specimens exposed to the synthetic slag.

The dashed lines of the above mentioned figures account for the stress increase due to an average loss
of the load bearing section by corrosion, which is
apparantly not sufficient to explain the measured
deterioration of mechanical properties.

All the results at T = 850°C obtained so far are
numerically listed in table 1. In fig. 6 the ratios
of rupture life or maximum flow stress measured in
air and under corrosion are plotted as a function
of strain rate. Whereas in the case of IN 939 no

significant influence of corrosion on time to rupture could be found at high strain rates, IN 738 LC
showed an almost constant reduction of about 30%,
which appears to be independant of the applied
strain rate. The maximum stress ratio

$\sigma_{max.corr.} / \sigma_{max.air}$ however decreases with decreasing strain rates, reaching a value of about 50% for
IN 738 LC at a strain rate of 2×10^{-9} s^{-1}.

4. Discussion

On the basis of the present test results, it can
generally be concluded that both nickel base alloys
IN 738 LC and IN 939 suffer from a reduction of flow
stress and time to rupture under hot corrosion environments. At high strain rates this effect becomes
less significant in the case of IN 738 whereas IN 939
shows no influence within the scatterband of reprodu-
cable test results. One possible explanation may be
the superior corrosion resistance of IN 939 together
with the relative short exposure time during which
corrosion attack can occur.

This may be comparable to similar observations in
the field of aqueous stress corrosion cracking investigations where at high strain rates the mechanical
properties are dominating compared to the relatively
small rate of chemical attack. Otherwise it is known
from these experiments, that a maximum of mutual
interaction between aqueous corrosion and mechanical
deformation does exist at some critical strain rate [8]
promoting brittle fracture of the protective oxide
films and stress corrosion crack propagation. If the

strain rate is too slow, oxide film repair is able
to keep pace with the rate of the corrosion processes
Thus ductile failure of the specimen will be the dominating factor as outlined above with experiments
 at very high strain rates.

In the present work metallografic post-examination
of specimens revealed heavy internal suldifidation
attack already after short exposure time of a few
hours. Evidently the hot corrosion simulating environment was too severe that any protective oxide
scale could have been formed during the early stages
of this test. Therefore no critical strain rate could
be established as discussed before. Further experiments are planned using preoxidized specimens to investigate this phenomenon.

From metallography no stress-induced acceleration
of hot corrosion attack was found. Strained as
well as unstrained specimens exhibited comparable
depth of penetration. Otherwise the marked reduction in rupture life and flow stress cannot be
explained on the basis of purely loss of load
bearing section as mentioned above. Even denuda-
tion of chromium from the subsurface due to corrosion cannot account for the reduced rupture
lives.

One possible explanation may be the formation
of ductile grain boundary coverings due to diffusion of sulfur rich contaminants, as recently
determined by Schmitt-Thomas et al by means of
Auger spectroscopy[5].This could not be established
because these investigations are still outstanding in

this work.From metallography of failed specimens the following degradation mechanism can be anticipated:

- subsurface chromium-and γ'-depletion, preferably at grain boundaries due to hot corrosion (fig. 7).

- subsequent internal sulfidation preferably along grain boundaries (fig. 7).

- creep crack initiation at the internally oxidized subsurface layer (fig. 8).

- creep crack propagation preferably into the notchlike depleted grain boundary zones (fig. 9).

- oxidation of sulfides accelerates sulfidation attack at the crack tip during straining and leads to deeper penetration of sulfur (fig. 10).

These sulfidation processes in advance of the intergranular cracks weaken the regions adjacent to the grain boundaries by extensive depletion of the alloying elements chromium and/or titanium which result in accelerated creep crack propagation.

Further research work has to be done, especially on the latter phenomena of accelerated creep crack propagation under hot corrosion environments to corroborate the outlined mechanisms.

5. Acknowledgement

This work has been supported by the German Ministry of Research and Technology within the national research and development program "Corrosion and Corrosion Protection".

6. References

1. H. W. Grünling, B. Ilschner, S. Leistikow, A. Rahmel, M. Schmidt: Werkstoffe und Korrosion 29 (1978) pp. 691 - 703

2. J. R. Nicholls, J. Samuel, R. C. Hurst, P. Hancock: International Conference on the Behaviour of High Temperature Alloys in Aggressive Environments, October 1979, Petten, The Netherlands

3. M. Schmidt, A. Rahmel: European Symposium on the Interaction between Corrosion and Mechanical Stress at High Temperatures, May 1980, Petten, The Netherlands

4. H. Huff, F. Schreiber: Werkstoffe und Korrosion 23 (1973) pp. 370 - 377

5. Kh.-G. Schmitt-Thomas, H. Meisel, H. J. Dorn: Werkstoffe und Korrosion 29 (1978) pp. 1 - 9

6. K. H. Kloos, J. Granacher, H. Demus: European
   Symposium on the Interaction between Corrosion
   and Mechanical Stress at High Temperatures,
   May 1980, Petten, The Netherlands

7. J. A. Goebel: Proceedings of the First Confe-
   rence on Advanced Materials for Alternative
   Fuel Capable Directly Fired Heat Engines,
   August 1979, Castine, Maine, USA pp. 473-488

8. R. N. Parkins, F. Mazza, J. J. Royuela, J. C.
   Scully: Werkstoffe und Korrosion 23 (1972)
   pp. 1020 - 1029

Table 1: Constant-Strain-Rate Test Results ($T = 850\,°C$)

Alloy	$\dot{\varepsilon}$ (s^{-1})	$\sigma_{max,\ air}$ (Nmm^{-2})	$t_{f,\ air}$ (h)	$\sigma_{max,\ corr}$ (Nmm^{-2})	$t_{f,\ corr}$ (h)	$\dfrac{\sigma_{max,\ corr}}{\sigma_{max,\ air}}$	$\dfrac{t_{f,\ corr}}{t_{f,\ air}}$
IN 738 LC	10^{-7}	385	106	345	78	0,90	0,74
	5×10^{-8}	338	202	270	137	0,80	0,68
	2×10^{-8}	293	194	222	158	0,76	0,81
	10^{-8}	318	512	275	343	0,86	0,67
	2×10^{-9}	199	3014	100	2130	0,50	0,71
	10^{-9}	140	7768	122	still running	0,87	still running
IN 939	10^{-7}	361	79	365	83	1,01	1,05
	5×10^{-8}	342	224	342	244	1,00	1,09
	10^{-8}	303	566	279	454	0,92	0,80

Fig. 1: Constant strain rate test equipment

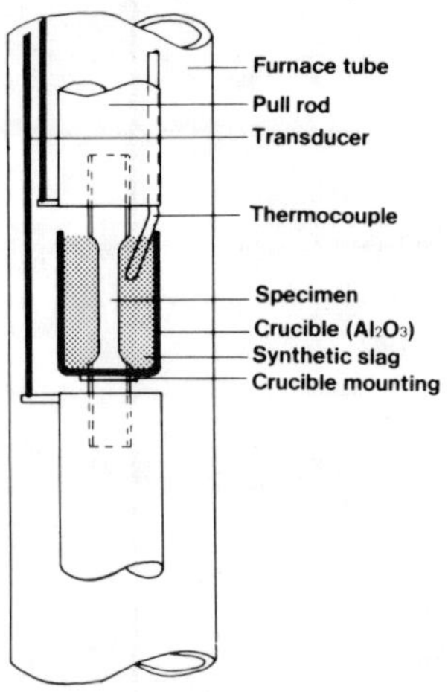

Fig. 2: Specimen mounting in CSR test machine

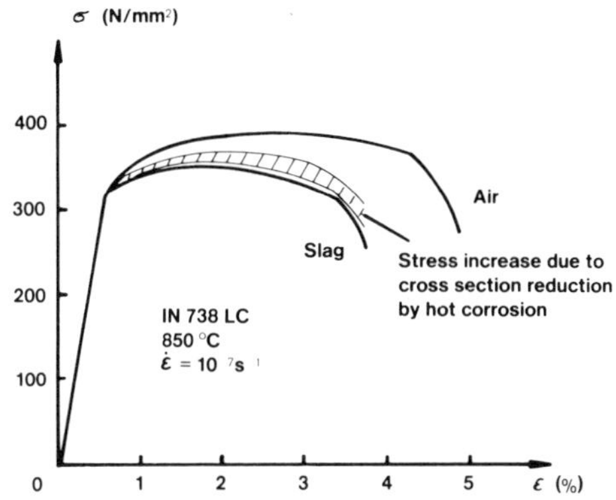

Fig. 3: Stress strain diagram of IN 738 LC with high strain rate

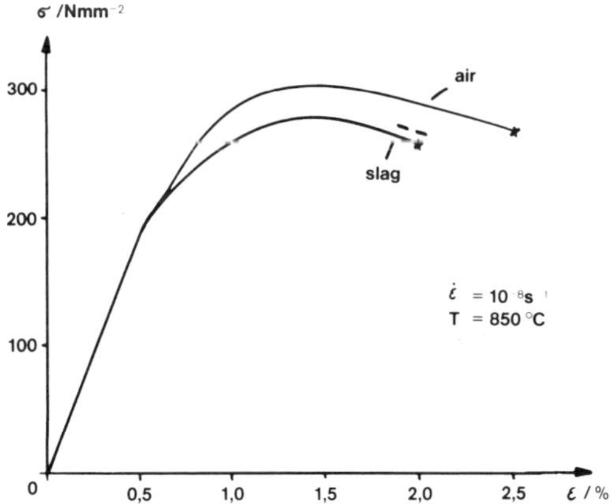

Fig. 4: Stress strain diagram of IN 939 with medium strain rate

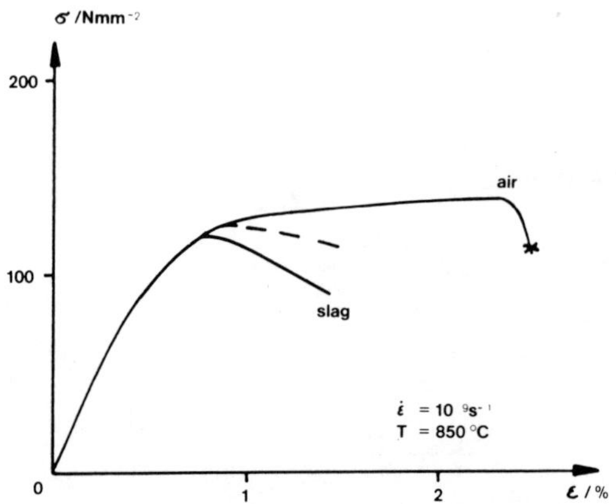

Fig. 5: Stress strain diagram of IN 738 LC with
low strain rate

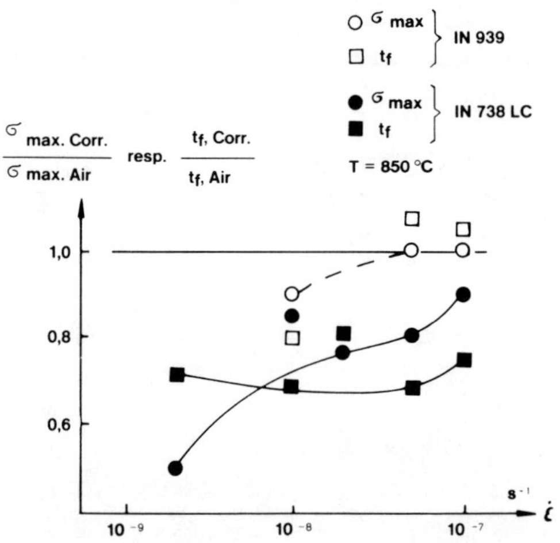

Fig. 6: Reduction of flow stress and fracture
life in hot corrosive environment

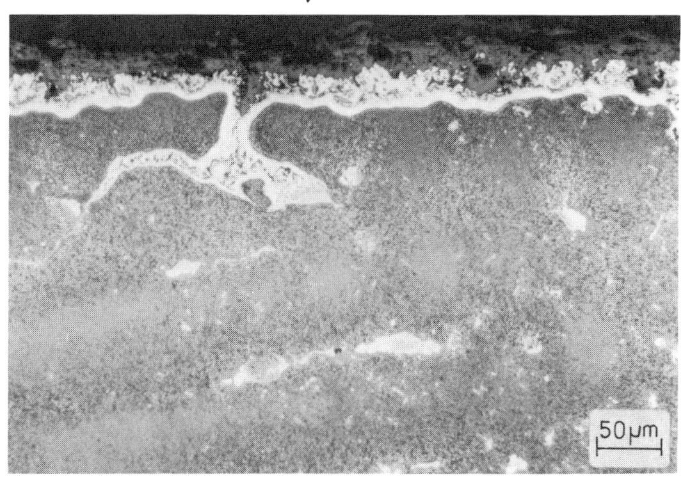

Fig. 7: Internal sulfidation of the denuded subsurface region (IN 738 LC, 10^{-8} s^{-1} 850°C, 343 h)

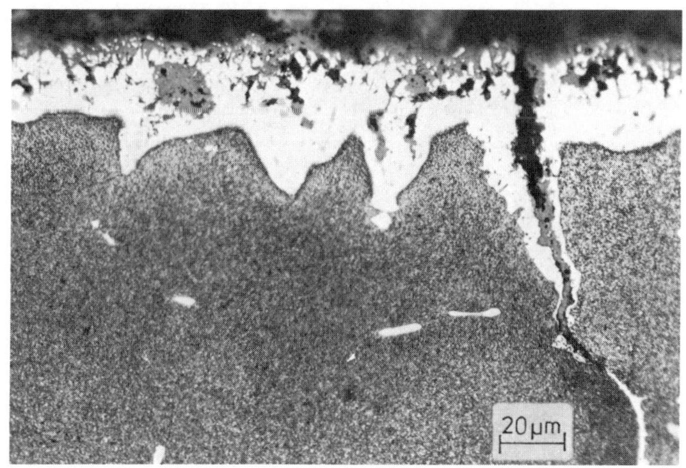

Fig. 8: Crack initiation at internal oxide products (IN 939, 10^{-8} s^{-1}, 850°C, 454 h)

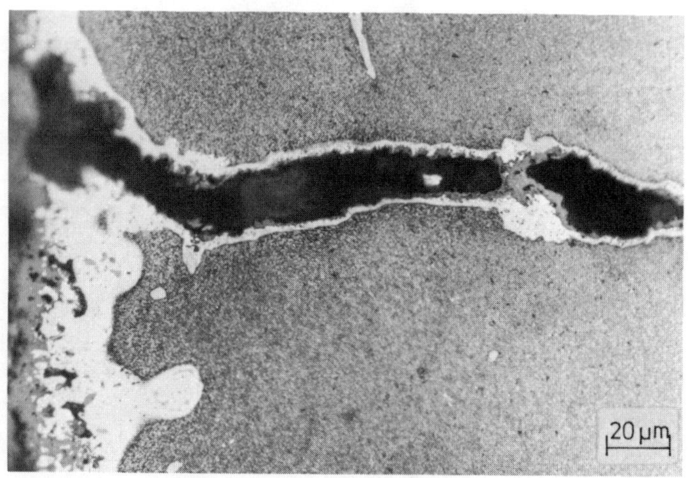

Fig. 9: **Intergranular** creep crack propagation under hot corrosion environment (IN 939, 10^{-8} s^{-1}, 850°C, 454 h)

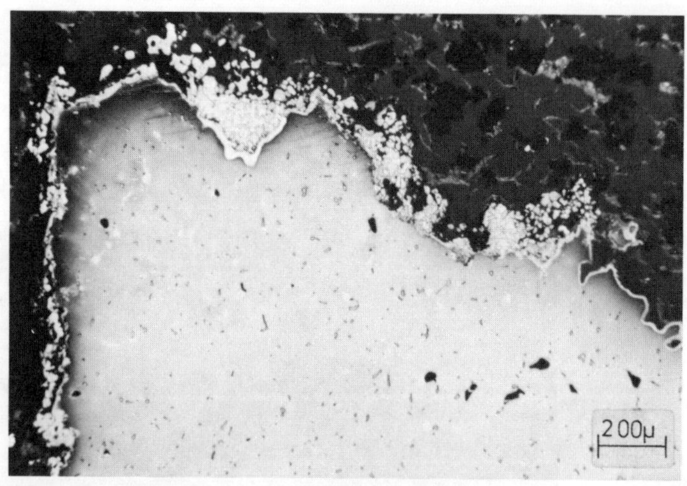

Fig.10: Ruptured specimen with typical heavy sulfidation attack (IN 738 LC, 2 x 10^{-9} s^{-1}, 850°C, 2130 h)

DISCUSSION

V. Guttmann: In all flow curves that you have presented, the common feature can be observed, that the first difference in the flow behaviour for the two different atmospheres occurs just when plastic deformation starts. Could you explain the reason for this effect?

W. Hartnagel: The reason for this effect is inherent in the used experimental technique. Before starting the test with the constant strain rate generated by the electric motor, the specimens have been manually loaded to a stress niveau near to elastic limit. This technique was necessary to load specimens in a shorter time and it avoids deviations of strain rate due to elastic deformation of the test machine.

H.W. Grünling: The papers presented in session 1b, especially concerning hot corrosion and sulfidation influence on creep and stress rupture properties seem to show very good agreement in results even when different test procedures were used. The most important result is that there is obviously no real effect on the basic creep mechanisms. Losses in rupture time and creep elongation are to be explained by means of facilation of creep crack formation and the enhancement of creep crack propagation. In the case of severe sulfidation and subsequent oxidation γ' and chromium depletion of the alloy due to sulphide formation is producing weak zones in to which the crack can more easily propagate.
Also, in the case of halide interaction, I do not believe that real creep processes are affected. In this case again, as also was pointed out by B. Ilschner, crack initiation influences and propagation mechanisms play the dominant role.

Microstructural aspects of fatigue crack propagation of cast nickel-base superalloys at 850°C in various environments

W. Hoffelner   M.O. Speidel [*]

Brown Boveri Research Center, CH-5405 Baden,
Switzerland
[*] now with: Eidgen. Techn. Hochschule, CH-8006 Zürich
Switzerland

SYNOPSIS

The fatigue crack propagation behaviour of the cast nickelbase superalloys IN 738 LC and IN 939 was studied at 850°C in various environments i.e.: vacuum, air and a service simulating sulphidizing atmosphere. While at medium ΔK-values the propagation rates in air and in the sulphidizing environment are twice as high as in vacuum, in the near threshold region the crack propagates faster in vacuum than in air. This behaviour will be explained as an interaction of environmentally induced crack accelerating and crack retarding effects.

INTRODUCTION

In a previous investigation[1] the influence of environment on the fatigue crack propagation rates of the cast nickelbase superalloys IN 939 and IN 738 LC at 850°C at medium cyclic stress intensity range ΔK (region of Paris law) has been studied. The used environments were vacuum, air and a service simulating sulphidizing atmosphere with and without chloride additions. The main results were:

o The crack propagation rates in air are about twice as high as in vacuum.
o The sulphidizing atmosphere shows no significant influence compared with the air values.
o Chloride additions change the fracture path from transgranular to intergranular with heavily branched cracks making an accurate determination of the propagation rates impossible.

o   From the drop of the crack propagation rates at low ΔK lower
    threshold values in vacuum could be expected than in air.

The aim of the present investigation is to study the near threshold
behaviour more in detail and to explain at least qualitatively the
influence of the environment on the fatigue crack propagation
rates.

EXPERIMENTAL

As specimen material the cast nickelbase superalloys IN 939 and
IN 738 LC have been used after standard heat treatment. These alloys consist of an austenitic matrix and coherent fcc γ'-particles
of the composition Ni_3 (Al, Ti) in 2 classes of size responsible
for the precipitation hardening. The main additional phases are
MC-type particles of composition (Ti, Ta, Nb) C occuring trans-
and intergranular and the chromium rich $M_{23}C_6$-carbides along the
grain boundaries. Fig. 1 shows the microstructure of both alloys.
From cast to size blocks, DCB-typ specimens were machined to the
dimensions previousely discussed[1].
Cyclic deformation was performed in a hydraulic closed loop machine
of 5 ton's capacity. As environments vacuum, air and a service
simulating sulphidizing atmosphere were used. A more detailed description of the testing facilities and the chemical composition
of the sulphidizing atmosphere can be found in literature[1].

EXPERIMENTAL RESULTS

Fig. 2 shows the fatigue crack propagation rates as a function of
the cyclic stress intensity range ΔK in various environments.

The well known stages of fatigue crack propagation could be found:

$$\Delta a / \Delta N \to \infty \qquad\qquad \Delta K \approx \Delta K_c \quad \text{(cyclic fracture toughness)}$$

$$\log \ \Delta a / \Delta N \propto \log \Delta K \qquad \frac{\Delta a}{\Delta N} = C \cdot (\frac{\Delta K}{E})^m \quad \text{(Paris law)}$$

$$\Delta a / \Delta N \to 0 \qquad\qquad \Delta K \approx \Delta K_o \quad \text{(fatigue threshold)}$$

In vacuum the region of Paris law can be described rather well
using $C = 5,1 \cdot 10^6$ and $m = 3,5$. The propagation rates are a little
bit higher than predicted by Speidel[2]. In aggressive environments
crack propagation rates of about two times higher, could be found
in this range, but no significant difference between air and sulphidizing atmosphere can be seen. But at the same R-value the fatigue threshold ΔK_o is considerably higher in air than in vacuum.

This behaviour which basically could be found also in steel[3] can
be explained on the basic of an interaction between crack accelerating effects and crack retarding effects.

Such effects, which are already discussed in literature[4,5,6] are
listed in tab. 1 where the arrows refer to acceleration ($\uparrow$) or retardation ($\downarrow$). How such effects can occur in the alloys IN 738 LC

and IN 939 is shown in figs. 3 - 5.

Fig. 3 shows the dissolution of γ'-particles around the crack.
This dissolution weakens the material and can lead to higher crack
growth rates. On the other hand, as mentioned by Scarlin[6], this
soft zone can be easily deformed, leading to crack tip blunting
and therefore increased ΔK_o-values.

Fig. 4 is a cross sectional view of the fracture path at low ΔK-
values ($\approx$ 10 MN.m$^{-3/2}$) in vacuum and in air showing transgranular
branched cracks in air and no crack branching in vacuum. In many
cases grain boundaries could be detected as starting point of
these branches in air and in the sulphidizing environment. At low
ΔK-values the crack only follows these grain boundaries for a
short distance and then propagates transgranularly as in fig. 5.
With increasing ΔK and also with decreasing frequency the amount
of interdendritic and intergranular fracture increases.
If chlorid is present the fracture path changes completely from
transgranular to intergranular with pronounced intergranular crack
branching as shown in fig. 6 reducing the actual ΔK at the crack
tip[7] leading to very low propagation rates of the fatigue cracks.

DISCUSSION

As shown in more detail in fig. 7 for air and vacuum, the shape
of the crack propagation curves of fig. 2 can be described by
superposition of two effects. The one leads to higher propagation
rates in air than in vacuum, and the other leads to a reduction of
the actual ΔK at the crack tip in the aggressive environment.

Considering crack branching to be responsible for this reduction
of ΔK, as a first approximation the measured ΔK-values have to be
reduced by $1/\sqrt{2}$ to get the actual ΔK at the tip of the crack[7].
This reduction leads to the dotted line in fig. 7. At medium ΔK-
levels this curve represents fatigue crack propagation rates of
about 6 times higher than in vacuum. This would be in a good agreement with investigations of the influence of air on the fatigue
crack propagation rates at elevated temperatures of wrought
Nimonic 105 which is not prone for crack branching[5]. In the near
threshold region, an obvious additional retarding effect, as crack
tip blunting in the γ'-free soft zones or crack closure effects,
has to be taken into consideration. Grain boundaries crossing the
fracture path and embrittlement of interdendritic regions could
account for crack branching while softening of the material at the
crack tip and prevention of crack rewelding could be the mechanisms
responsible for the acceleration of fatigue crack propagation rates
in air. The influence of sulphidizing atmosphere could be treated
in this way. Probabely the higher propagation rates in this environment are balanced by more pronounced crack branching.

CONCLUSIONS

Propagation of fatigue cracks in cast nickelbase-superalloys in aggressive environments at high temperatures is governed by a rather complicated interaction of environmentally induced chemical and mechanical effects leading to crack acceleration as well as crack retardation.

Their superposition can result in a strongly ΔK dependent influence of environment on the crack growth rates as shown for the alloy IN 738 and IN 939. The fatigue threshold values are lower in vacuum than in air, but at higher ΔK-values the fatigue crack rates are 2 times faster in air than in vacuum.

This behaviour could be explained mainly as combination of changes in strength in front of the crack tip and crack branching in the aggressive environment.

ACKNOWLEDGEMENT

The authors would like to thank Mr. R. Baumann for helpful assistance on metallographic investigations.

This work was financed in part by the Swiss Federal Government and performed within the framework of COST - action 50.

LITERATURE

[1] W. Hoffelner, M.O. Speidel, International Conference on the Behaviour of High Temperature Alloys in Aggressive Environments, Petten, The Netherlands, Oct. 15.-18., 1979, Conf. Proc. in print.

[2] M.O. Speidel in High Temperature Materials in Gas Turbines, Ed. P.R. Sahm, M.O. Speidel, Elsevier, Amsterdam 1974, p. 207.

[3] R.P. Skelton, J.R. Haigh, Mat. Sci. Eng. 36, 1978, 17 - 25.

[4] M.O. Speidel, R.B. Scarlin in International Conference "Gefüge und Bruch", Leoben, Austria, 25. Nov. 1976, p. 163.

[5] R.B. Scarlin, International Conf. on Fracture 1977 ICF 4, Conf. Proc. Vol. 2, Waterloo Press 1977, p. 849.

[6] R.B. Scarlin, in ASTM STP 675, ASTM 1979, p. 396.

[7] M.O. Speidel, The Theory of Stress Corrosion Cracking in Alloys, Nato Sci. Aff. Div., Brussels 1971, p. 346

Environmentally induced effect	Influence on the material	Influence on fatigue crack propagation rates
dissolution of phases	change the mechanical properties at the crack tip	↑ ↓
	weakening of material	↑
	crack tip blunting	↓
crack branching	reduction of actual ΔK at crack tip	↓
grain boundary attack	weakening of grain boundaries	↑
	intergranular crack branching	↓
oxid layer	prevents rewelding	↑
	prevents resharpening of crack tip	↓
	hindrance of dislocation movement	↓
crack closure	reduction of actual ΔK at crack tip	↓

Tab. 1:   Possible influences of the environment on fatigue crack propagation of cast nickelbase superalloys at high temperatures.
(↑ crack acceleration, ↓ crack retardation)

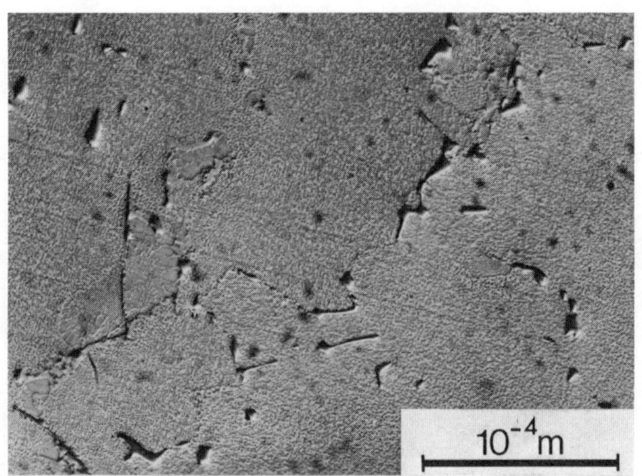

IN 738 LC

IN 939

Fig. 1:   Microstructure of the nickelbase superalloys IN 738 LC
and IN 939 showing blocky MC-carbides, fine $M_{23}C_6$ carbides at grain boundaries and the coherent γ'phase as
mottled background (light micrographs, glyceregia etching).

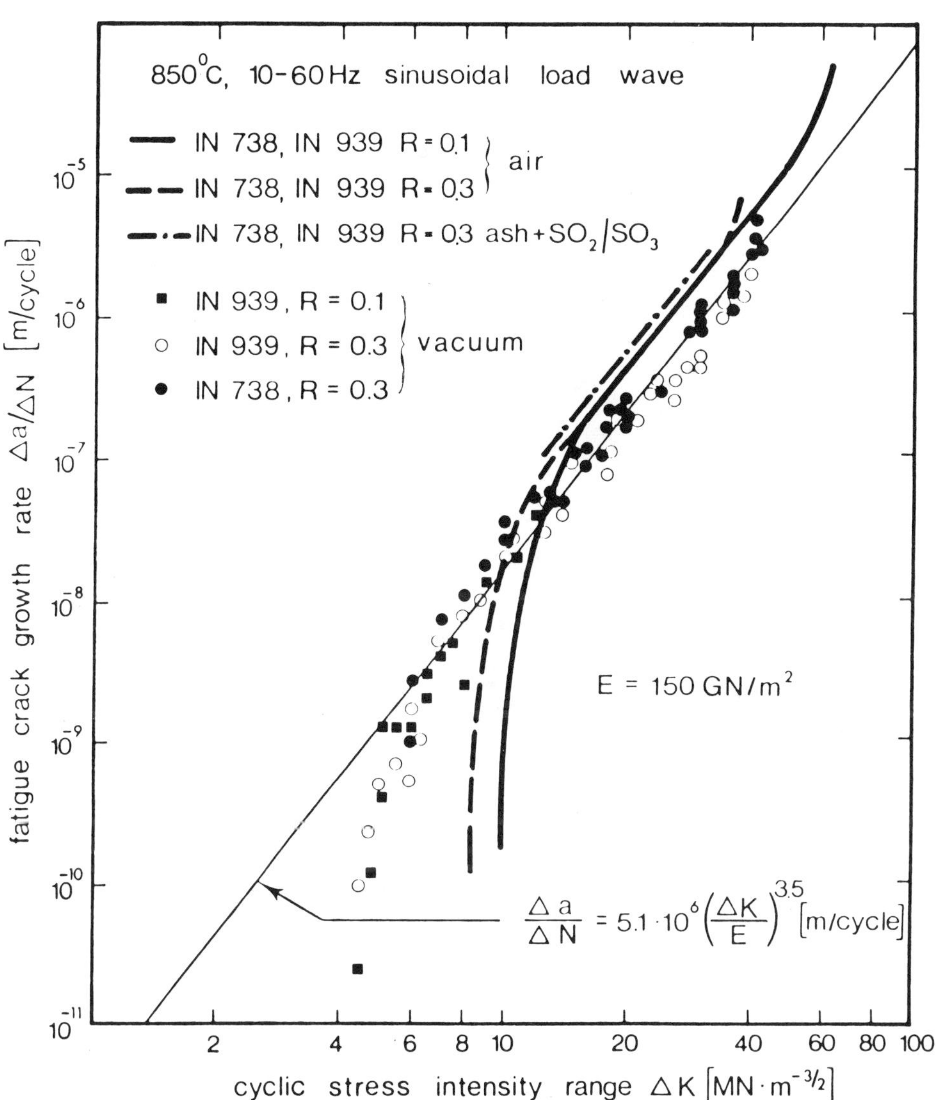

Fig. 2: The influence of environment on the fatigue crack propagation rates of IN 738 LC and IN 939 at 850°C.

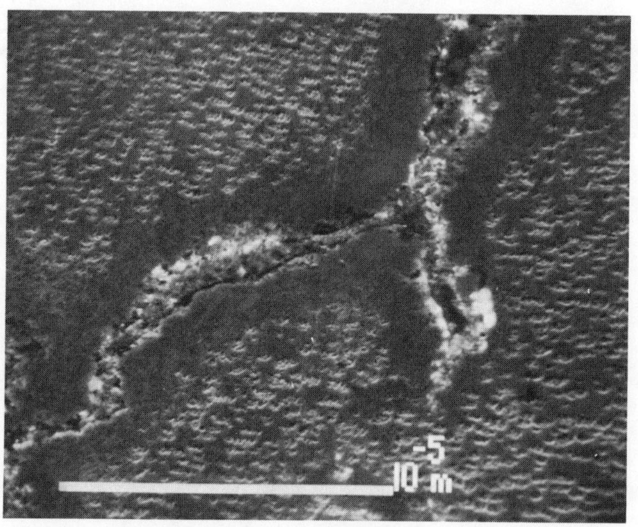

Fig. 3:   SEM micrograph of a fatigue crack in IN 738 LC tested
           in air showing dissolution of γ'-particles around the
           crack tip.

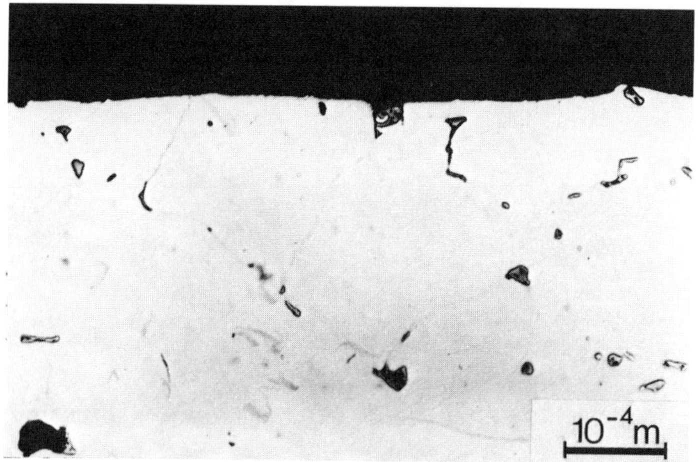

vacuum

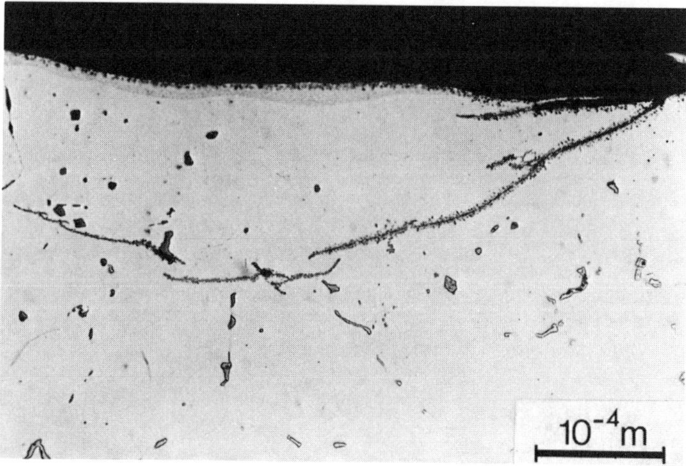

air

Fig. 4: Fracture path of a fatigue fracture in IN 939 tested in vacuum and in air. Pronounced crack branching can be seen at the air tested sample.

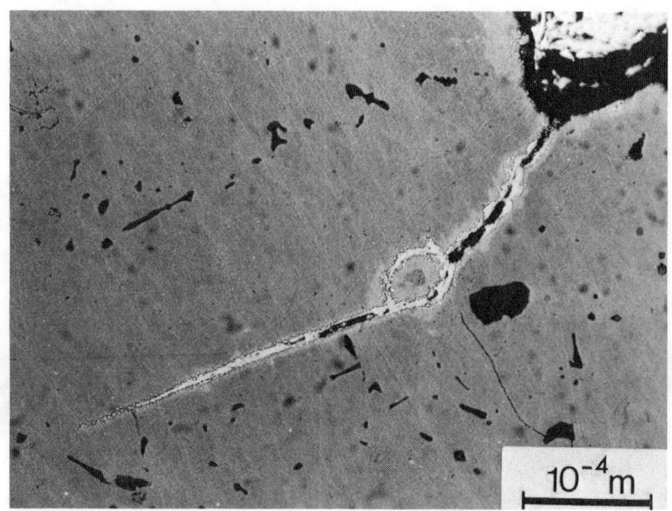

Fig. 5:   Light micrograph of a fatigue crack branch in IN 738 LC
          tested in air + SO_2/SO_3. The crack starts intergranular
          and changes to transgranular.

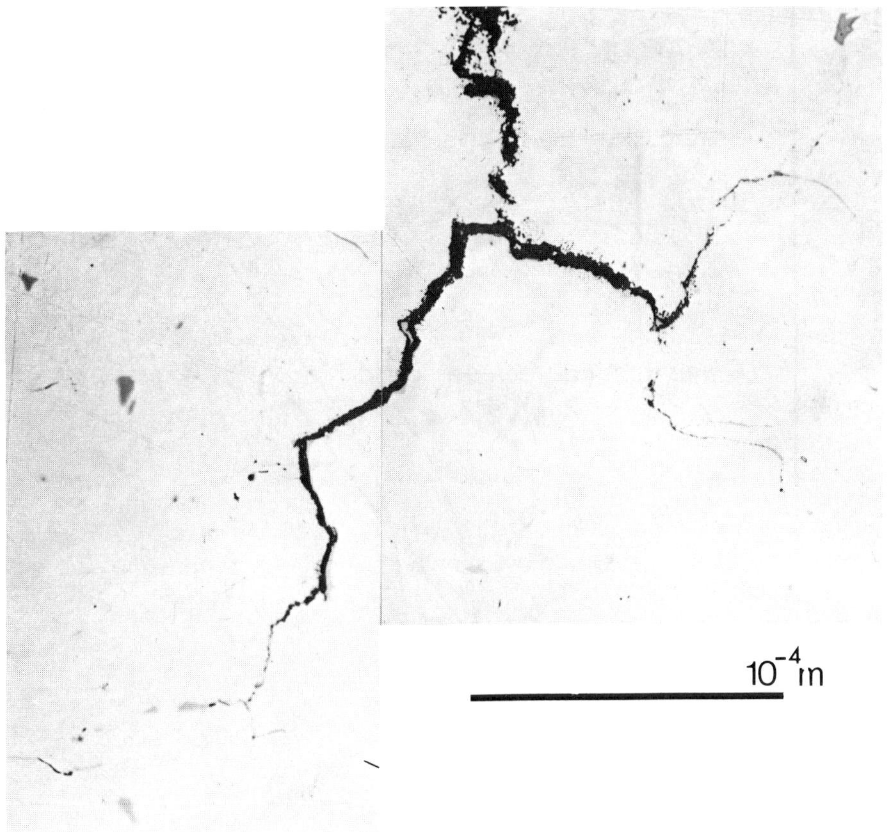

Fig. 6: The shape of the fatigue crack in IN 738 LC tested in sulphidizing atmosphere with chloride additions.

W. *HOFFELNER* and *M.O. SPEIDEL*

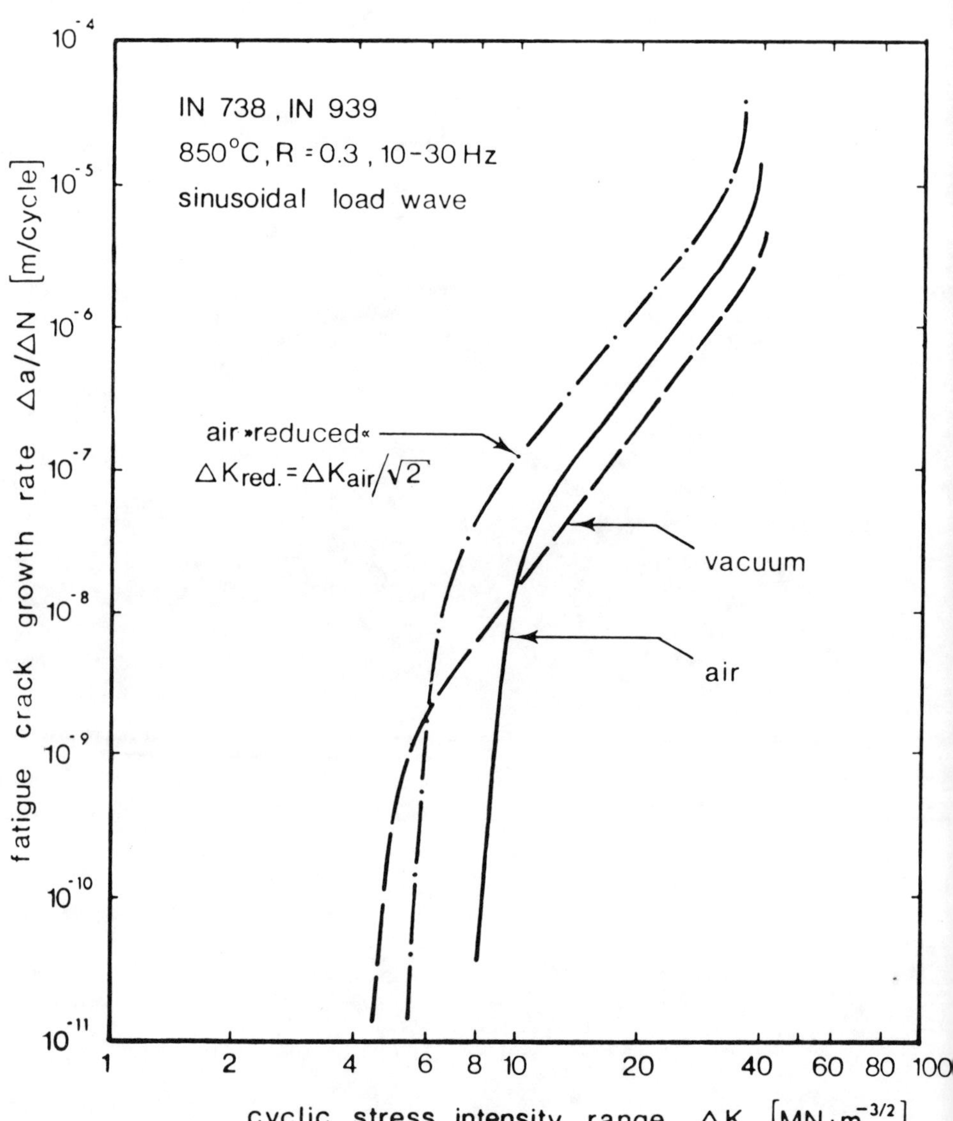

Fig. 7: The influence of environment on the fatigue crack propagation rates of IN 738 LC and IN 939 at 850°C.

DISCUSSION

B. Ilschner: This question is related to the interpretation given
by you for enhanced crack propagation caused by microstructural
changes such as dissolution. My point is that γ' dissolution is
a time dependent rate process which should be rather insensitive
with respect to the ΔK value. Therefore, one could expect that
at high crack propagation rates the γ' dissolution effect is much
less pronounced than at low $\Delta a/\Delta N$. In contrast, your curves
show a parallel shift (i.e. a constant factor) of the "air" as
"vacuum" curves over a large range of crack propagation rate.

W. Hoffelner: The main part of our investigations was concerned
with low crack propagation rates in order to get an explanation
for these low threshold values in vacuum. In all our metallurgical
investigations we found this dissolution of gamma prime phase
at the crack tip. For more details the knowledge of the kinetics
of the gamma prime dissolution along the highly deformed, reactive
surface in the vincinity of the crack tip would be required.
However, as pointed out before, fatigue crack growth at high
temperature is a very complex superposition of a lot of mechanism,
partly crack accelerating, partly crack retarding. As far as the
slope of the crack propagation is concerned, maybe the double
logarithmic plot is not sensitive enough to show small effects.

T. Ericsson: Crack propagation rate should depend on frequency.
Low frequency giving more time for corrosion. Could you elaborate
on the frequency effect you have observed.

W. Hoffelner: As presented by myself at the last meeting in
October in Petten on corrosion, at a particular ΔK value, the
crack propagation rate increases with decreasing frequency.
But it could be shown that this increase is mainly related to
creep crack growth and not to corrosion.

THE EFFECTS OF SUBSTRATE CREEP ON THE STRUCTURE OF OXIDE SCALES

FORMED ON Fe-9%Cr STEELS

P.C. Rowlands, M.I. Manning and J. Soo

Central Electricity Research Laboratories,
Leatherhead, Surrey, U.K.

SYNOPSIS

Fe-9%Cr steel specimens have been subjected to combined oxidation/
carburisation and axial creep at $600^{\circ}C$ in $1\%CO-CO_2$ atmospheres.
Transmission electron microscopy and optical metallography have
been used to study the effects of strains of between 0.3 and 7%
imposed over 4300h on the oxide scale microstructures. The two
layers of the duplex scales formed have been found to deform in
different ways. The Fe_3O_4 outer layer consists of large columnar
crystals (10X1.5μm) overlying a band of equiaxed crystals ($\sim$2μm).
The Fe_3O_4 layers have not fractured even at the highest strain
rates. The equiaxed crystals possibly deform by a process of continuous recrystallisation/oxidation and retain a homogeneous structure. The columnar crystals accommodate the strain by the formation
of pores predominantly within the oxide crystals. Dislocation and
stacking fault arrays are associated with the pores which grow
larger with increasing strain and have volume fractions approximately
proportional to the total strain. The inner, chromium containing
spinel layers, which have a very much finer grain size ($\sim$0.15μm)
deformed by different processes. At the lower strain rates, a form
of grain boundary sliding occurred, with subsequent infilling of the
grain boundary voids with Fe_3O_4. At the highest strain rate,
periodic fractures occurred through the inner layer normal to the
applied strain direction, which also filled with Fe_3O_4 crystals.
At intermediate strain rates outer layer - type laminations within
the inner layer were sometimes observed in the plane of the oxidised
surface. These observations help to explain factors affecting the
adherence and integrity of oxide scales formed on boiler tubes under
the influence of operational strains.

INTRODUCTION

The tests described in this paper were primarily conducted to
measure the effects of oxidation and carburisation on the
mechanical properties of 9Cr1Mo steels and this aspect of the work
has been reported elsewhere[1]. However, for high pressure pipework
and structural components subject to creep or other deformation at
high temperatures, it is possible that the deformation of the metal
may adversely effect the protective oxide scales. Thus strain of
the metal substrate might cause exfoliation of the oxide or stress
enhanced oxidation both of which could reduce the component life.
In this paper, the effects of imposed creep strains on the structure of the oxide scales grown during the tests is reported.

EXPERIMENTAL DETAIL

The compositional analysis of the steel used is given in Table 1.
Hollow tubular specimens of the form shown in Fig. 1 were machined
from 25mm dia bar to give a tubular gauge length section of 15mm
OD, 11mm ID and of 25.4mm in length. The specimens were normalised
at 1000°C for 0.5h and tempered at 700°C for 0.5h. The bore of the
test length was polished to a 600 grit SiC finish.

Gas line couplings were welded onto the capsules, the insides being
flushed with argon during welding to prevent oxidation of the inner
surfaces. Longitudinal loads were applied to the specimens using
hollow loading bars which allowed gas access, to give the stresses
shown in Table 2 and on Fig. 2 of between 46-110MNm^{-2} for the five
tests.

The specimens were enclosed in radiant electrical furnaces and maintained at 600°C with flowing CO_2 -1%CO - 625vppm H_2O at 4.1MNm^{-2}
pressure on the inside and with air on the outside. The stresses
were sometimes altered during the tests (Fig. 2) in order to
accumulate the desired creep strains at the end of the 4300h test
period. Elongation of the test sections during the tests were
measured using linear voltage - displacement transducers and recorded automatically.

After the tests, sections were cut from the gauge lengths for optical and transmission electron microscopy. For optical metallography, both circumferential and axial sections were prepared
from the gauge length and a section from the thicker, 'shoulder'
region was also prepared. The sections were mounted by hot pressing
in diallyl resin and polished to 1μm diamond finish. Oxide thickness measurements were made from both the strained gauge lengths
of the specimens and from the unstrained shoulders.

The TEM foils were prepared by a combination of mechanical pre-

paration and ion-beam thinning using the techniques described
by Manning and Rowlands[2]. At least two foils were prepared from
each specimen.

RESULTS

Optical Metallography

Visual examination of the specimens after testing showed no evidence of scale spalling or cracking. The oxide surfaces on the
bore were uniform and featureless.

From the metallographically prepared sections, measurements of the
oxide thicknesses measured on the gauge length and the thicker
'shoulder' region of each specimen were made as shown in Table 3.
Every value in the Table represents the mean of at least 15 readings at each selected location. The Table also gives the standard
deviations of the measurements.

By examination of carefully polished metallographic sections it
was evident that there were no cracks in the outer layer oxides of
any of the specimens. Careful etching of the scale microstructures
in 50^V/o HCl/H_2O showed that the outer layer scales consisted of
two distinct regions (Plate 1a). In a 15µm thick band closest to
the inner layer, the scale consisted of equiaxed crystals, while
the rest of the scale exhibited long columnar grains (approx.
10x1.5x1.5µm) normal to the scale surface. In addition small pores
are apparent in the columnar outer layer scale region but are much
less noticeable in the inner equiaxed regions of the Fe_3O_4 scale.

The etched sections revealed less about the inner layer scales whose
grain size is below optical resolution. However, occasionally, bands
of outer - type oxide scale were observed within the inner - layer
scales. These bands were normal to the specimen surface and about
1-2µm thick and 20-30µm long. They were present in the gauge length
regions of samples strained to 0.3 and 0.7% but only occurred in
localised regions where the scale was thicker. In the most highly
strained specimen, strained to 6.7%, a search was made optically
for small cracks in the inner layer normal to the specimen surface
which had been revealed in TEM. These were found, and are seen in
Plate 1 close to the interface between the inner and outer layers
where they are widest. These cracks were seldom more than a micron
wide even close to the outer layer and had a spacing of ∿300/cm
measured on an axially prepared specimen.

Transmission Electron Microscopy

Transmission electron micrographs of the outer layer scales clearly
showed the existence of numerous small pores in the columnar regions
of outer layer scales. These were generally much less common in the
equiaxed Fe_3O_4 crystal region.

Plate 2 shows representative micrographs of the pores in the columnar regions for each specimen. It was apparent that the total
number of pores was similar in each sample although they progressively increased in average size with increasing total specimen strain.
It was impractical to make a rigorous estimation of the volume per
cent porosity in each sample due to the non-uniformity of thickness
of the ion-beam thinned specimens. Nevertheless, at least two
samples were prepared from each specimen, and as large an area as
possible was examined, to minimise the element of subjectivity in
determining representative micrographs. Based on TEM micrographs
such as Plate 2, the mean diameter of the pores was estimated for
each specimen and in Fig. 3 the values obtained are plotted against
$(strain)^{1/3}$. As the foils were judged to be $\sim$0.12µm thick, and the
pore diameters could be estimated from the micrographs, a mean
volume density of $\sim$10^{11} pores cm^{-3} was calculated.

Diffraction contrast from dislocations or stacking fault arrays
were almost invariably associated with each pore. Dislocations
tended to extend from the pores to neighbouring features in the
foils such as grain boundaries and other pores (Plate 3). This behaviour was equally evident in the specimens strained to 0.33% and
7% strain (Plate 3).

The equiaxed region of the outer layer closest to the inner layer
was pore free with a low dislocation density. The inner layer/
outer layer interface was always distinct and free of any porosity
(Plate 4a).

The inner layer scales consisted of equiaxed grains of $\sim$0.15µm
diameter although the first row of inner layer crystallites appear
columnar and of a characteristic length (Plate 4a). This regular
behaviour soon disappears deeper in the inner layer although the
scale just below the outer layer often appeared striated on a fine
scale (Plate 4d).

On the 0.33% strained specimen, bands of outer-layer type crystals
were apparent (Plate 4b). These bands were parallel to the surface
and were typically 1-2µm thick with a columnar crystal morphology.
Plate 4c shows a feature observed in the inner layer scale of the
specimen strained to 0.7%. The inner layer crystallites appear to
have broken apart with magnetite filling the adjoining spaces. No
clearly identifiable inner layer degradation was observed in the
specimens strained to 1.35 and 2.5%.

Only on axial sections of the specimen strained to 7% were through-
scale cracks apparent in the inner layer. TEM and optical images
of these cracks close to the inner oxide/outer oxide interface are
shown in Plate 1. In the TEM image, the crack is seen to be filled
with outer layer Fe_3O_4 crystals. These contain many pores, the

columnar crystals having their long axes across the crack rather
than normal to the surface as in the outer layer Plate 1(b). The
outer layer scale just above the inner layer crack is defect-free
with a small equiaxed grain directly above it in a region clearly
subjected to very high local deformation rates.

DISCUSSION

Oxide Morphology and Microstructure

There is often some uncertainty about whether microstructural
features observed in oxide scales using optical microscopy or TEM
methods are introduced during the preparation stages. In this study,
the microstructural features which were large enough to be detected
optically have been observed both in ion-beam prepared TEM foils
and in diamond polished optical microsections. Thus there appears
little doubt that the extent and nature of porosity and microcrack-
ing in the oxide scales have been unambiguously identified by the
combination of the two techniques. It is thus clear that the Fe_3O_4
outer layer is able to withstand the applied straining rates of
up to $1.5 \times 10^{-5} h^{-1}$ and to accumulate deformations of 7% without
fracture.

A theoretical analysis has been made of the maximum deformation
rates sustainable by iron oxides[3] utilising available creep data
for bulk iron oxide samples. This analysis predicts that 60µm
thick Fe_3O_4 layers, as formed in these tests, should tolerate
straining at rates an order of magnitude higher than those employed.
It is likely that there were localised regions in the outer oxide
scales where deformation rates were sometimes considerably higher
than the applied creep rates. Such regions would exist above the
cracks in the inner layer as in Plate 1b. A plausible interpretation of Plate 1b is that at some earlier stage in oxidation, when
the inner layer crack formed, it propagated into the outer layer.
Subsequently, however, the outer layer crack largely healed, the
inner layer crack was bridged by Fe_3O_4 crystals and a new equiaxed
Fe_3O_4 grain recrystallised in the outer layer above the crack.
After the inner layer had cracked in this way, further deformation
of the layer would occur by the cracks widening (Plate 1b). This
would result in high localised strain being applied to the Fe_3O_4
crystals within the cracks causing the observed porosity, although
fresh oxide deposition at this site could minimise the requirements
for creep accommodation. It is presumably true that at any point
within the scale, but increasingly towards the metal interface,
fresh oxidation is avaialble to facilitate crystal growth or to
nucleate fresh grains to fill volume created by straining. Plate 1b
provides some evidence then that continuous recrystallisation
assisted by oxidation may be occurring in the equiaxed region of
the outer layer scale and within gaps opened up by the strain
within the creep resistant inner layer. Certainly in the equiaxed

crystal regions, deformation is accomplished without the development of significant porosity, although within the inner layer cracks, the Fe_3O_4 crystals are columnar and contain significant porosity.

The larger part of the outer layer scale consists however of columnar grains ($\sim$10 x 1.5 x 1.5μm). In this region, deformation is accompanied by the growth of pores within the columnar region as shown in Plate 2. The majority of the pores appear within grains rather than at grain boundaries, although it is possible that subsequent grain boundary migration might have occurred since pore nucleation. The density of pores appears similar in the variously strained specimens, but they increase in size with increasing strain. To test the hypothesis that pore growth accommodates a significant part of the applied strain estimates of the mean pore diameter are plotted against the cube root of the strain in Fig. 3. Large errors are associated with the estimates of mean pore diameters due to the non-uniformity of the thin foils precluding statistical evaluation of their sizes and distribution. The density of pores was determined from the electron micrographs by estimating the mean number of pores per unit area, N_A, and calculating the mean number per unit volume from $N_V = N_A/(h+D)$ where h is the foil thickness of $\sim$ 0.12μ and D is the measured value of the larger void diameters in the section. N_V was also assessed from optical micrographs using $N_V = N_A/D$. Both optical and TEM were in agreement, both methods giving values of $\sim$ 10^{11} pores cm^{-3} for all the sections examined. Taking a mean value of 10^{11} cm^3, a line can be drawn on Fig. 3 for the mean pore diameters required to accommodate all the strain. Another line is drawn for the pore diameters required to accommodate ^{1}/10th of the applied strain by volume creation in pores.

In all the samples, dislocations associated with the pores could always be found under certain imaging conditions (Plate 3). Dislocations associated with pores have been reported in other studies of fine grained ceramic materials. Very similar micrographs were obtained by Tighe[4] in hot pressed alumina and extruded MgO. Deformation microstructures of fine grained ceramics were also investigated by Heuer et al[5] who noted that the ability of pores to pin dislocations appears to be a general phenomena. Evidence for grain boundary sliding was also found in deformed alumina where small triple point voids were observed[5] in thin foils. It is not clear whether the pores should pin dislocations and thus hinder deformation as suggested by Heuer et al[5], or whether, by acting as dislocation sources and sinks, they facilitate deformation. In this work, it is at least apparent that the pores grow in size with increasing plastic strain with little fresh nucleation which, must indicate the ability of the dislocations to transport the matter around. For the pores to grow both ionic species must be transported and this leads to the interesting speculation that dislocation motion could transport oxygen through compact outer layer oxide scales.

Oxide Deformation Modes

Three separate deformation modes of the inner layer scales were observed. In the sample strained to 0.7%, the small inner layer crystallites appear to have parted in the applied strain direction with formation of further outer layer type oxide in the interstices normal to the metal surface. In the sample strained to 0.33%, inner layer fractures of a different nature were observed. Fractures within the inner layer in the plane of the specimen surface were filled with columnar outer layer type crystals. Where these failures had occurred, the inner layer scale was thicker than in adjacent unfailed regions. These failures give rise to ∿1μm thick Fe_3O_4 laminations within the inner layer which are morphologically very similar to outer-layer type inclusions seen in transmission micrographs of breakaway oxide scales[6] on 9Cr steels after prolonged oxidation in high pressure CO_2 environments. The third failure mode, observed in the sample strained to 6.7%, of cracks through the inner layer normal to the applied strain direction is more akin to the brittle fractures occurring during rapidly applied straining of iron oxides.[7] The cracks were widest at the outermost part of the inner layer which is consistent with the oldest part of the oxide having accummulated the greatest strain. These observations provide new evidence for the creep resistant nature of the inner layer scales on 9Cr steels. This is in spite of their very much smaller grain size which should facilitate creep. The increased creep resistance of this layer is probably due to the addition of chromium and other alloying elements to the spinel phase M_3O_4. Laboratory measurements have shown that Cr diffusivities in Fe_3O_4 are at least $X10^3$ less than self diffusion of iron and this is expected to be reflected in creep behaviour[8].

Because of the small areas examined from each specimen following TEM foil preparation, it is possible that further examinations might have revealed evidence of more than one deformation mode acting in certain specimens. No clear evidence for any structural degradation of the inner layer was apparent in the specimens strained to 1.35 and 2.5% strain. Although this may be due to the restricted areas examined, it is notable that these specimens showed the lowest overall scale thicknesses (Table 3).

Effect of Strain on Oxidation

The effects of the microstructural degradation discussed above on the oxidation rate have been assessed by comparing mean oxide thicknesses from the gauge length of each specimen with that from the shoulder assuming it to be virtually unstrained. The mean oxide thicknesses on the shoulder were 109μm for the 0.33, 1.35 and 6.75% strain tests which were all conducted on the same machine and ∿119μm on the two tests conducted on another machine which may account for the minor variation of thickness of unstrained regions.

To facilitate comparison of strain accelerated oxidation behaviour
of each specimen, the percentage increase in oxide layer thicknesses
of the strained region over the shoulder region have been tabulated
for each specimen in Table 4. The inner layer thicknesses show the
larger percentage increases of the strained regions. The high
values at .33% strain are associated with the formation of outer
layer type laminations within the inner layer (Plate 4b). Rather
surprisingly, the strain acceleration factor subsequently falls at
progressively higher strain rates from $1.7 \rightarrow 8 \times 10^{-6} h^{-1}$. The outer
layer is actually slightly thinner for the specimen exhibiting
grain boundary sliding of the inner layer (0.7% strain). Only at
the highest strain rate with a total strain of 6.75% is this trend
reversed. This specimen shows a considerable increase in overall
oxide thickness of 45% most of which was due to inner layer thickening by 63% (Table 4).

Oxide thicknesses were carefully measured at many locations to
obtain the means and standard deviations. Simple statistical tests
show that the measured changes in Table 4 are significant at the
99% confidence level. The shoulder temperatures were found to be
only $1^{\circ}C$ lower than the $600^{\circ}C$ controlled at the gauge length with
the large radiant electric furnaces employed, and thus the thickness increases of the strained portions cannot be explained on
this basis.

The proportionately larger increase in inner layer oxide thickness
compared with the outer layer clearly indicates that one effect of
applied strain is to enhance oxidant ingress. This enhancement is
more likely to be due to the degradation of the inner layer microstructure rather than to the increase in pore diameter, since the
outer layer is generally considered to be sufficiently porous to
support inner layer growth[9]. The observed grain boundary sliding
and subsequent infilling with magnetite is one possible cause of
enhanced oxidant ingress.

The strain rates employed in these tests greatly exceed those to
which 9Cr steels would be subjected in the CO_2 coolant environment
of AGR boilers. Consequently, the relatively small enhancement of
oxidation in the gauge region of the strained specimens is
encouraging. A speculative explanation of the lower oxide thicknesses of intermediate strain rates (Table 3) could consider that
inherent compressive stresses in the scale during growth were
partially relieved by the tensile stresses applied to the inner
layer. Further work would be required to discover whether this
effect was reproducible. At the highest strain rate of 1.6×10^{-5} a
definite oxidation enhancement was associated with magnetite
filled scale cracks through the inner layer.

These observations also have some relevance in considering the
failure modes of $2\frac{1}{4}Cr1Mo$ superheater and reheater tubes operating

at approximately 600°C in fossil fired plant. Here, strain rates
approach those applied in these tests and the inner layer of the
scales formed on both the steamside and flue gas side of the tubes
have been found to exhibit excessively thick oxide scales with
laminated inner layer scale morphologies. A mechanism proposed to
account for this accelerated oxidation[10] assumes that inner layer
laminations are caused by straining of the inner layer scales due
to the radial tube distension as it creeps under internal pressure.
Observations obtained in this study have confirmed the two assumptions of this model[10] - that outer layer magnetite is easily de-
formable and that inner layer scales containing chromium are considerably more creep resistant.

These tests, although conducted on tubular specimens did not reproduce the straining conditions of pressurised tubing. The capsules
were pulled uniaxially with the internal pressure being a very minor
stress, whilst pressure tubes distend mainly radially as the hoop
stresses are twice the axial stresses. It was proposed[10] that radial
tube distension favoured laminated scale formation because it creates
space for the formation of a new layer. In these tests, it is most
likely that the tube diameter would actually contract during the
test because of the Poisson contraction and this might well affect
the nature of inner layer scale failures.

In summary, the application of TEM methods to study the deformation
microstructures of oxides strained during mechanical testing gives
additional information about creep and fracture processes that com-
plements optical microscopic investigations. An improved understanding of the mechanical properties of oxide scales can shed new
light on the resistance of protective scales of sustained deformation in service and may help in interpreting the role of scale
breakdown in the accelerated oxidation processes to which ferritic
alloy steels are prone.

CONCLUSIONS

1. 9Cr steels have been strained during oxidation at 600°C at
    strain rates up to $1.6 \times 10^{-5} h^{-1}$ and to total strains of 6.75%
    and the resulting scale microstructures examined optically and
    by transmission electron microscopy.

2. Small, but significant, increases in oxide thicknesses were ob-
    served at all the strains from 0.33 to 6.75% when compared with
    unstrained areas. The highest strain resulted in a 45% increase
    in oxide thickness. Proportionately larger increases in inner
    layer oxide thickness compared with the outer layer were observed.

3. No scale spalling or total fracturing of the scales occurred.
    The outer layer of the usual duplex oxide layer was found to
    consist of small equiaxed Fe_3O_4 crystals ($\sim$2µm dia) close to
    the inner layer and longer columnar crystals towards the gas/

oxide interface. Inner layer grain sizes were $\sim 0.15\mu$m.

4. The equiaxed Fe_3O_4 crystal layers accommodated the applied deformation without development of any related microstructural features. The columnar Fe_3O_4 layer accomplished the deformation with the growth of voids within the scale.

5. The total volume of the creep voids was approximately related to total strain as $(\varepsilon)^{1/3}$.

6. In all cases, dislocation networks extended from the void surfaces to adjacent microstructural features.

7. The inner layer scales were not able to deform without some degradation. Three deformation modes were observed. (1) At lower strain rates inner layer crystallites exhibited a form of grain boundary sliding with Fe_3O_4 forming in the grain interstices. (2) Thin (1μm – 2μm) thick bands of outer – layer type were observed within the inner layers and are similar to those formed during breakaway oxidation. (3) Finally, at the higher strain rates, cracks normal to the metal surface were observed in the inner layer scales. These subsequently filled with outer-type crystals.

8. A comparison was made between optical and TEM microstructures. With careful preparation, the same pores and cracks could be observed using both methods.

9. The fracture observations thus obtained support the view that 100μm thick Fe_3O_4 layers can accommodate strain rates up to $2\times10^{-5}h^{-1}$, but that the inner layer scales are not able to withstand even the lower strain rates without some microstructural degradation. However, a progressive increase in oxidation rate was only occasioned the highest strain rate employed.

REFERENCES

1. J. Soo and P.C. Rowlands "The effects of combined oxidation and carburisation on the mechanical properties of Fe-9Cr/Mo steels". 1980, Proc. Conference on 'Environmental Degradation of high temperature materials', Isle of Man 1-3 April.

2. M.I. Manning and P.C. Rowlands "A method of preparing TEM foils from thick oxides and metal/oxide interfaces". Submitted to Br. Corr. Journal.

3. M.I. Manning, "Limits to the integrity of oxide scales growing on curved surfaces and distending metal substrates". This publication.

4.   N.J. Tighe "Microstructure of Fine-grain Ceramics" Chapter 7
     pp 109-133, 1970 "Ultrafine-grain Ceramics" New York,
     Syracuse University Press.

5.   A.H. Heuer, R.M. Cannon and N.J. Tighe "Plastic Deformation in
     Fine-Grain Ceramics" Chapter 16 pp 339-365, 1970 PUltrafine-
     grain Ceramics", New York, Syracuse University Press.

6.   P.C. Rowlands and M.I. Manning, to be published.

7.   M.I. Manning and S.A. Richardson "Strain-tolerance of steam
     grown oxide scales" CEGB Report RD/L/N25/78.

8.   J.D. Hodge, J. Electrochem. Soc., 1978, 125, No. 2, 55C.

9.   Hampton, J., Rowlands, P.C. and Teare, P.W., CEGB report
     RD/L/N49/78, 1978.

10. M.I. Manning and D.B. Meadowcroft "Effects of Tube creep
     strains on laminated scale formation in ferritic pressure
     tubing" 1980. CEGB Report RD/L/R2012.

ACKNOWLEDGEMENTS

This paper is published by permission of the Central Electricity
Generating Board.

TABLE 1

Chemical Composition of the Steel Used $[^w/o]$

C	Si	Mn	Cr	Mo	S	P	Cu
0.11	0.33	0.38	8.7	0.95	0.007	0.013	0.09

TABLE 2

Summary of CO_2 Creep/Oxidation Tests

M/C Test No.	Stress [MPa]	Final Strain %	Total Time [h]	Strain Rate $[h^{-1}]$
1	46	0.33	4300	7.7×10^{-7}
2	70	0.72	4300	1.7×10^{-6}
3	90	1.35	4300	3.1×10^{-6}
4	80/110	2.50	4300	8×10^{-6}
5	100/120/100	6.75	4300	1.6×10^{-5}

TABLE 3

Mean oxide thicknesses for strained and unstrained sections of specimens

Strain %	Oxide Thickness μm					
	Unstrained Shoulder			Strained gauge length		
	Total	Outer layer	Inner layer	Total	Outer layer	Inner layer
0.33	109±10	57±4	52	137±10	65±7	72
0.72	118± 5	64±3	54	133±20	60±5	73
1.35	109±13	57±6	52	119±13	61±5	58
2.50	120±13	62±5	58	124±19	64±6	60
6.75	108±16	56±8	52	157±10	72±5	85

TABLE 4

Percentage change in oxide thickness formed on
strained gauge length compared with unstrained shoulder

Strain %	Percentage increase in oxide thickness		
	Inner layer	Outer layer	Total
0.33	38.0	14.0	25.7
0.72	35.0	6.6	12.7
1.35	11.5	7.0	9.2
2.50	3.4	3.2	3.3
6.75	63.0	28.0	45.0

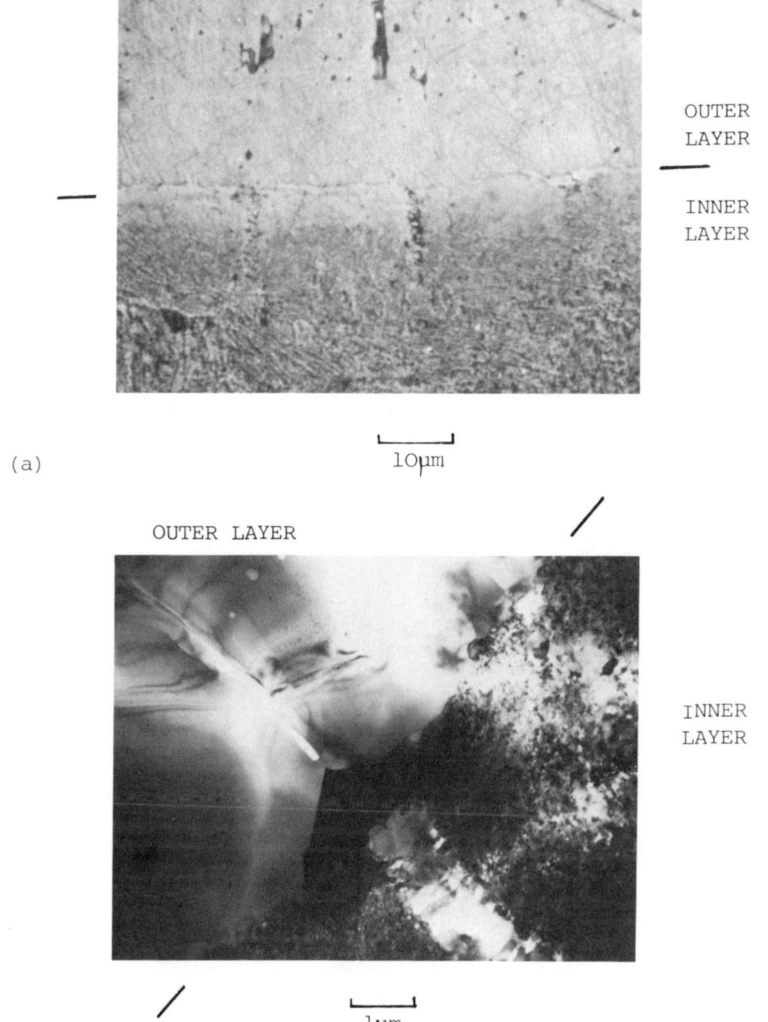

(a)

(b)

Plate 1.   Cracks through inner layer on specimen strained to
            6.7%

            (a)   Optical micrograph
            (b)   T.E.M. micrograph (200 kV)

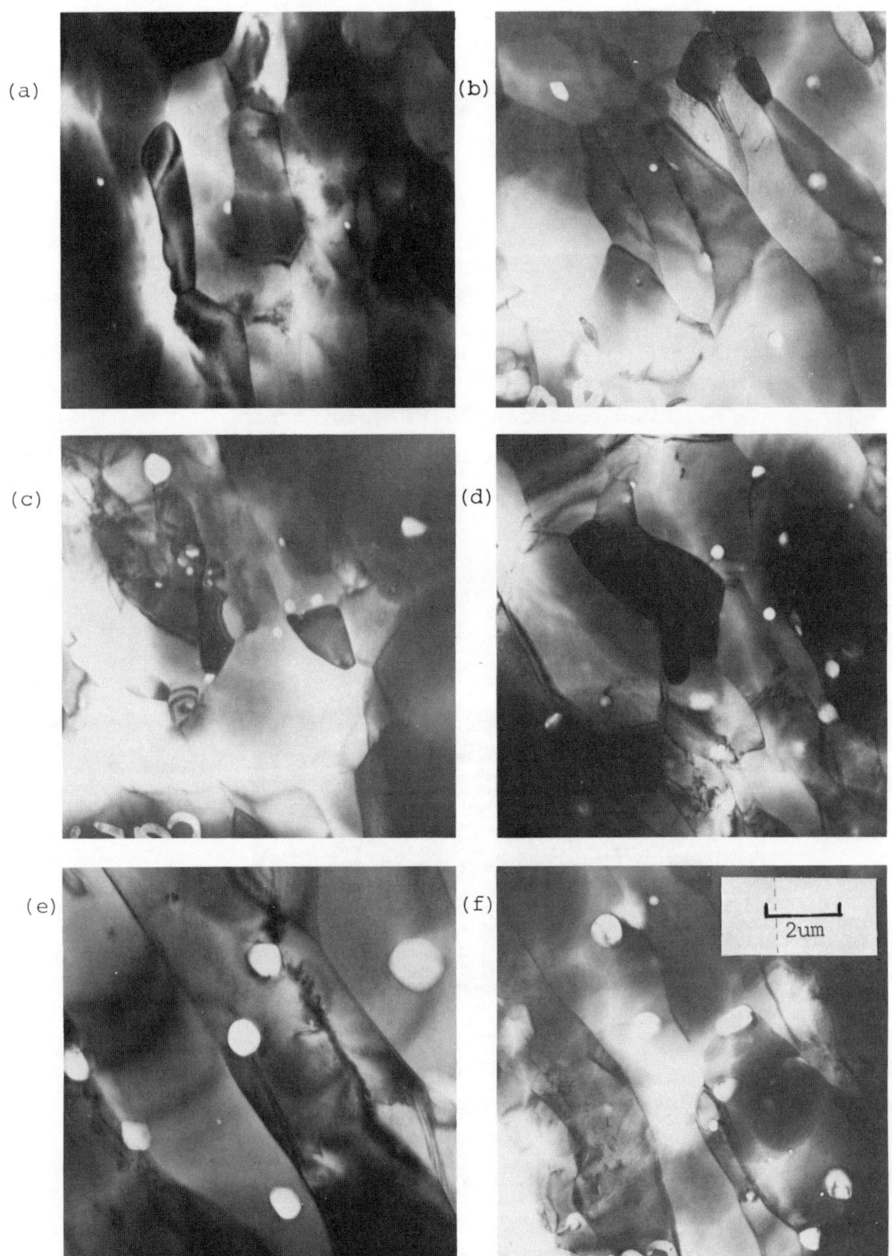

Plate 2. Pores in the columnar region of the outer layer scales
          of the specimens strained to 0%,(a);0.33%,(b);0.7%,(c);
          1.35%,(d);2.5%,(e); and 6.7%(f).

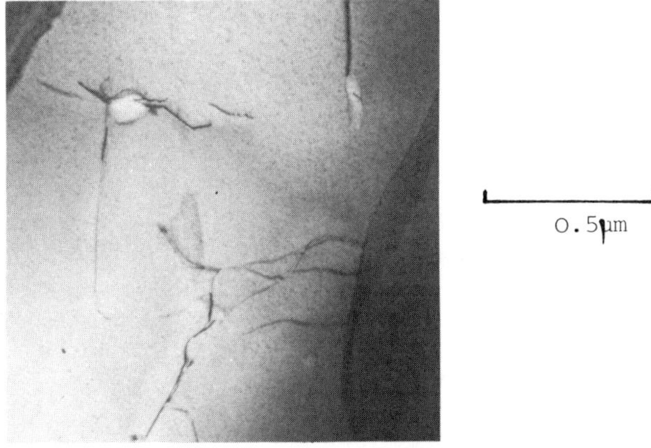

(a) Sample strained to 0.33%   (1000kV)

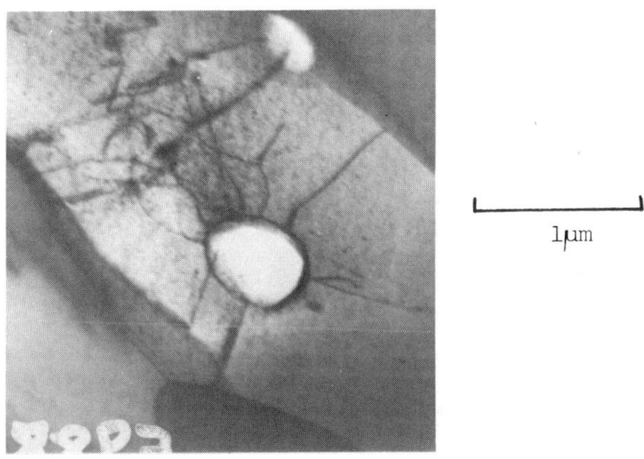

(b) Sample strained to 6.7%   (200kV)

Plate 3. Dislocations associated with pores in outer layer
          scales at both the lowest (a) and the highest (b)
          strain rates employed.

(a) |—————| 1µm

(b) |———| 1µm

(c) |—————| 0.5µm

(d) |——| 1µm

Plate 4.   Inner Layer Scale Morphologies.

   (a) Inner/outer oxide layer interface on sample strained
       to 0.33%.   (1000kV)
   (b) Outer layer type laminations within the inner layer
       region on sample strained to 0.33%.   (200kV)
   (c) Fe_3O_4 filled gaps between the inner layer crystallites
       on the sample rowstrained to 0.7%   (200 kV)
   (d) Striated inner layer region on sample strained to
       6.7%.   (200 kV)

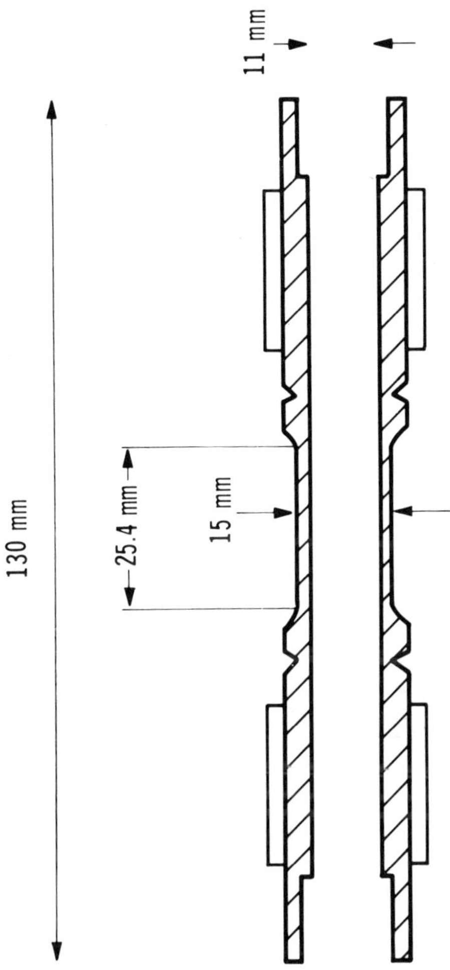

Fig 1. Tubular 9Cr1Mo specimens used for studying creep/oxidation interactions in 4.1 MNm^{-2} CO_2.

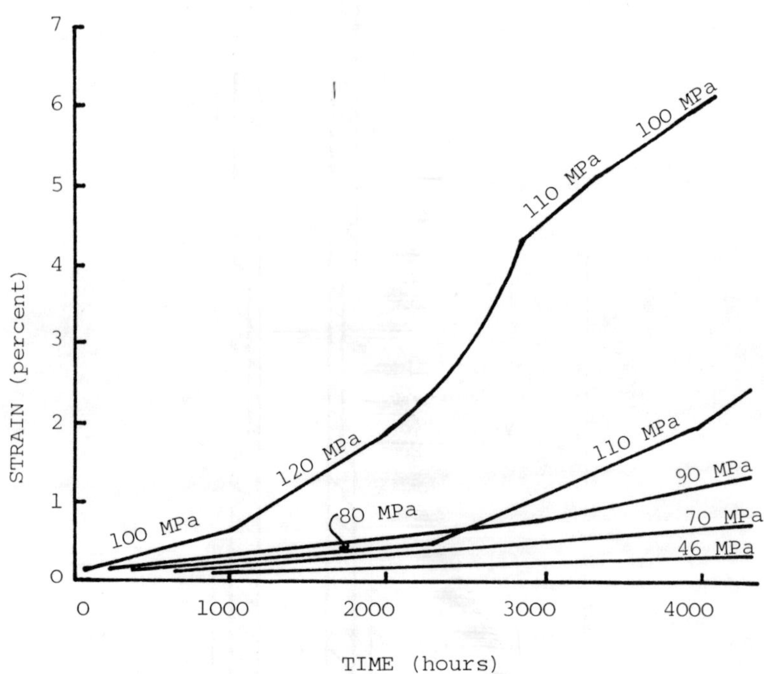

Fig 2.   Strain/time curves for the tubular 9Cr1Mo
test specimens loaded at 600°C at the
stresses indicated.

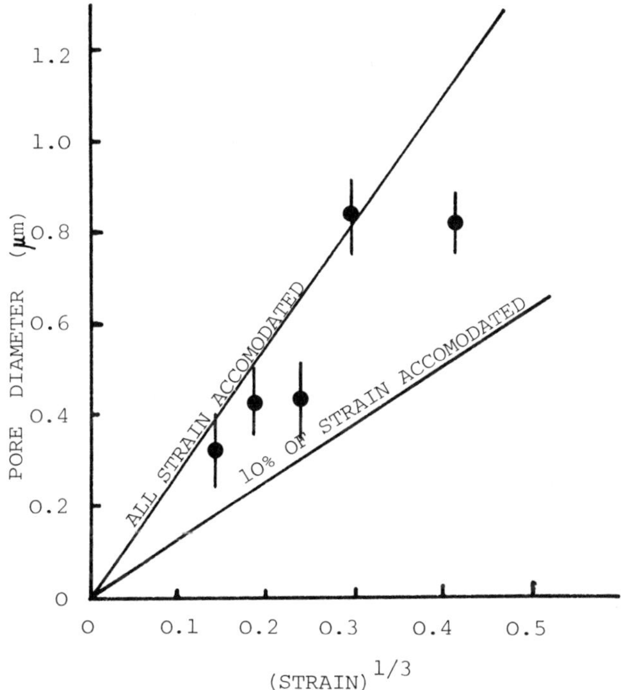

Fig. 3. Measurements of the outer layer pore diameters as
a function of the cube root of the total applied
strain. The lines drawn represent the corresponding
strain accomodation by volume creation assuming a
density of pores of 10^{11}/cm^3

DISCUSSION

<u>A. Palazzi</u>: Iron is a metal that unfortunately shows us passivation; oxydation proceeds through diffusion of vacancies.

In order to have an iron base metal that withstands corrosion at high temperature, we add chromium and aluminium whose property is to form diamond like oxides i.e. spinels.

Spinels are growing epitaxially and are well rooted on the metal base because they are non-stoichiometric compounds.
Properties of FeO, Fe_2O_3, Fe_3O_4 are of little use if the importance of spinel intermediate layer is not fully dealt with.

<u>M.I. Manning</u>: It is to be expected that the structure of an oxide will only directly influence the adherence of the layer (via lattice correspondencies across the metal/oxide interface) when the grain size of the oxide is large compared with the layer thickness. The inner layer scales formed on the ferritic steels used in this work are very fine grained with at least 10^3 diameters across the layer thickness.

<u>R. Rolls</u>: One of your illustrations showed void deformation in the Fe_3O_4 outer scale layer. What is believed to be the process for void formation? Is it by vacancy condensation?

Is it known for the temperature of your studies by what deformation mechanisms the creep of Fe_3O_4 occurs. It would be interesting to know if this is assited by the vacancy flux in the Fe_3O_4, which might explain the degree of deformation observed, especially if the void formation implies a saturation of vacancies in the Fe_3O_4.

<u>M.I. Manning</u>: In these tests, the deformation of the oxide was imposed by creep of the metallic substrate and consequently the stress in the oxide is unknown, and as I don't think cavity growth as a consequence of creep in Fe_3O_4 has yet been considered analytically, I can not be specific about the deformation mechanism. Our observations clearly show, however, that all the creep voids are associated with dislocations, and that the total void volume is roughly proportional to the applied strain.

INFLUENCE OF CREEP STRESS ON THE MORPHOLOGY OF
OXIDE SCALES DEVELOPED ON Ni-20 Cr-1,4 Si ALLOY

P. FAURE, B. PIERAGGI, F. DABOSI

Laboratoire de Métallurgie Physique, ERA 263, ENSCT

118, Route de Narbonne - 31077 Toulouse Cédex (France)

SYNOPSIS

This study was undertook in order to determine the influence of creep stress on the structure and morphology of oxide scales developed on preoxidized specimens. The oxidation temperature of creep specimens are between 950 and 1150° C ; the oxidation time is between 50 and 100 hours. After that preoxidation treatments, the creep tests were carried out at 800° C under constant load.

Metallographic examinations, X-ray diffraction and EPMA on unstrained specimens and unstressed parts of creep specimens revealed in both cases the oxide scales are formed by an external scale of NiO and $NiCr_2O_4$, a continuous layer of Cr_2O_3 and a subscale of SiO_2. If the oxidation temperature is below 1000° C, the silica subscales are continuous with only a slight intergranular penetration. On the contrary, above 1000° C, the silica subscales are discontinuous and the intergranular formation of silica becomes more important.

The same observations made on the stressed parts of creep specimens showed two main modifications of the morphological features of the oxide scales. At preoxidation temperatures below 1000° C, an internal oxidation of chromium is observed, it preferentially takes place on grain boundaries. At preoxidation temperatures above 1000° C, an important increase of the internal and intergranular formation of silica is observed. These observations can explain the effect of preoxidation treatments on creep rate and fracture morphology.

1. INTRODUCTION

Previous study of the oxidation kinetics and mechanisms of Ni-20 Cr-1.4 Si alloy [1] allowed to show the influence of silicon upon the structural and morphological features of oxide scales developed during the oxidation of that alloy. This alloy, used as heating element in electrical furnace, may be subjected to stresses during its service life. So, this work presents some results about the effect of stress upon the morphology of oxide scales formed during oxidation of that Ni-Cr alloy.
It is divided into three parts, respectively related to :
- The structure and morphology of oxide scales developed on unstressed specimens.
- The influence of oxide layers upon creep properties
- The morphology of oxide scales after creep test.

2. MATERIALS AND EXPERIMENTAL METHODS

The composition of studied materials is given in Table 1. Specimens cut from 1.5 mm thick sheet allowed to determine the structural and morphological features of oxide layers in unstressed conditions. The creep specimens were machined from a wrought billet with 80 x 80 mm^2 section.

All the specimens cut from these materials, were vacuum annealed for 4 hours at 1035° C ; after that annealing treatment, the two studied materiels are characterized by an equiaxed structure whose grain size is about 100 µm.

Rectangular specimens were cut from the thick sheet, they have an average area of about 4 cm^2. They were oxidized in pure oxygen at temperature between 880 and 1150° C and for oxidation time between 1 hours and 100 hours.

The oxide morphology and structure developed at various stages of oxidation was examined using optical metallography, X-ray diffraction, scanning electron microscopy and electron microprobe analysis.

Figure 1 gives the dimensions of creep specimens, the specimen axis is parallel to the major axis of wrought billet. These creep specimens were preoxidized in the same way that rectangular specimens ; the preoxidation treatments were carried out at temperature between 950 and 1150° C for time between 50 and 100 hours.

The creep tests were carried out in air at 800° C under constant load corresponding to a stress equal to 28 or 33 MPa. After creep tests, stressed and unstressed parts of specimens were examined in the same way that rectangular specimens. Creep rupture surfaces were also analysed.

3. RESULTS

3.1. Structure and morphology of oxide layers developed on unstressed specimens.

Oxide layers developed on unstressed specimens at temperature below 1100° C are well adherent, spallation of external oxide layer was only observed when the scales are formed above 1100° C.

X-ray diffraction shows that in the whole temperature range, the scales contain NiO, Cr_2O_3 and SiO_2 oxides and $NiCr_2O_4$ spinel. NiO and $NiCr_2O_4$ are the major constituants of the outer scale. Tridymite allotropic modification of SiO_2 oxide is always detected even from the early stage of oxidation. However, the structure and morphology of these oxide layers vary with time and temperature.

Figure 2 shows the typical distribution of the major constituants in the oxide layers formed during the first stage of oxidation and Figure 3 shows the concentration profile of Ni, Cr and Si through the scale. These observations shows the scale is divided into three subscales, the outer subscale contains NiO and $NiCr_2O_4$, the medium subscale is formed by a continuous layer of Cr_2O_3 and the inner subscale by a thin and quasi-continuous layer of SiO_2.

The same features are observed on oxide scales formed below 1000° C however great the oxidation time. For long time oxidation, it was only observed a slight intergranular oxidation of silicon in the vicinity of the scale.

As it can be seen from Figure 4, the SiO_2 inner layer does not remain continuous when the oxidation temperature is higher than 1000° C and the oxidation time higher than 5 hours. In this case, the SiO_2 inner layer is irregular and discontinuous. In other part, the intergranular oxidation of silicon is much more important as it is showed from Figure 5.

The same features are observed from specimens oxidized at temperature higher than 1100° C but as already noticed, spallation of outer scale leads to the observation of a Cr_2O_3 external scale. Figure 6 shows the grain boundaries of metallic substrate are underlyed by a thicker layer of Cr_2O_3 oxide.

In other part, it was checked that, in the whole temperature and time range of this study, the oxide scales developed on unstressed creep specimens, shows the same morphological and structural features that those developed on flat specimens. The only difference between these two types of specimen is the better adherence on creep specimens of oxide scale formed above 1100° C. This is probably due to the cylindrical symetry of creep specimen which leads to a different repartition of thermal stresses.

3.2. Influence of oxide layers upon creep

The creep tests were performed under a load which leads to relative large creep rate. So, the duration of these creep tests is less than 500 hours under a stress of 28 MPa and 250 hours under a stress of 33 MPa. The preoxidation treatment is characterized by the equivalent weight gain per unit $\Delta m/S$ expressed in mg.cm^{-2} and deduced from the previous kinetic studies [1].

The results are reported on Figure 7 which shows the variation of secondary creep rate $\dot{\varepsilon}_S$ versus $\Delta m/S$. In spite of their dispersion, these results show that the secondary creep rate decreases when $\Delta m/S$ remains lower than about 0.5 mg.cm^{-2} or 0.3 mg.cm^{-2} according to the stress value ; above these $\Delta m/S$ values, the dispersion increases.

For the lower values of $\Delta m/S$, the influence of preoxidation seems to be independent of temperature and time. On the contrary for the higher $\frac{\Delta m}{S}$ values, the temperature and the time of preoxidation treatment appears to be more important as it can be seen from figure 8 and 9 which show the influence of preoxidation temperature and time.

In other part, it should be noticed that the preoxidation treatment influences the time to rupture. However, it is not possible to define clearly this influence owing to the important dispersion of these results.

3.3. Morphology of oxide layers after creep test

In a first stage, it was verified from metallographic examinations that the morphology of oxide scales lying on unstressed parts of test specimens is not altered after creep test. So, the alteration observed on stressed parts may be assigned to influence of creep stress.

Depending on preoxidation temperature, two sorts of alterations are observed :
- One is characterized by the formation of internal Cr_2O_3 when the preoxidation treatment is made below 1000° C (Figure 10).
- The other, for preoxidation treatments above 1000° C, leads to an increase of internal and intergranular oxydation of silicon (Figure 11).

However, these morphological alterations are mainly observed on test specimens which lead to the lower rupture life and the higher creep rate.

Examination of rupture surface shows that the influence of preoxidation treatments upon creep properties can be related to that intergranular and internal oxidation. That is very clear from Figure 12 which shows the emergence of SiO_2 particles on rupture surface. More, cracking of oxide layers is often observed (Figure 13) ; these cracks can act as initiation for alloy cracking when the oxide scales are adherent. This figure shows also that the oxide layers developped above 1100° C on cylindrical specimens are adherent.

4. DISCUSSION AND INTERPRETATION

The observed influence of silicon upon the structure and morphology of oxide scales are in agreement with previous works [2-4] ; the discussion of the growth mechanism of these scales has been made elsewhere [1]. However, our study allowed to determine more precisely the variation with time and temperature of the morphological features of these scales.

These observations allows to explain the influence of preoxidation treatments upon creep properties :

- For preoxidation treatments characterized by a low $\Delta m/S$ value, the scale is adherent and the silica subscale is continuous. So, the decrease of creep rate may be related to an increase of the total strenght of the alloy-scale "composite" due to the high strenght of the scale.

- For preoxidation treatments characterized by an higher $\Delta m/S$ value, the situation is more complex owing to the dispersion of results. However, it may be assumed that this dispersion is related to the extent of intergranular oxidation. Indeed, when the preoxidation is performed at temperature leading to the formation of a continuous scale of SiO_2, the initial intergranular oxidation is slight and the creep rate decreases. On the contrary, when the preoxidation treatments is performed at high temperature, it leads to a more important intergranular oxidation, this intergranular oxidation reduces creep resistance and leads to an increase of creep rate. More, the internal oxidation of Cr and Si during creep test increases this effect. These observations allow to understand the increase of creep rate with temperature (Figure 8) and time (Figure 9) of preoxidation treatments.

As the creep test temperature was low in regard to the preoxidation temperature, it may be assumed that no further oxidation occurs during creep tests. So, the internal oxidation of Cr and Si which was observed after creep tests can be assigned to the influence of stresses about diffusion processes through the oxide scale. More, the effect of stresses upon diffusion processes related to oxide scale growth has been recognized [5] though there are only few experimental evidence [6-8].

However, it may be assumed that cracking of oxide scales can also lead to an increase of oxygen activity beneath the oxide scale which can favour the observed internal oxidation. So, the observations made during this study must be completed in order to obtain more quantitative results before to give a more detailed and complete interpretation of the observed effect of stresses upon the morphology of oxide scales.

Acknowledgements : The authors are indebted to Dr. M. BRUN of So-
ciété Turboméca for the use of Laboratory facilities throughout
the course of this study and Acieries d'Imphy which supplied the
studied alloys.

REFERENCE

1. P. Faure, 3nd Cycle Thesis (1979) Toulouse
2. D.L. Douglass, J.S. Armijo, Oxid. Metals, (1970) 2, 207.
3. A. Takei, K. Nii, Nippon Kinzobu Gakkaishi (1971) 40, 32.
4. A. Takei, K. Nii, Nippon Kinzoku Gakkaishi (1971) 40, 696.
5. A.T. Fromhold Jr. in "Stress effects and the oxidation of
    metals", ed. J.V. Cathcart, 3-74, 1973, New-york, The Metal-
    lurgical Society of AIME.
6. J. Stringer, Corrosion Science, (1970) 10, 513.
7. J.V. Cathcart, in "High Temperature gas-metal Reactions in
    mixed environments, eds. S.A. Jansson, Z.A. Foroulis, 63-83,
    (1973), New-York, The Metallurgical Society of AIME.
8. P. Hancock, R.C. Hurst, in Advances in Corrosion Science and
    Technology, 1-84, (1974), New-York, Plenum Press.

Table I
Composition of studied alloys

	C	Mn	S	Fe	Si	Cr	Ni	Total
1.5 mm thick sheet.	0,020	0,02	0,002	0,37	1,36	20,07	78,11	99,95
80 x 80 mm^2 wrought billet	0,033	0,07	0,004	0,22	1,37	19,95	78,30	99,95

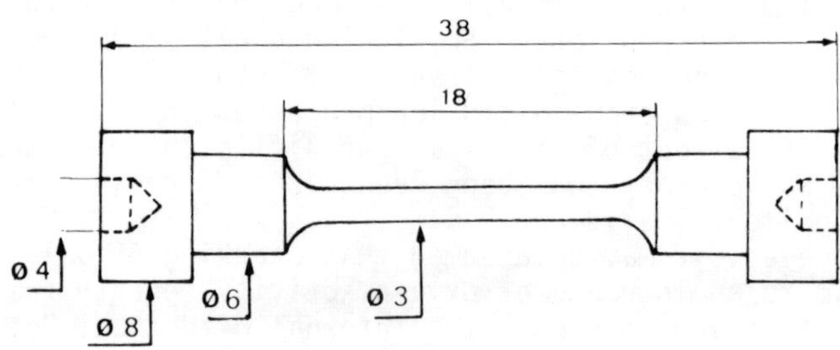

Figure 1 - Creep test specimen.

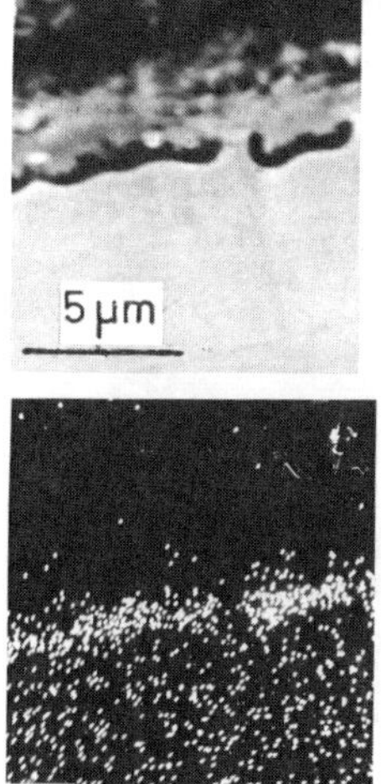
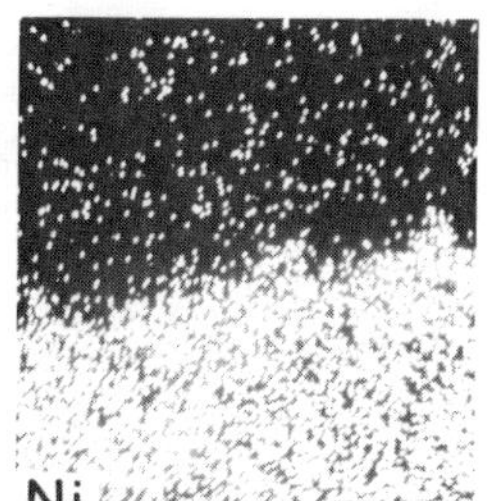

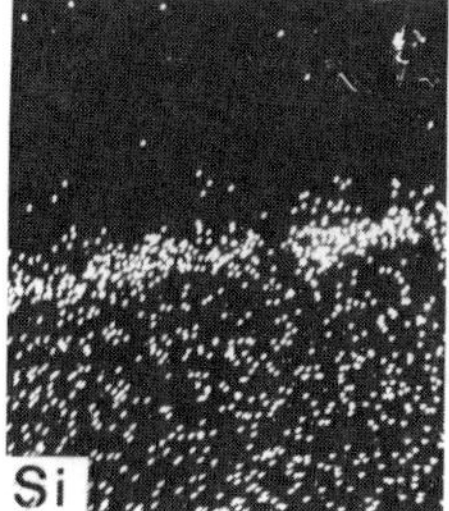

Figure 2 - Distribution of Ni, Cr, Si in oxide layers formed on specimens oxidized at 1015° C during 5 hours.

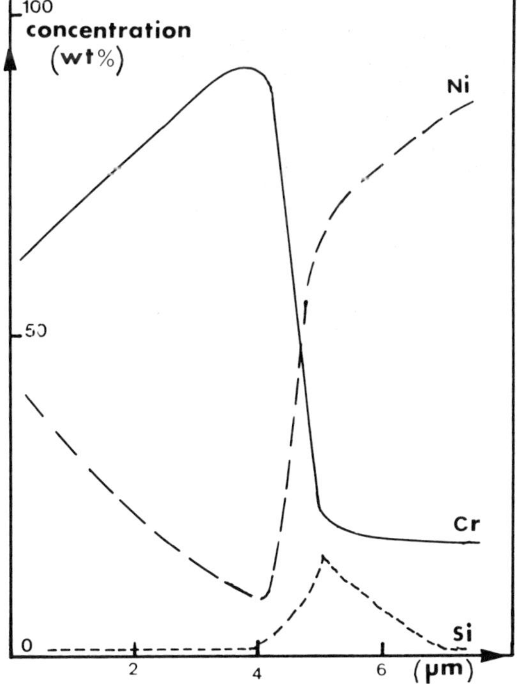

Figure 3 - Concentration profile across oxide layers formed in the above conditions.

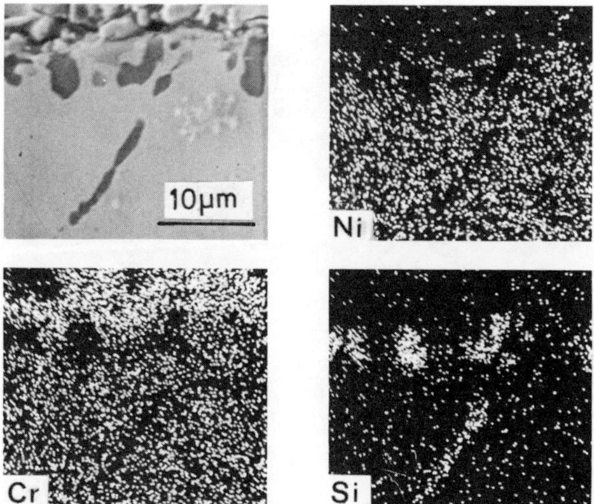

Figure 4 - Distribution of Ni, Cr, Si in oxide layers formed on specimens oxidized at 1015° C during 75 hours.

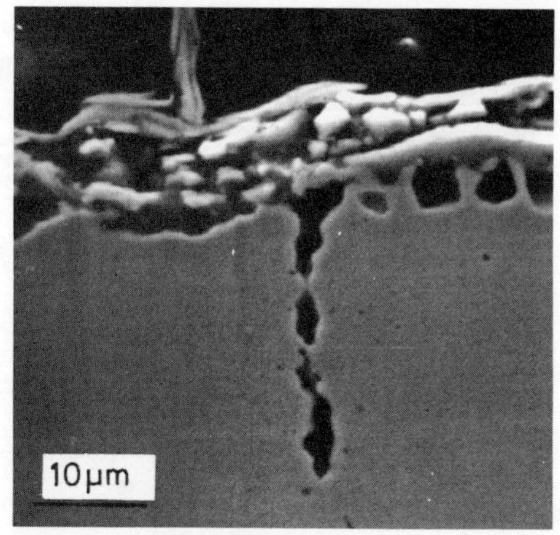

Figure 5 - Morphology of oxide layers formed in the same conditions as above.

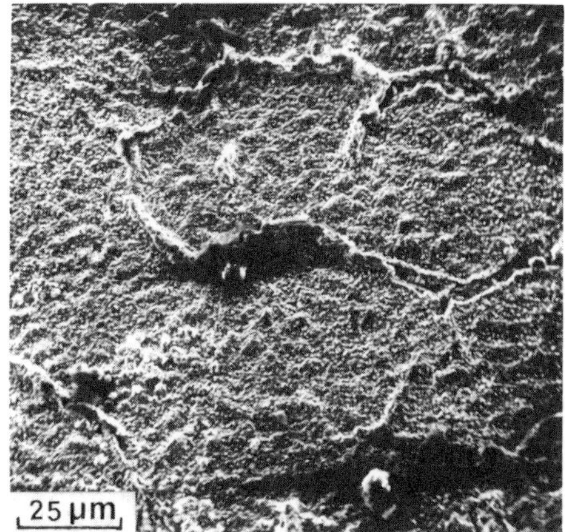

Figure 6 - Surface of a specimen oxidized at 1130° C during 5 hours.

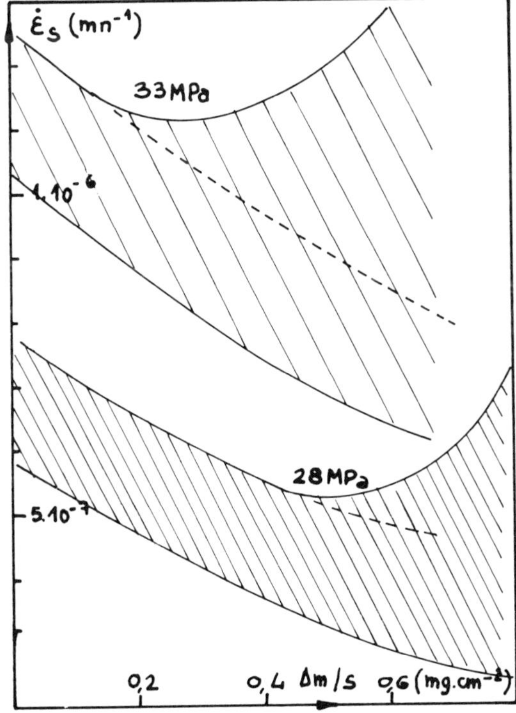

Figure 7 - Variation of secondary creep rate $\dot{\varepsilon}_S$ versus $\Delta m/S$.

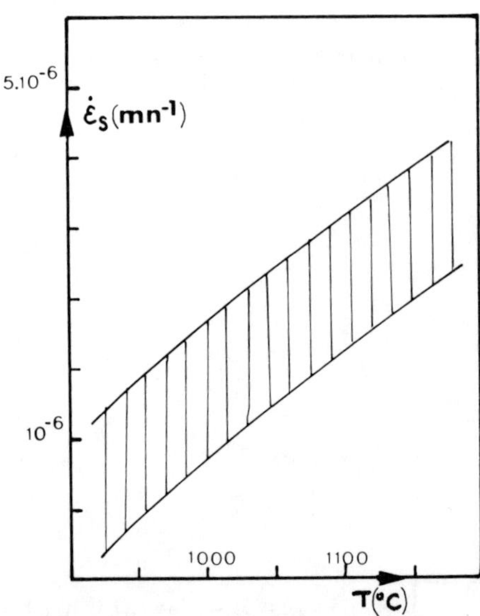

Figure 8 - Variation of $\dot{\varepsilon}_S$ versus preoxidation temperature for isochronal (t = 50 h) preoxidation treatments.

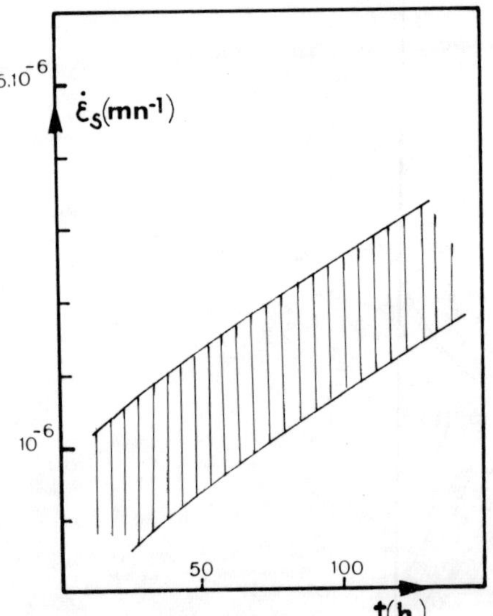

Figure 9 - Variation of $\dot{\varepsilon}_S$ versus preoxidation for isothermal (T = 975° C) preoxidation treatments.

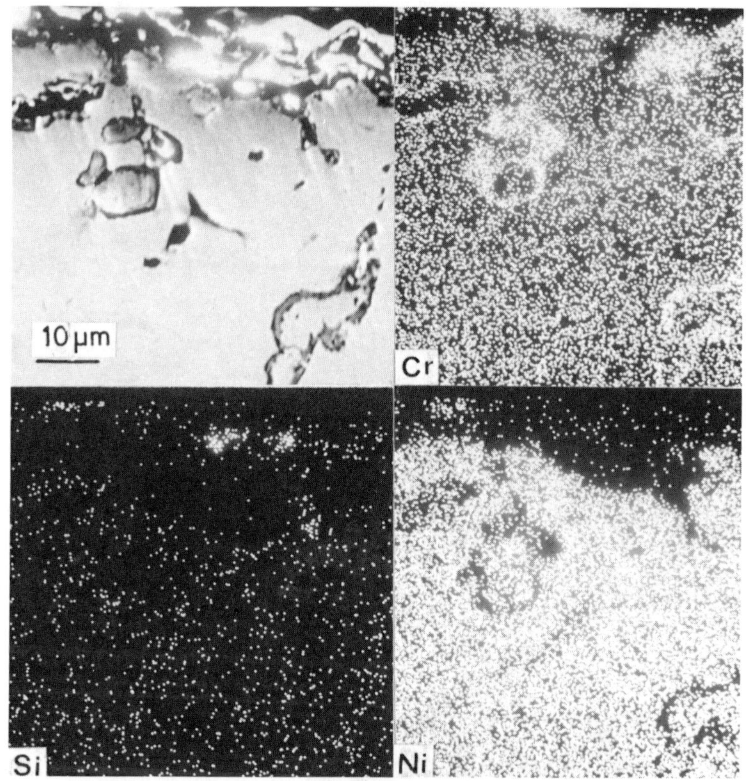

Figure 10 - Internal oxidation of Cr observed from stressed parts of a creep specimen preoxidized 95 hours at 975° C.

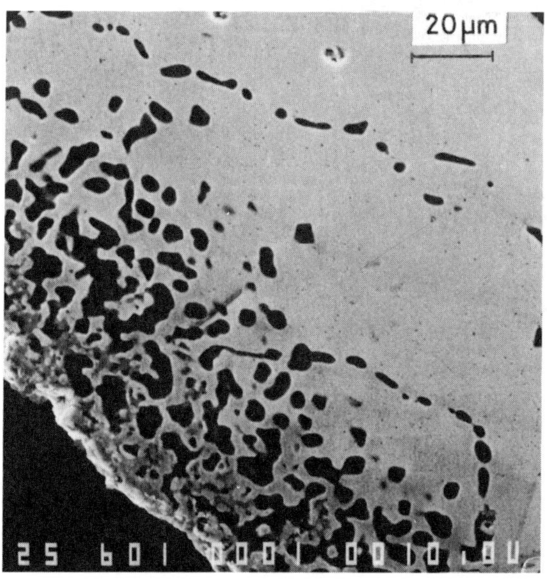

Figure 11 - Internal oxidation of Si observed from stressed parts of a creep specimen preoxidized 50 hours at 1050° C.

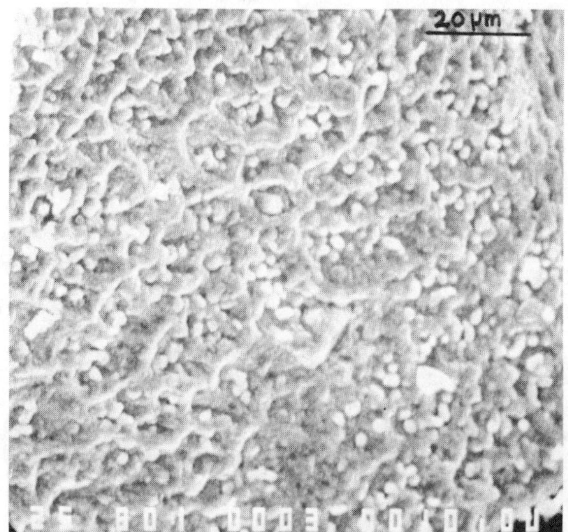

Figure 12 - Rupture surface of a creep specimen preoxidized 50 hours at 1130° C.

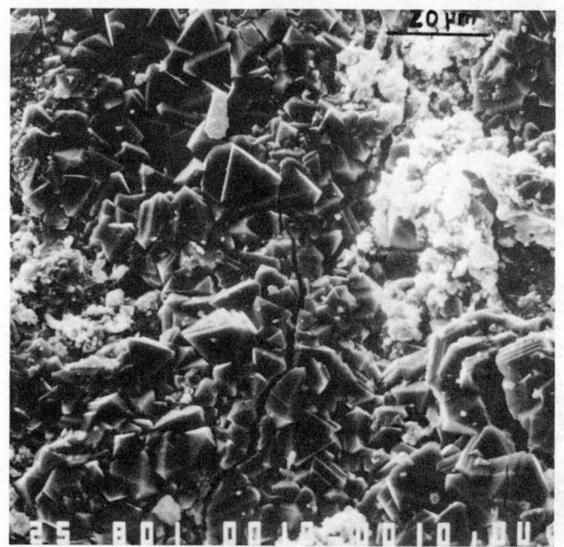

Figure 13 - Cracking of oxide layer formed on a creep specimen preoxidized 50 hours at 1130° C.

LIMITS TO THE INTEGRITY OF OXIDE SCALES GROWING ON CURVED SURFACES AND DISTENDING METAL SUBSTRATES

M.I. Manning

Central Electricity Research Laboratories,
Leatherhead, Surrey, U.K.

SYNOPSIS

Utilising an elastic strain energy criterion for oxide scale failure, the tolerance of scales to strains applied slowly during oxidation is analysed. It is found that scale integrity is particularly sensitive to the creep properties of the constituent oxide phases. Using creep data that are available for the three iron oxides, FeO, Fe_3O_4 and Fe_2O_3, the limiting deformation rates predicted for maintained scale integrity have been compared with laboratory investigations of oxidation during deformation. It is shown that oxidation on curved surfaces also imposes calculable strains on scales. If such conformational strains and those due to oxidation at edges and corners of specimens are not accommodated by creep, departures from protective oxidation kinetics may be observed. The oxidation of wires is proposed as a method of studying the creep behaviour of growing oxide scales and investigating the role of strain in promoting transitions in oxidation kinetics.

INTRODUCTION

There are so many factors that could contribute to interactions between stress and high temperature corrosion that there is a danger that the perceived complexity of the problem may hinder identification of the major underlying trends. Unless observations obtained by different workers on different systems can be compared using simple guidelines some of the experimental information is wasted. In this paper fracture considerations for oxide scale failure are used to predict the effects of mechanical strains on oxide scale integrity and consequently on oxidation kinetics.

Although additional factors not considered in this analysis undoubtedly influence strain oxidation interactions, the scale integrity criteria described do enable an appraisal of experimental
observations obtained for the iron oxide scales considered as an
example. The analysis indicates the order of magnitude of applied
strain rates that are likely to disrupt the integrity of scale
layers containing wustite, FeO, magnetite, Fe_3O_4 and haematite,
Fe_2O_3. The predicted effects depend sensitively on the scale thicknesses and on the differing creep properties of the constituent
oxide phases.

Calculable straining rates may be imposed during the oxidation of
curved surfaces, particularly at the edges and corners of laboratory
oxidation samples. In addition, the oxidation of high curvature
surfaces such as wires provides a means of investigating the role
of scale creep behaviour in affecting scale integrity under strain
and promoting transitions in oxidation kinetics.

SCALE FAILURE CRITERIA IN TERMS OF INSTANTANEOUS ELASTIC
STRAIN LEVELS

The strains that oxides can tolerate have been considered in some
detail in recent work at this laboratory[1]. To summarise the results,
the strain tolerance in compression was obtained by comparing the
strain energy stored in the scale with the energy required to generate two new surfaces by delamination, and is expected to be given
by

$$\varepsilon_c = \sqrt{\frac{B\gamma}{hE}} \qquad \qquad \dots (1)$$

where ε_c is the critical elastic strain level for spalling, γ is
the fracture surface energy of the delaminating interface, B is a
factor of about 4, E is the elastic modulus and h is the thickness
of the oxide layer. It was shown[1] that γ/E had a similar value for
many oxides of about 4×10^{-11}m. Failure of scales in tension was
found to occur in a similar way with an inverse dependence on the
square root of the scale thickness[1].

These findings were summarised in the form of an oxide failure mode
map as shown in Figure 1. In this map, oxide scale integrity is
maintained in the region where

$$\varepsilon_c < \left| \sqrt{\frac{1.6 \times 10^{-4}}{h}} \right| \qquad \qquad \dots (2)$$

where h is the scale thickness in microns. Outside this region,
scale fracture and spalling are anticipated and protective
oxidation kinetics can no longer be guaranteed.

These limits to the elastic strains that oxides can withstand (2)
are the starting point for the analyses which follow. By allowing
for creep relaxation in oxide scales being strained during oxidation, the elastic strain level can be calculated and compared
with (2). First the effects of straining during oxidation without
creep relaxation will be assessed.

LINEAR STRAINING DURING OXIDATION WITHOUT CREEP RELAXATION

If a creep resistant oxide scale is linearly strained during oxidation, the elastic strain level varies through the scale thickness depending upon the oxidation kinetics. If constant straining
at a rate $\dot{\varepsilon}_a$ is applied and parabolic oxidation kinetics are assumed,
the failure criterion of the preceding section can be used. To
generate the equivalent of (1) the stored energy in the scale is
calculated assuming it to be strain free at the instant of formation,
but to accumulate subsequent strain at the applied rate. If a
creep resistant oxide has grown parabolically on a uniform metal
surface to a thickness h in time t with rate constant Kp

$$h^2 = Kp_t \qquad \qquad \ldots (3)$$

Considering a small thickness of this layer, dh', at a distance,
h', from the oxide formed originally it will have age

$$\frac{(h^2-h'^2)}{Kp} \qquad \qquad \ldots (4)$$

The strain at this location, h', is given by

$$\varepsilon_h' = \frac{(h^2-h'^2)}{Kp} \dot{\varepsilon}_a \qquad \qquad \ldots (5)$$

With an elastic modulus for the oxide of E, the energy stored in
unit area of the scale in the thickness, dh', can be found and
integrated to give the total energy storage per unit area of the
whole scale

$$\frac{4}{15} \frac{E \dot{\varepsilon}_a^2}{Kp^2} h^5 \qquad \qquad \ldots (6)$$

If it is instead assumed that all the strain was applied at the end
of the oxidation period rather than continuously during oxidation
a stored energy of

$$\frac{1}{2} \frac{E \dot{\varepsilon}_a^2}{Kp^2} h^5 \qquad \qquad \ldots (7)$$

is obtained. Thus linear straining during parabolic oxidation

results in 8/15th of the energy storage that would occur if all
the strain were applied at the end of the period. Equating the
strain energy with the energy to produce new surfaces as in (1)
the critical scale thickness at failure is given by

$$h = \left[\frac{15}{4} \frac{Kp^2}{E} \frac{}{2} \right]^{1/5} \qquad \dots (8)$$

LINEAR STRAINING DURING OXIDATION WITH CREEP RELAXATION

A simplified and usual format for expressing creep data is used
in this paper. This needs some justification. Creep data are
normally collected on bulk oxide samples and analysed for their
stress and grain size dependence in order to elucidate details of
the diffusion mechanisms. In this paper, the order of magnitude
of the creep strain is the main concern and the creep relation is
best re-arranged in terms of the elastic strain for comparison with
the scale failure criteria discussed earlier. Consequently, a
simplified format for expressing creep data has been employed.
This ignores grain size or stoichiometry effects on creep rate but
yields the order of magnitude of the creep relaxation rate. A
linear, secondary creep relationship can thus be written in the
form

$$\dot{\varepsilon} = K \varepsilon^n \qquad \dots (9)$$

where ε is the elastic strain in the oxide, (the more usual form is
$\dot{\varepsilon} = k'\sigma^n$ where σ is the stress).

So, considering again an oxide scale constrained to extend at an
imposed strain rate, $\dot{\varepsilon}_a$, in a time interval Δt, the oxide will have
been additionally strained by $\dot{\varepsilon}_a \Delta t$. However, in this time interval,
strain relaxation due to oxide creep will have occurred tending to
reduce the elastic strain level. During the time interval Δt, the
change in elastic strain level $\Delta \varepsilon$ is thus

$$\Delta \varepsilon = (\dot{\varepsilon}_a - K\varepsilon^n) \Delta t \qquad \dots (10)$$

The elastic strain level in the oxide will be given by the solution
of this equation for the particular conditions. It will be a function of position through the oxide thickness. For constant $\dot{\varepsilon}_a$ and
for n integral, this equation is soluble for a given position in
the oxide. Even in these cases, the expression for ε as a function
of position is too complex to allow analytical evaluation of the
stored energy. Although numerical methods are best pursued where
a specific solution is sought, in this paper the limit situations
for (10) are assessed.

The first limit solution of (10) is where creep relaxation is negligible (K=0) and elastic strains build up progressively. The failure criterion is given by (8).

The second case is where, after a period where the elastic strain builds up, the applied strain rate becomes just balanced by secondary creep such that

$$\dot{\varepsilon}_a = K\varepsilon^n \qquad \qquad \ldots (11)$$

If this steady state is achieved on a timescale which is short compared to the total oxidation time, the strain is uniform through the scale and the failure criteria is readily calculated from (1) and (11).

$$\dot{\varepsilon}_a = K\left[\frac{B\gamma}{hE}\right]^{n/2} \qquad \qquad \ldots (12)$$

The range of applicability of each limit solution has been calculated[2] but they are omitted here.

CREEP DATA FOR IRON OXIDES

Wüstite

Reppich[3] investigated the compressive creep of both polycrystalline and single crystal FeO in the temperature range 650-1100°C. Fitting an equation to this data for polycrystalline FeO gives, after some readjustment of units,

$$\dot{\varepsilon} = 1.83 \times 10^{27} \varepsilon^{3.7} e^{-41920/T} h^{-1} \qquad \qquad \ldots (13)$$

where ε is the elastic strain in the oxide and T is in °K. The equation was obtained in terms of strain by substituting $E\varepsilon=\sigma$ where σ is the stress and E is the elastic modulus of the oxide. Lacking any specific value of E for FeO, a value of $4 \times 10^{10} Nm^{-2}$ was used. This value was that measured[1] for Fe_3O_4 at 600°C.

Magnetite

A creep relationship for Fe_3O_4 derived from the work of Crouch[4] is

$$\dot{\varepsilon} = 5.85 \times 10^{19} \varepsilon^3 e^{-29697/T} h^{-1} \qquad \qquad \ldots (14)$$

using the value of E of $4 \times 10^{10} Nm^{-2}$.

Haematite

The only creep data available for Fe_2O_3 do not extend below 770°C. Crouch[5] found

$$\dot{\varepsilon} = K e^{-40400/T} \sigma^2 \qquad \ldots (15)$$

and at $770^\circ C$ and $\sigma = 10^7 Nm^{-2}$ the measured creep rate was $2 \times 10^{-10} s^{-1}$. Using an elastic modulus[1] for Fe_2O_3 of $8 \times 10^{10} Nm^{-2}$ gives

$$\dot{\varepsilon} = 3.07 \times 10^{18} e^{-40400/T} \varepsilon^2 h^{-1} \qquad \ldots (16)$$

PREDICTED OXIDE THICKNESSES FOR MAINTAINED SCALE INTEGRITY
UNDER LINEAR APPLIED STRAINING

The creep relationships discussed above have been used with (12) to predict the limiting oxide thicknesses that will just withstand steady straining at various constant rates assuming the steady state creep approximation considered earlier. The results for FeO, Fe_3O_4 and Fe_2O_3 are shown together in Figure 2. To retain clarity in the figure, the predicted lines have been shortened, so that they do not overlap, although in principle they extend to cover the whole field shown. Construction of this diagram contains the assumption that the scales are not thickening appreciably during the applied straining. If they are, the elastic strain will vary across the scale thickness and the scales would fail at slightly larger total scale thicknesses than those shown in Figure 2. In addition, failures predicted from Figure 2 would only occur after the elastic strain had built up to the failure levels and this would introduce a finite incubation time before failures occurred after the start of linear applied straining.

OXIDATION ON CURVED SURFACES

Oxidation of curved surfaces imposes calculable strain rates on oxide scales that are directly related to the rate of oxidation. In this case consider that an amount of metal corresponding to a thickness, dh_{ml}, is ixidised in time, dt, to form an oxide thickness ϕdh_{ml} where ϕ is the volume ratio of oxide formed to metal consumed taking into account any porosity in the oxide. Careful consideration must be given to where the metal comes from and where the oxide forms. Suppose that a fraction V of the metal is effectively supplied from the bulk of the metal as a result of vacancy dispersion within the metal and that the rest of the metal (a fraction 1-V) is supplied from the metal surface immediately below the oxide. Let a fraction α of the oxide form at the outer surface of the scale and a fraction $(1-\alpha)$ form at the metal/oxide interface. If scale integrity and contact with the metal are to be maintained, the scale will have to be displaced outwards relative to the metal by an amount

$$(1-\alpha)\phi dh_{ml} - \tfrac{3}{4}(1-V) \, dh_{ml} \qquad \ldots (17)$$

The bulk of the already formed scale is thus moving away from a reference point within the metal substrate at a rate

$$\frac{dh_{ml}}{dt} \left[\phi(1-\alpha) - (1-V) \right] \qquad \dots (18)$$

On a flat surface, this displacement is unimpeded and ideally oxidation would proceed on a microscopic level in such a way that stresses associated with oxidation would be minimised by maintaining a planar oxidation front. However, on a cylindrical surface oxidising with smooth microplanar interfaces, the strain rate imposed on the scale parallel to the metal/oxide interface due to the requirement of translating the bulk of the previously formed scale is

$$\frac{d\varepsilon}{dt} = \frac{M}{R} \frac{dh_{ml}}{dt} \qquad \dots (19)$$

where

$$M = (1-\alpha)\phi - (1-V) \qquad \dots (20)$$

and R is the radius of curvature of the surface. Thus a hoop strain builds up as oxidation proceeds.

CREEP RESISTANT SCALES (ELASTIC APPROXIMATION)

If it is assumed that stress relaxation within the growing scale can be neglected, then there is a relationship between the elastic strain in the scale and the metal loss. The oxide which grows when the metal loss is h_{ml} will have developed a strain M/R $(H_{ml}-h_{ml})$ when the metal loss becomes H_{ml}. The increase in strain energy stored in unit area of the oxide associated with an increase in metal loss from h_{ml} to $h_{ml}+dh_{ml}$ is thus

$$dJ = \tfrac{1}{2}\phi \, dh_{ml} \, \frac{M^2}{R^2} \left[H_{ml}-h_{ml} \right]^2 E \qquad \dots (21)$$

whence integrating

$$J = \frac{M^2 E \, H_{ml}^3}{6R^2} \qquad \dots (22)$$

hence, the limiting metal loss H_{ml} before scale failure is

$$H_{ml} = \left[12 \frac{\gamma}{E} \frac{R^2}{M^2\phi} \right]^{1/3} \qquad \dots (23)$$

ALLOWANCE FOR CREEP RELAXATION (BY STEADY STATE CREEP APPROXIMATION)

Allowances can be made for stress relaxation during the oxidation
of curved surfaces using the same arguments as before. Since creep
relaxation is a time dependent process, the time dependence of the
oxidation rate must be specified before the two effects can be combined. Assuming parabolic oxidation kinetics (3), and using (19)

$$\frac{d\varepsilon}{dt} = \frac{M}{2R} \cdot \frac{Kp_{ml}}{h_{ml}} \qquad \ldots (24)$$

where h_{ml} is the metal wastage due to oxidation and Kp_{ml} is the
parabolic rate constant for metal loss. Making allowance for
creep relaxation (9), (24) becomes

$$\frac{d\varepsilon}{dt} = \frac{M}{2R} \frac{Kp_{ml}}{h_{ml}} - K\varepsilon^n \qquad \ldots (25)$$

This equation has no general solution and it is necessary to specify
the creep rate exponent to obtain solutions. Even so, the stress
and elastic strain vary across the scale and (25) is best solved
numerically for particular cases. However, utilising the steady
state creep approximation,

$$\frac{M}{2R} \frac{Kp_{ml}}{h_{ml}} = K\varepsilon^n \qquad \ldots (26)$$

a critical radius of curvature factor (2R/M) can be deduced for
oxide scales of particular thicknesses to remain adherent.

Re-arranging and using (1) for the critical strain the adherence
criteria becomes

$$\left[\frac{2R}{M}\right]_{crit} = \frac{Kp_{ml}}{K} h_{ml}^{(n/2-1)} \left[B\gamma/E\right]^{-n/2} \qquad \ldots (27)$$

This equation, which is independent of h_{ml} if the creep stress exponent is 2 has been solved for the cases of iron oxide formation.
For the case of duplex scale formation on alloy steels M=0 so no
effects of curvature on oxidation are anticipated. For the case
of single layer growth on pure iron M=2 and therefore curvature
effects are anticipated. A value of Kp_{ml} for iron oxidation was
obtained by fitting a relation to data for mild steel given in
Armitt et al[1].

$$Kp_{ml} = 2.44 \times 10^{16} e^{-30000/T} \; \mu m^2 \, h^{-1} \qquad \ldots (28)$$

Thus

$$\left[\frac{2R}{M}\right]_{crit} = \frac{2.44 \times 10^{16}}{K} \, e^{-30000/T} \, h_{ml}^{(n/2-1)}$$

$$(1.6 \times 10^{-4})^{-n/2} \qquad \ldots \ (29)$$

Utilising the creep relations already given for FeO, Fe_3O_4 and Fe_2O_3, in Fig. 3 are plotted the predicted minimum radii of curvature factors (2R/M microns) that single layer iron oxide scales would be expected to withstand at various metal losses. Since M=2 for the case considered, the radius of curvature, R, can be read off directly. Also marked on the figure is a line labelled sample edges and corners. This was constructed on the basis that the curvature present at initially sharp edges and corners of oxidised iron or mild steel specimens is approximately given by the metal loss (see Plate 1).

DISCUSSION AND COMPARISON WITH AVAILABLE DATA

In experimental studies reported at this conference[7], Rowlands, Manning and Soo strained the scales forming on 9Cr steels at 600°C during oxidation. The highest strain rate employed was $1.6 \times 10^{-5} h^{-1}$ and the wholly magnetite outer layer scales accommodated this deformation rate readily, attaining final strains of ∿7% without fracture. During the tests the scales grew to thicknesses up to 160μm (80μm for the Fe_3O_4 layer). These observations accord with the predictions of Fig. 1 as the strain rates employed are well below the critical strain rates for fracture of magnetite at 600°C. Berchtold, Sockel and Ilschner[8] strained mild steel during air oxidation at 600°C at strain rates between 10^{-5} and $5 \times 10^{-3} h^{-1}$. Under the conditions employed, FeO was the major scale constituent, although Fe_3O_4 and Fe_2O_3 would also be minor constituents. In these tests gross scale cracking occurred at the higher strains resulting in the formation of thick scales, some of which approached 400μm in thickness[8]. A value for the critical straining rate for maintained scale integrity of $3.6 \times 10^{-5} h^{-1}$ was reported at scale thicknesses of about 100μm. This value corresponds closely with the predicted critical strain rate for FeO at 600°C shown in Fig. 2. No data are available to test the predicted values for Fe_2O_3 which is considerably more creep resistant. For the inner layer mixed spinel scales formed on 9Cr steels, Rowlands et al[7] found they were not able to stand the strain rates applied (10^{-6}-$1.6 \times 10^{-5} h^{-1}$) and fractures had occurred at the higher strains. However, cracks so formed had filled with Fe_3O_4 and these were not externally evident below the uncracked outer Fe_3O_4 layers, but were found only by detailed metallography. Thus, although no creep data are available for the chromium containing inner layer oxide scales formed on alloy steels, they are clearly considerably more creep

resistant than the magnetite of the outer layers. Whilst M, defined by (20), is zero for the normal two-layered growth of oxide
scales on iron based alloys, and no effects of curvature on oxidation are anticipated for alloy steels, Fig. 3 is applicable, however, to the case of the formation of single layered scales growing on iron for which M=2, with all the scale growth assumed to be
at the gas/oxide interface. The figure shows that any iron oxide
phase would be unable to form a single layer without fracture at
the high curvatures associated with edges and corners for which the
curvature has been assumed to be equal to the metal loss due to
oxidative 'blunting' of the corner as shown in Plate 1. This shows
how, in accord with Fig. 3, the single layer scale formed on the
flat surfaces of a sample of a rimming steel during oxidation in
1%CO/CO_2 at 500°C has not been maintained at the specimen edge
where two-layer growth has initiated. This is shown in Plate 1
where the inner layer scale can be seen at the specimen edge.
Because two-layer growth has an M value of O, a change in oxidation
mode has obviated the requirement for excessive strain accommodation
at the edge. This illustrates that a transition in oxidation mode
is a possible alternative to scale failure in high curvature regions.
Fig. 3 also includes data for FeO and Fe_2O_3, although these phases
do not normally form single layer scales on iron. However, it
would be predicted that for creep resistant oxides like Fe_2O_3, radius
of curvature effects on oxidation would be apparent, even at radii
of curvature large compared with the oxide thickness. Consequently
such effects could be studied by the oxidation of wires of varying
radii. In fact, due to the sensitivity of such predictions to the
creep properties assumed for the scale in Fig. 3, it is apparent
that inferences regarding creep properties during oxidation might
be obtained from measurements of curvature effects on scale integrity
on other oxidation systems.

In summary, general analyses have been presented in this paper for
the effects of linearly applied strain and the effects of surface
curvature upon oxide scale integrity. These analyses have been used
with available creep data for iron oxides to predict these effects
for the oxidation of steels. For the example of iron oxide formation,
where experimental data do exist they are in line with theoretical
expectation. There is no experimental data on the creep properties
of the mixed spinel phases forming the inner scale layer on oxidised
alloy steels, although they are apparently creep resistant.

Where creep data for other systems are available and can be expressed
as in (9), figures equivalent to Fig. 2 and Fig. 3 could be constructed for other oxidation systems using the analysis presented.

CONCLUSIONS

1. An elastic strain energy criterion for oxide scale failure can
    be used to assess the limiting strain rates that oxide layers
    should withstand.

2. Using available creep data for iron oxides, the effects of
    linear straining during oxidation and of surface curvature on
    the mechanical response of the scales have been predicted.

3. Where experimental data are available for iron oxides, they are
    in agreement with prediction.

4. The analyses presented here can be applied to any oxidation
    system for which oxide creep data are available.

5. The creep properties of oxide scales during oxidation could be
    deduced from adherence observations on oxidised wires.

ACKNOWLEDGEMENT

This paper is published by permission of the Central Electricity
Generating Board.

REFERENCES

1. J. Armitt, D.R. Holmes, M.I. Manning, D.B. Meadowcroft and
    E. Metcalfe, 'The spalling of steam grown oxide from super-
    heater and reheater tube steels', 1978, EPRI Report FP 686.

2. M.I. Manning and D.B. Meadowcroft, 'Effects of tube creep
    strains on laminated scale formation in ferritic pressure
    tubing', 1979, CEGB Report RD/L/R2012.

3. B. Reppich, 'Plastische Verformung von Eisen-II-Oxid' Phys.
    Stat. Sol., 1967, $\underline{20}$, pp 69-82.

4. A.G. Crouch, 1979, Private Communication.

5. A.G. Crouch, 'High temperature deformation of polycrystalline
    Fe_2O_3', J. American Ceramic Society, 1972, $\underline{55}$ (11) pp 558-563.

6. O. Kubascheswi and B.E. Hopkins, 'Oxidation of metals and alloys',
    1962, London, Butterworths.

7. P.C. Rowlands, M.I. Manning and J. Soo, 'The effect of substrate
    creep on the structure of oxide scales formed on Fe-9%Cr steels',
    this publication.

8. L. Berchtold, H.G. Sockel and B. Ilschner, 'The influence of
    deformation on the oxidation of mild steel', International
    Conference on the Behaviour of High Temperature Alloys in
    Aggressive Environments, 15-18 October 1979, Petten (NH),
    The Netherlands.

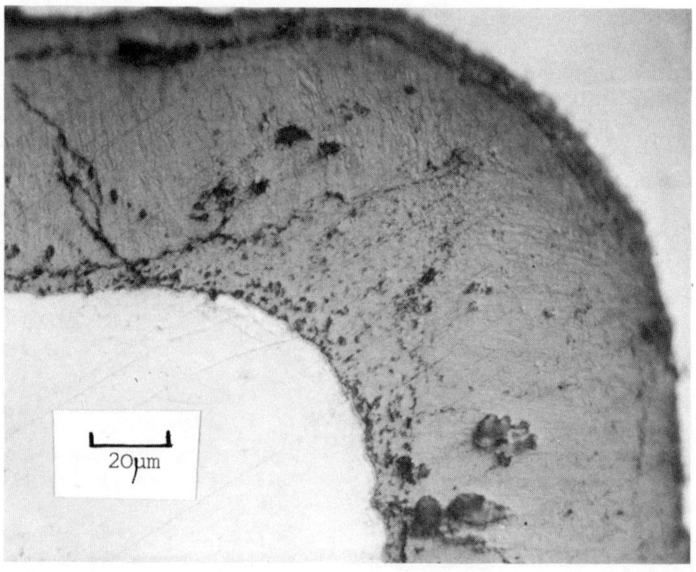

Plate 1.   Metallographic section through the edge of a mild
           steel sample oxidised for 1500h at 500°C in
           1%CO/CO$_2$ at 40 bar with 1000 vpm H$_2$O. The scale
           has been etched in 50 v/o HCl/H$_2$O to show up the
           two layer structure present only at the corner.

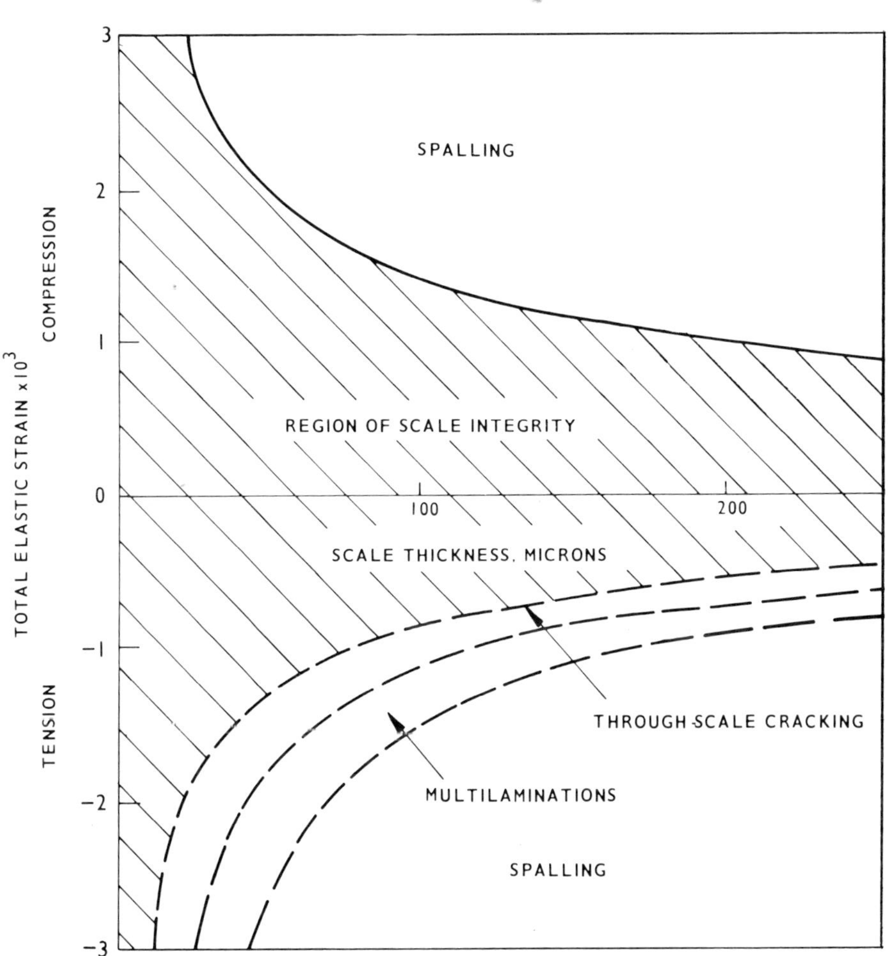

Fig 1. Oxide failure mode map. Oxide scale integrity is maintained within the shaded area, but oxide failures occur at greater strains[1].

                    M.I. MANNING

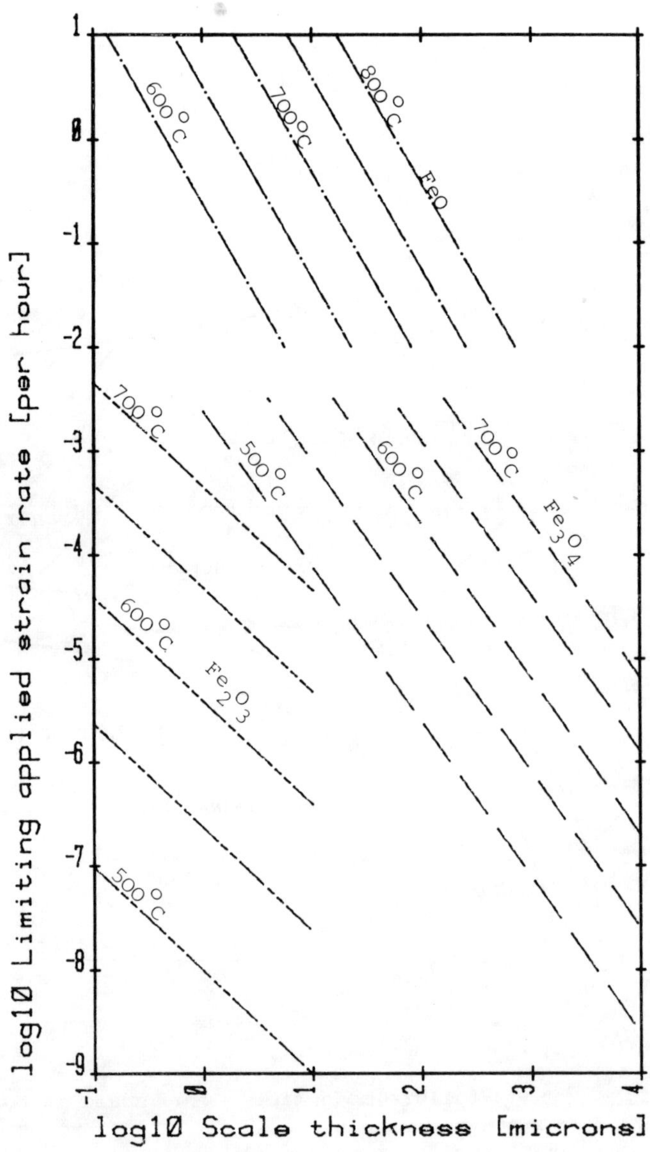

Fig 2. Limiting strain rates that iron oxide
layers should withstand without failure

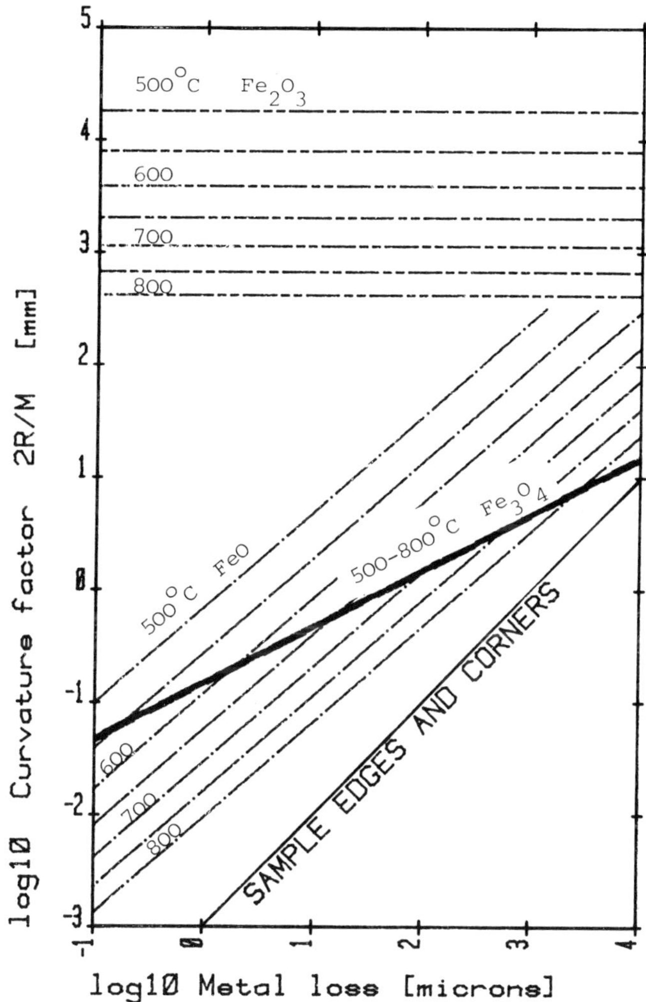

Fig 3. Integrity limits for single layer oxides growing on curved iron surfaces

DISCUSSION

T. Ericsson: It is well known that high stresses can occur in growing oxide scales. Very often the stresses are compressive in the scale. How is this taken account for in the analysis of Manning and Bell, or can the stresses be neglected in their materials.

M.I. Manning: Growth stresses are not generally observed for duplex scales growing on steels, even when the scales are relatively thick. I have proposed that the reason for this is that the scale displacement vector, M. (defined in the paper) is zero for two layer growth. Growth stresses due to surface curvature are expected, however, during the laminated scale growth discussed by Forrest and Bell as M=1 in this case.

P. Kofstad: From available data on creep of oxides it is well known that the creep rate is dependent on the partial pressure of oxygen in equilibrium with the oxide, i.e., on the non-stochiometry (defect concentrations) under various conditions. Over a growing scale, there is a large oxygen potential gradient. How do these aspects affect the overall model presented in the paper?

M.I. Manning: In the interests of a simplified treatment, the analyses in this paper have excluded consideration of the effects of oxygen potential and grain size on the creep behaviour of oxide scales. However, the experimental creep data available for iron oxides generally include measurements of these effects and, consequently, the analyses could be extended to take account of the variation of oxygen potential and grain size across the thickness of an oxide scale.

THE RELATIONSHIP BETWEEN THE DEFORMATION OF SUBSTRATE
AND SCALE IN THE MECHANISM OF MULTILAYER OXIDE FORMATION
ON LOW ALLOY FERRITIC STEELS

J.E. Forrest and P.S. Bell

Central Electricity Research Laboratories

Leatherhead, Surrey, U.K.

<u>SYNOPSIS</u>

Low alloy ferritic steels normally form a duplex (inner layer
spinel/outer layer magnetite) scale when exposed to low $p(O)_2$
oxidizing environments at high temperatures, e.g. steam or carbon
dioxide. Under certain conditions this protective scale can break
down to give a scale consisting of many pairs of spinel and
magnetite layers. These multilayer scales can have several
deleterious effects on the components used in high temperature
generating plant, e.g. mechanical jacking of crevices, enhanced
oxidation rates or release of oxide debris. Growth of each
successive pair of layers occurs when the existing surface scale
breaks down allowing oxidant access to the metal/scale interface.
Two ways in which this occurs are:

(1) Chemical agents in the environment interfere with the growth
     of the scale e.g. breakaway corrosion in CO_2. Scale growth
     in crevices will continue leading to the generation of forces
     sufficient to deform the metal substrate (jacking).

(2) Mechanical strain can lead to disruption of the oxide scale.
     When strain causes the existing scale to become detached or
     where the scale remains adherent but is cracked, oxidant
     ingress can occur initiating the growth of a new layer of
     scale. Substrate deformation can arise from external stresses,
     the operation of components in the creep range, cooling strains,
     and heat flux strains. The combined response of substrate and
     scale will then determine the type of multilayer scale growth.

INTRODUCTION

Low chromium ferritic steels are used extensively in modern
generating plant for pressurized components operating with metal
temperatures up to nearly 600°C. The successful use of these
steels for the life of the plant (in excess of 100kh) depends on
their forming a protective oxide scale. However, the strains to
which these components are subject during various phases of plant
operation may result in the loss of protective oxidation kinetics
and the growth of a multilayer scale. Apart from external strains,
those arising from changes in temperature (thermal mismatch
strain) and internal pressurization (creep strain) are important.
Usually of lesser importance are the strains that result from
rapid changes in temperature (thermal shock strain) and the
application or removal of a heat flux. The problems that can arise
from the growth of thicker less protective multilayer scales
include: increased metal loss rates giving shorter component
lifetime, reduced efficiency of heat transfer surfaces (often with
a consequent increase in metal temperature), turbine erosion and
tube blocking if the scale spalls, and mechanical deformation of
surrounding components when scales continue to grow in confined
spaces. This paper examines some of the mechanisms which can lead
to the growth of multilayer scales with examples from plant
components and recent laboratory investigations.

DUPLEX AND MULTILAYER GROWTH

Scales grown on low alloy ferritic steels over a wide range of
temperatures up to 700°C consist of an inner layer iron-chrome
spinel and an outer layer of pure iron oxide. In air approximately
half the outer layer is haematite (Fe_2O_3), while in lower $p(O_2)$
oxidizing environments the outer layer is wholly magnetite (Fe_3O_4).
The growth of this type of scale is shown schematically in Figure
1(a). The oxidation reaction proceeds via the outward transport
of iron to the scale/gas interface to form the outer layer, and
the inward transport of oxidant to the metal/scale interface to
form the inner layer. The growth of a multilayer scale is shown
in Figure 1(b). The scale comprises several pairs of layers of
magnetite and iron-chrome spinel. The growth mechanism of each
new pair or layers is similar to that of a duplex scale except that
it occurs under the previously formed scale. A prerequisite for
the formation of a new pair of layers at the metal/scale interface
is that the partial pressure of oxidant at this site must rise to
a level at which magnetite is thermodynamically stable.

The kinetics of duplex scale growth are often approximately parabolic, with the rate determined by the thickness of inner layer
spinel. The overall kinetics of multilayer growth may be linear
if the growth of each new pair of layers is determined by the
thickness of inner layer adjacent to the metal/scale interface.

If some protection is afforded by the overlying layers the kinetics
will fall below linear, although the rate of scale growth should
always be greater than that of a simple duplex scale.

MECHANISMS OF MULTILAYER SCALE FORMATION

Figure 2 shows schematic representations of several mechanisms of
multilayer scale growth, each of which is discussed briefly below.

Loss of Inner Layer Protectivity (Figure 2a)
The formation of this type of multilayer scale is associated with
the presence of chemical agents other than oxygen. The build up
of chemical agent in the scale, e.g. carbon in CO_2 oxidation or
acidity in a hydrolysable chloride solution, reduces the
protective nature of the inner layer thus allowing an increase in
the oxygen partial pressure at the metal/scale interface. If this
pressure reaches the dissociation pressure of magnetite then new
duplex layers will start to grow at the metal interface underneath
the initially formed scale. Mechanical strain may not play a
significant role in causing the growth of new layers since they
form isothermally on specimens which are not subject to externally
applied strain.

Scale Detachment with Varying Degrees of Restraint (Figure 2b)
In this case the growth of a new pair of layers will take place
when the existing scale separates at the metal/scale interface to
expose the bare metal surface. The critical level of strain in the
scale required to cause separation may be produced by several
means. Two important examples are externally applied mechanical
deformation and cooling strains that result from mismatch in
thermal expansion of the scale and metal substrate. One factor
that controls the extent of growth of a new layer of scale is the
presence of any external restraint which acts on the scale and may
assist in bonding the magnetite layer of the newly formed duplex
scale to the inner layer of the scale which had detached, thus
cutting off the path for the supply of gaseous oxidant.

Oxidant Ingress via Through-scale Cracks (Figure 2c)
Under certain conditions of strain the scale will undergo tensile
cracking without detachment. A new duplex layer can then grow at
the exposed metal surface. When there is a sufficient density of
through-scale cracks a continuous layer will be formed under the
existing scale.

Oxidation in Volume Created by Vacancy Condensation (Figure 2d)
Under certain conditions the rate of inward diffusion of oxygen to
form the inner layer of a duplex scale is insufficient to balance
the outward diffusion of iron with the result that vacancies may
be left in the metal. In this case the vacancies remain in the
metal until a supersaturation is reached when they precipitate

rapidly at the metal/scale interface resulting in void formation
and the separation of regions of scale. The increased oxygen
partial pressure at the surface under the separated scale can lead
to the growth of a new duplex layer.

EXAMPLES OF MULTILAYER SCALE GROWTH

Scale Detachment Caused by Thermal Cycling

Figure 3 shows two specimens each consisting of several strips of
Fe-2¼Cr-1Mo steel bolted together at one end. Both specimens were
thermally cycled several times between 600°C and room temperature
(one week at temperature followed by furnace cooling). The left
hand specimen in Figure 3, exposed in a low oxygen partial
pressure $N_2/H_2/H_2O$ gas mixture, showed no significant change in
dimensions and was covered with an adherent duplex scale
consisting of an inner layer iron-chrome spinel and an outer layer of
magnetite with no haematite. The scale on the right hand
specimen, which had been exposed in air, consisted of several
detached layers of duplex scale equal in number to the number of
thermal cycles. The width of the stack at a given position away
from the clamping nut and bolt increased by the same amount with
each thermal cycle. The total change in width was negligible at
the clamped end, and increased towards the free end as shown in
Figure 4. Approximately half the thickness of each outer layer
was haematite, and the individual flakes of scale were curved with
the haematite layer forming the convex surface. The behaviour of
a typical scale is shown schematically in Figure 5 and the values
of stress and elastic strain energy in the component layers of the
scale in Table 1. Strain energy is induced in the scale on
cooling from the growth temperature due to the mismatch in
expansion coefficients. It has been shown[1] that only a very small
fraction of this strain energy (<1%) is released when the scale
initially detaches since the compressive force in the haematite
layer nearly balances the tensile forces in the magnetite and iron-
chrome spinel layers. Only when the whole scale bends can there be
an appreciable ($\sim$ 50%) release in the stored strain energy. There
are two consequences of this requirement for the scale to bend:
when there is a strong restraint acting perpendicular to the scale,
the energy that would be released by scale bending may not be
sufficient to overcome the work required to displace the restraint
and the scale will be prevented from detaching. On the other hand,
if the restraint is weak the energy released when the scale bends
can do work in displacing the restraint. Calculation shows that
the force exerted by the bending of a typical 50μm thick scale is
insufficient to stretch the clamping bolt, but is capable of
opening the stack of strips in regions of weaker restraint.

Scale Detachment caused by Externally Applied Mechanical Deformation

The arrangement shown in Figure 6 consists of a cylindrical
section Fe-2¼Cr-1Mo fatigue test specimen with a tightly fitting

Type 316 stainless steel collar. The strain amplitude was ± 0.5%
plastic deformation and the test was made in air at 600°C.
Application of this strain cycle caused detachment of the scale
and during the subsequent 60m hold time in tension a new layer of
scale grew in the small gap between the collar and test piece
(formed by the Poisson contraction of the latter). Reapplying the
compressive strain caused a small but permanent outward displacement of the collar, since the volume of the scale grown during
the hold time was greater than the volume of metal from which it
was formed (Pilling-Bedworth ratio ∿2). Examination of the
sectioned specimen shown in Figure 6 indicated that the number of
layers of scale corresponded to the number of fatigue cycles. We
call this type of process where the repeated application of a
small reversible plastic deformation in one component can accumulate a larger net strain in another component 'oxide ratchetting'.

Multilayer Growth Associated with Through Scale Cracking
A typical example of a multilayer scale which had grown on the
steamside (inner) surface of a reheater tube is shown in Figure 7.
Initially a duplex scale approximately 200μm thick was formed
before the transition to multilayer (laminated) growth. Usually
the inner spinel and outer magnetite layers of a duplex scale on
this steel are of approximately equal thickness; in this case the
extra thickness of the outer magnetite layer over the spinel layer
is caused by the continued transport of iron from the metal to the
steam/scale interface. Through-scale cracks are a common feature
of these scales. In the example shown in Figure 8 cracks have
passed through the same region of scale a number of times. On each
occasion a new crack extends down to the metal/scale interface
before filling with magnetite. This is shown more clearly in
Figure 9 where a new layer is growing at the metal/scale interface
and the crack has just started to fill with magnetite. Detachment
of the existing scale is not a requisite for this mechanism,
rather growth of a new duplex scale will be initiated at the
exposed metal surface at the bottom of the crack, and extend along
the interface as shown in Figure 9. The steps in the operation of
this mechanism are shown schematically in Figure 10. Tensile
strain arising from differential contraction of the tube and scale
during cooling, and creep strain resulting from the slow dilation
of the pressurized tube while at temperature, are important in the
operation of this mechanism. Similarities in the pattern of the
layers of scale on tubes from neighbouring regions of a boiler
indicate that specific events in operation such as shutdowns can
initiate the growth of a new layer. Creep dilation of the tube
causes the cracks to remain open on returning to operating
temperatures when they fill with magnetite.

Multilayer Growth Through Vacancy Condensation
The example of this type of multilayer growth, shown in Figure 11,
is taken from an Fe-2¼Cr-1Mo superheater tube. Although the

operation of this mechanism may not involve the interaction of
strain with the substrate or scale it provides an interesting
contrast with the previous through-scale cracking mechanism. In
this case there were no through-scale cracks, and the number of
layers formed by the operation of this mechanism was much smaller
than that formed by through-scale cracking, i.e. less than five
compared with frequently over one hundred. The upper micrograph
in Figure 11 shows voids at the metal/scale interface and also a
row of voids which have been left behind in the scale as a result
of the inward growth of the inner layer. Although the row of
voids would have covered approximately 80% of the interface, this
was insufficient to give the degree of scale separation required
to initiate the growth of a new layer. However, the gap indicated
in the lower micrograph of Figure 11 shows that complete
separation of the scale could occur (in the through-scale cracking
mechanism the layers were usually adherent).

<u>SUMMARY</u>

Several distinct types of multilayer scale on low alloy ferritic
steels can result from the interaction of mechanical strain on the
metal substrate and oxide scale. However, in high pressure CO_2 and
acid chloride solutions multilayer growth occurs without any
apparent substrate/ or scale/strain interaction. If such an interaction is a requirement for the growth of multilayer scales in these
two environments then it is most likely to be self generated as a
result of the scale growth process. Thus although the growth of
these scales in crevices may give rise to high stresses capable of
causing severe mechanical deformation of neighbouring components,
e.g. bolt stretching in the Magnox reactors and water tube
'denting' in PWR steam generators, such stresses may not be present
when the same multilayer scale grows on an exposed surface. Multilayer scales which grow as the result of fracture or detachment of
the scale can only give rise to similar high levels of mechanical
deformation in the case of oxide ratchetting, where the mechanical
work must come from the externally applied load. In the absence of
a ratchetting mechanism multilayer growth which results from scale
cracking or detachment can deform only a weakly restrained crevice,
and more often the degree of crevice restraint will determine
whether initial scale detachment will be energetically feasible.
Even where detachment or cracking does occur the growth of a new
layer of scale may not result in crevice displacement since new
scale will only grow to fill the free space in the crevice; once
this has happened the supply of oxidant into the crevice will be
cut off and scale growth will cease for all practical purposes.

	Haematite		Magnetite		Spinel		Total Energy
	σ	Q	σ	Q	σ	Q	
At growth temperature (600°C)	zero	zero	zero	zero	zero	zero	zero
After cooling to 0°C	+212.2	1.39	-98.3	0.55	-40.9	0.5	2.44
After detachment and uniform expansion	+195.8	1.19	-105.1	0.63	-45.1	0.61	2.43
At equilibrium after uniform bending	non-uniform	0.79	non-uniform	0.52	non-uniform	0.11	1.42

TABLE 1: Stress, σ(MN/m^2) and Stored Elastic Strain Energy, Q(J/m^2) for a Three-layer Scale, 5μm Haematite, 5μm Magnetite and 10μm Inner Layer Spinel

ACKNOWLEDGEMENTS

This work was carried out at the Central Electricity Research Laboratories and is published by permission of the Central Electricity Generating Board.

REFERENCE

1. P.S. Bell and J.E. Forrest, CEGB Report No. RD/L/R2009, 1980.

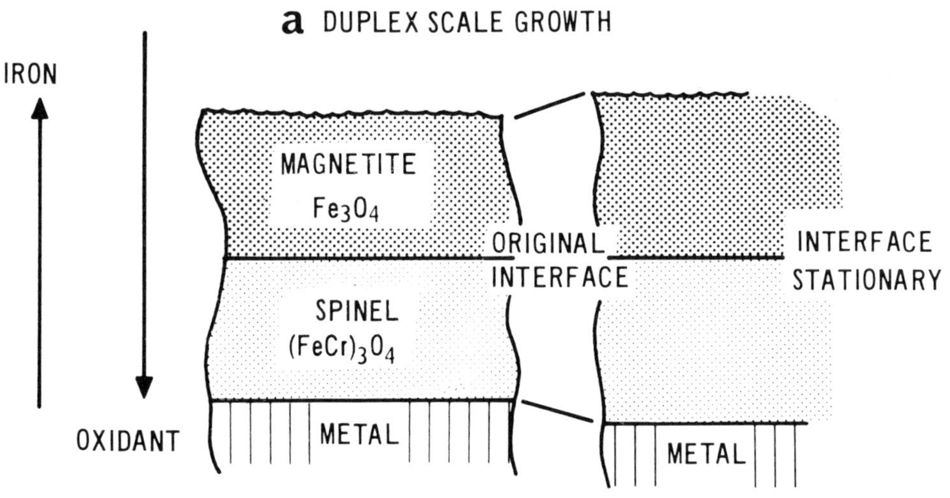

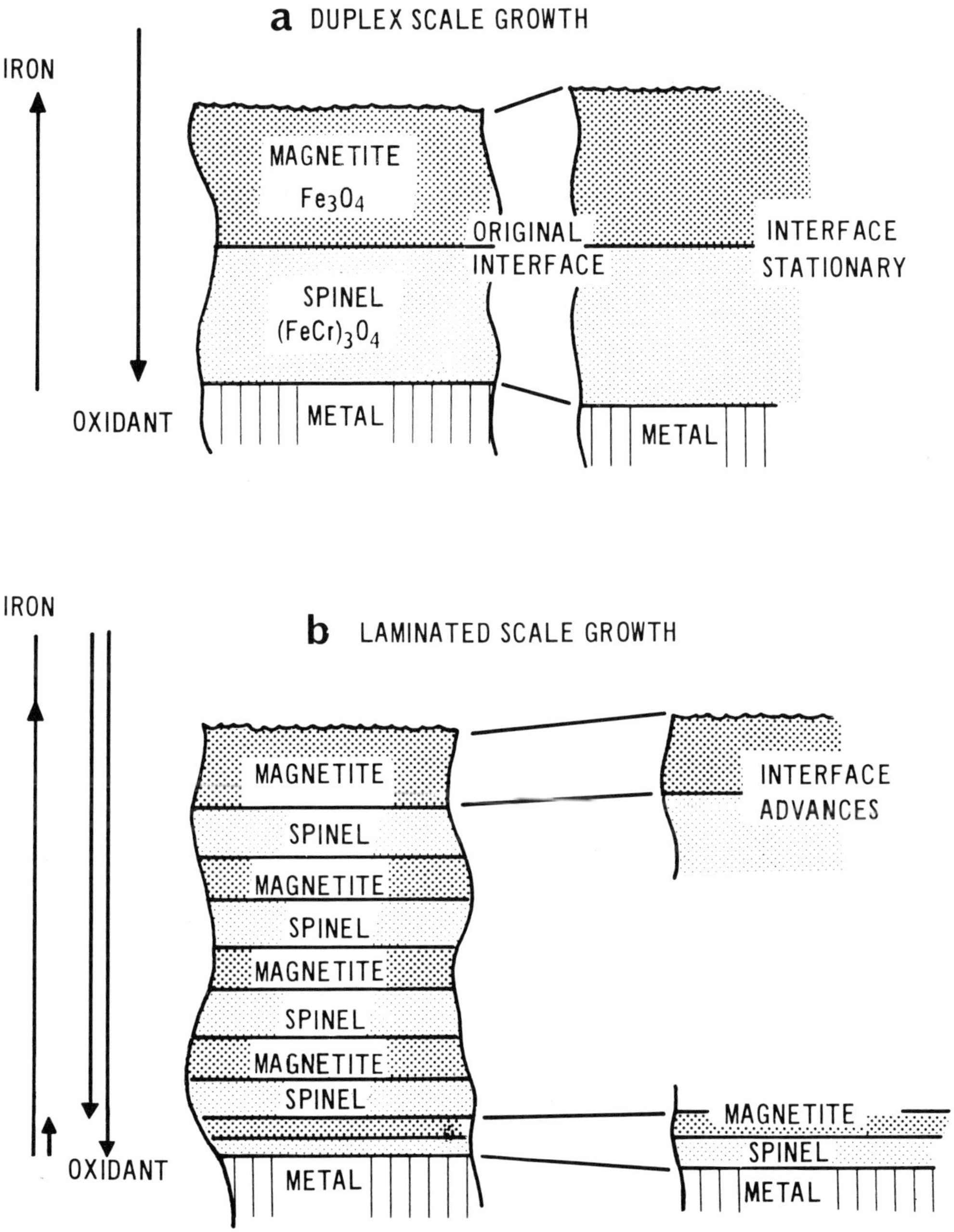

FIG.1:   Growth of Duplex and Laminated Scales on Ferritic Steels

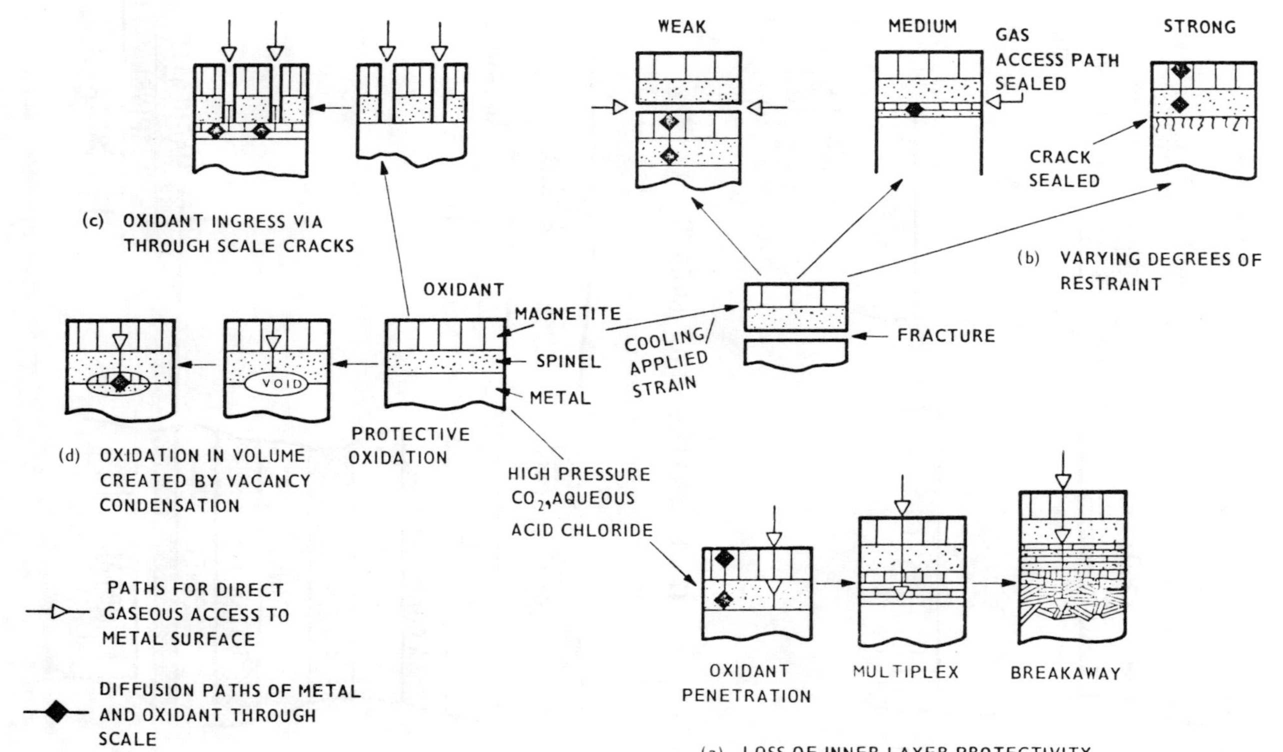

Fig. 2: Schematic Representation of the Mechanisms of Multilayer Scale Growth on Ferritic Steels

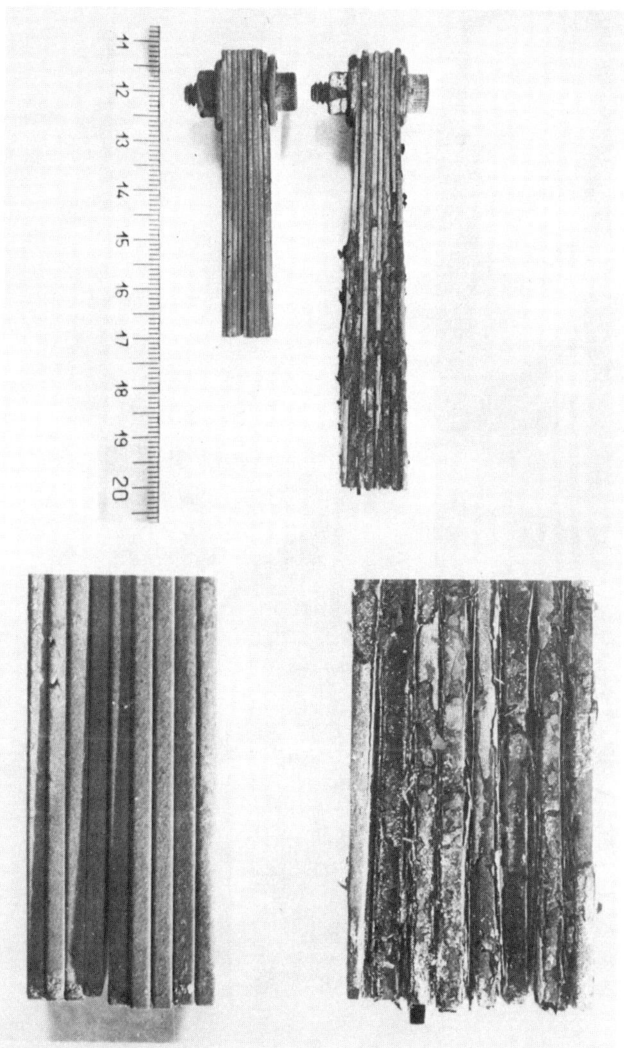

FIG.3: Stacks of Fe-2½Cr-1Mo after Thermally Cycling between 600°C and Room Temperature. Left Hand Specimen Oxidized in $N_2/H_2/H_2O$ Gas Mixture, Right Hand Specimen in Air

     J.E. FORREST and P.S. BELL

FIG.4: Section of Stack Specimen Thermally Cycled between 600°C and Room Temperature in Air. Note Curvature of Scale Flakes in Right Hand Micrograph

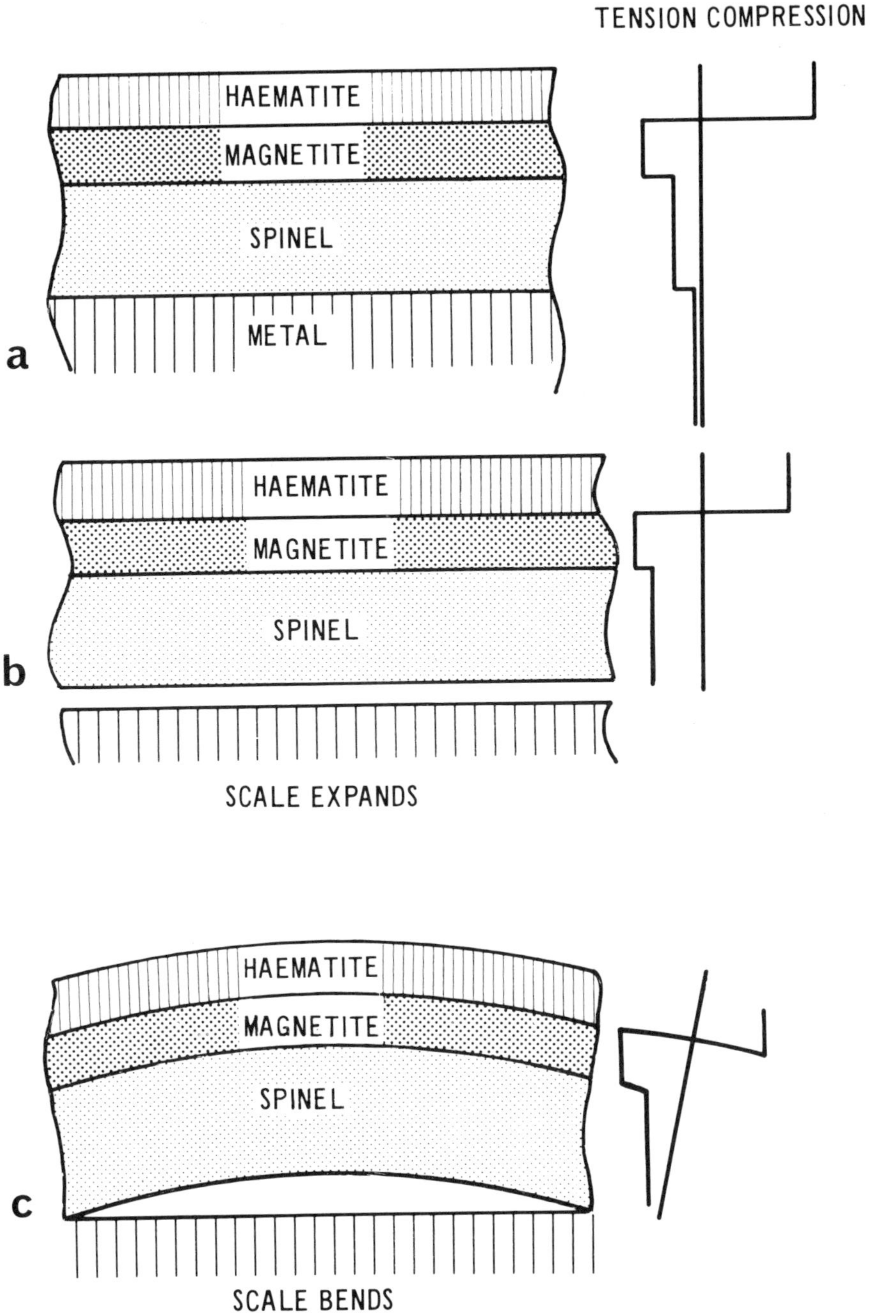

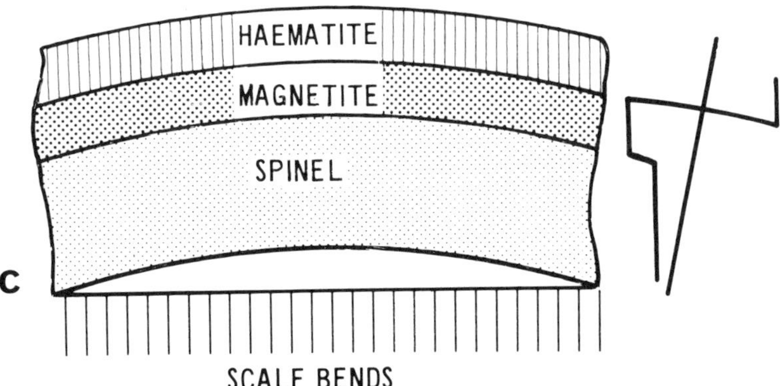

FIG.5: Schematic Representation of the Stresses in a Three Layer Scale Grown on an Fe-2¼Cr-1Mo Steel. (a) After Cooling to Room Temperature, (b) Scale Separated at metal/scale Interface (c) Scale Bent.

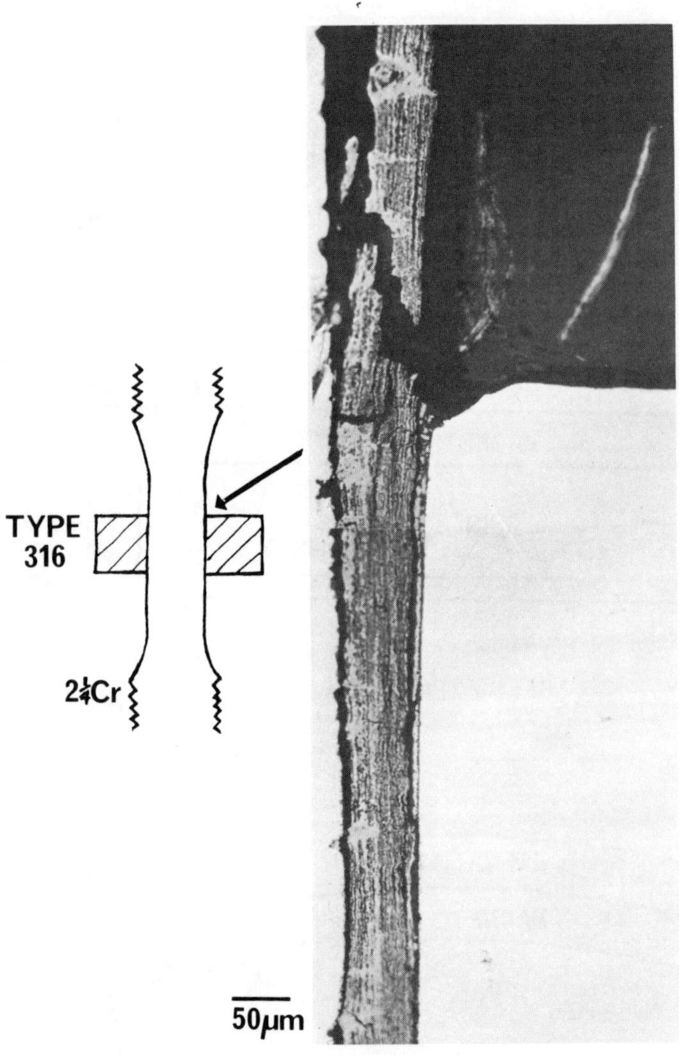

FIG.6: Diagram showing Arrangement of Fatigue Specimen and Collar, and an Optical Micrograph of the Multilayer Scale which has grown in the Crevice

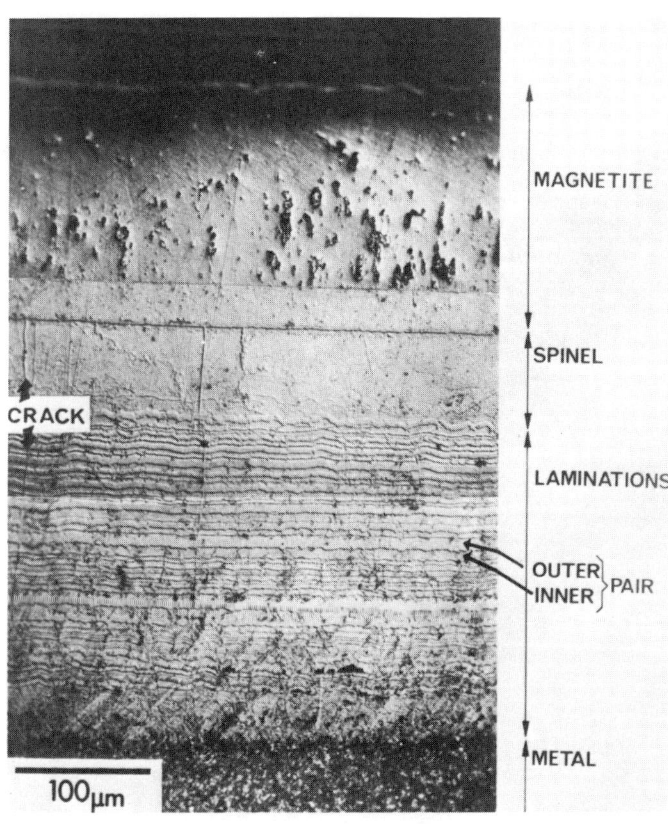

FIG.7: Optical Micrograph of a Multilayer Scale Grown on the Steamside of an Fe-2½Cr-1Mo Reheater Tube

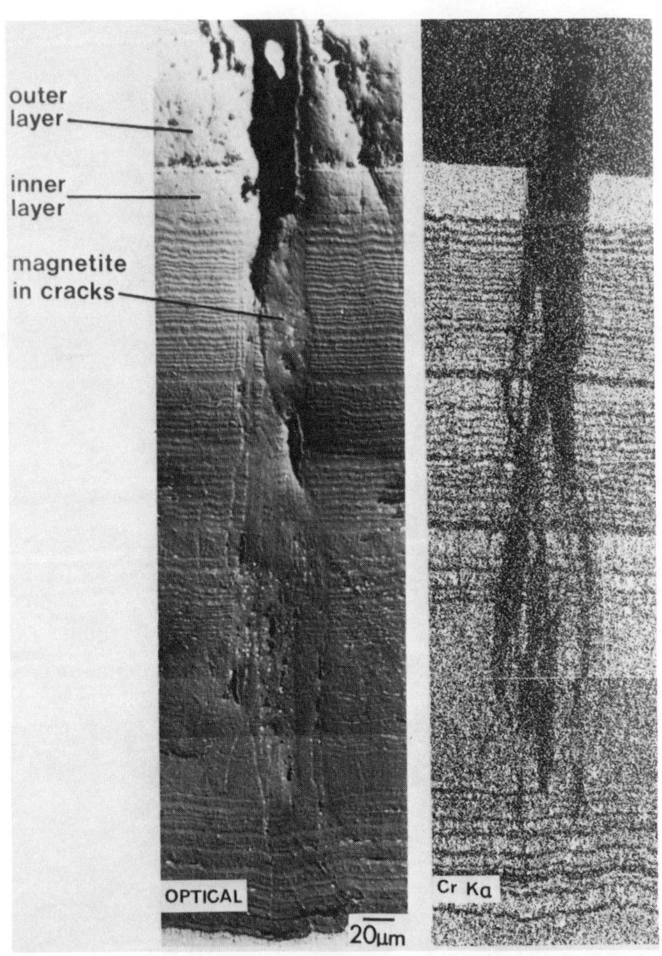

FIG.8:   Optical Micrograph and Elemental Chromium distribution of
a Multilayer Scale in a Region showing Repeated Cracking

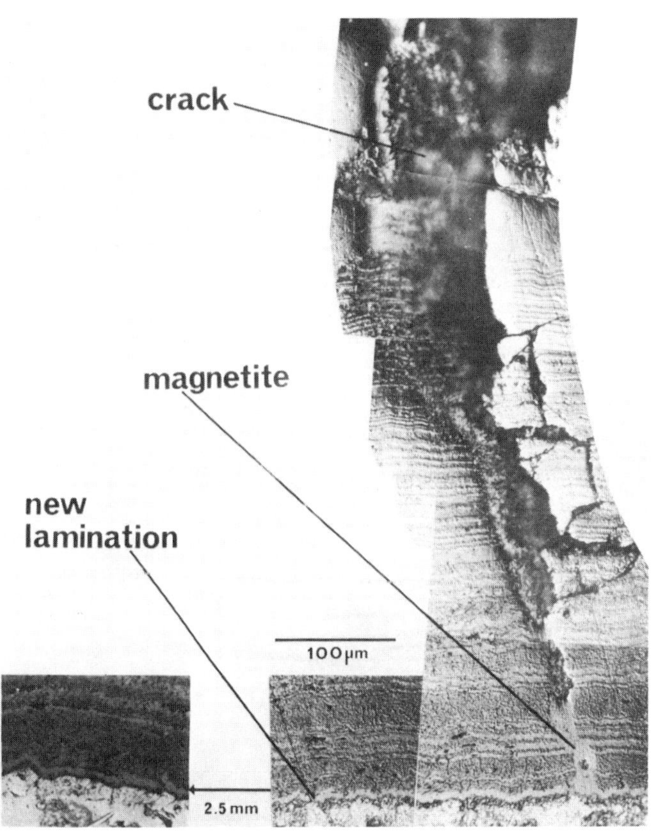

FIG.9: Optical Micrograph of a Multilayer Scale showing Crack Partially Filled with Magnetite and New Duplex Layer Growing at the Metal/Scale Interface

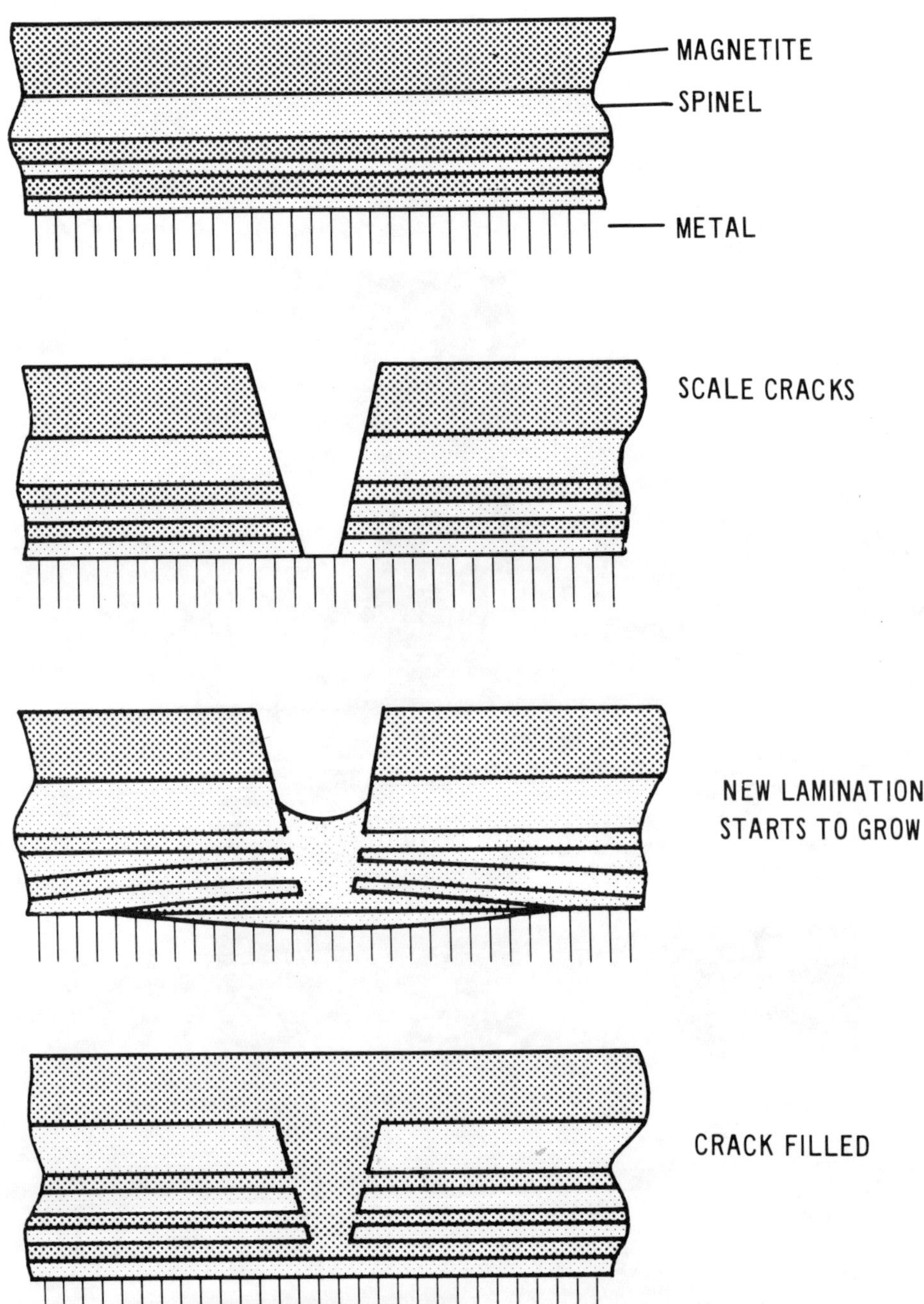

FIG.10: Schematic Illustration of the Stages in the Growth of Multilayer Scales by the Process of Through-scale Cracking

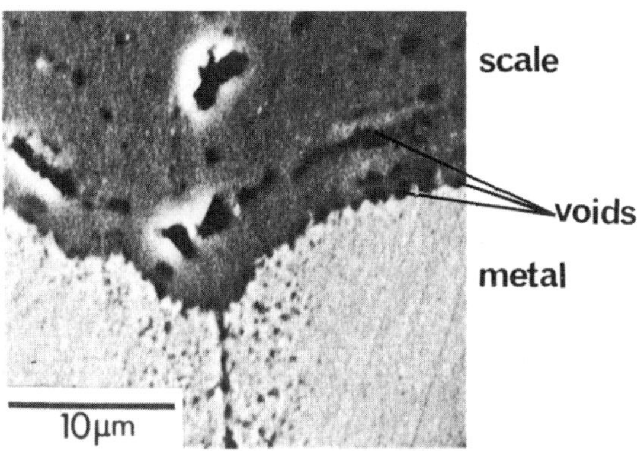

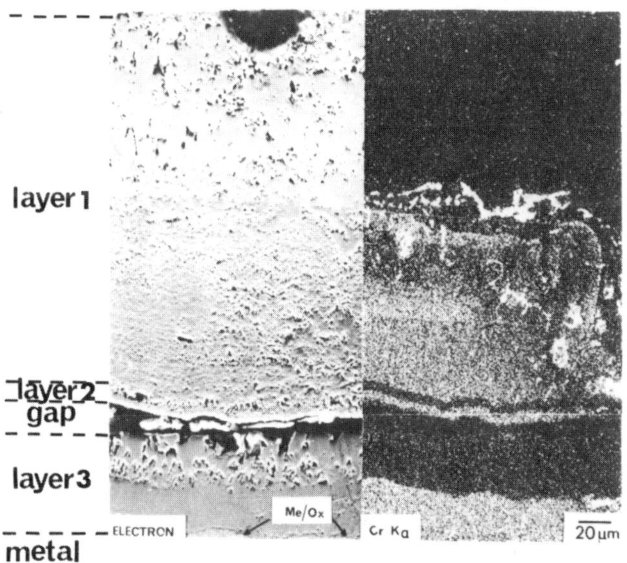

FIG.11: Scanning Electron Micrograph (top) shows Rows of Voids at
the Metal/Scale Interface. Lower Pair of Micrographs
shows Multilayer Scale and Elemental Chromium
Distribution. Note Separation of One Pair of Layers.

DISCUSSION

H.J. Grabke: In your paper, you presented a schematic drawing in which it was indicated that different tensile and compressive stresses occur in the oxide layers: haematite, magnetite and spinel. How did you get the information on these stresses and are there quantitative values available?

P.S. Bell: The stresses in the haematite, magnetite and spinel layers arise from the differential thermal contraction of the layers and metal substrate on cooling from the scale growth temperature.   Values of these stresses can be calculated from the thermal expansion coefficients and Young's moduli of the phases in the scale and metal substrate.   It is assumed that the scale is stress free at the growth temperature.   Details of the calculation of the cooling stress and the bending behaviour of the scales after detachment from the metal are given in the report by Bell and Forrest referenced at the end of this paper.

I.R. McLauchlin: I would like to ask Bell if he found any effect of bolt tightening torque on oxidation behaviour at interfaces in his bolted carbon samples.

P.S. Bell: The specific effect of bolt tightening torque on the oxidation behaviour in crevices was not studied in this work. However it was observed that when the oxidant was either air or water vapour then the crevice became sealed when scale on opposite faces of the crevice impinged. The magnetite did not continue to grow within the crevice but was deposited only at free surfaces. Inner layer spinel did continue to grow within the crevice although the rate of growth was very low due to the long oxidant access path through the magnetite which sealed the crevice, and since inner layer spinel effectively grows in the volume vacated by the oxidized iron there was no stress generation. It is noted that this behaviour is different from that which may be observed when the oxidant is a CO_2 based gas or an aqueous acid chloride solution, when growth of both magnetite and spinel may continue inside a crevice giving rise to stress generation and "jacking" of the crevice.

STRESS-INDUCED INTERGRANULAR CRACKING
OF NICKEL-CONTAINING ALLOYS IN
PETROCHEMICAL PROCESS EQUIPMENT

P.A. Collins, Z.A. Foroulis and M.J. Humphries

Exxon Research and Engineering Co.

Florham Park, NJ   07932

Synopsis

The intergranular stress-induced corrosion cracking of a nickel-base alloy (Inconel 600) and of an iron/nickel alloy (Incoloy 800) in service in petrochemical process units is described. Corrosion-cracking of the nickel-base alloy shroud occurred when it was exposed in a nitriding environment (predominantly NH_3) containing small amounts of H_2S and oxygen, in the temperature range of 350-510°C. Electron Probe microanalysis was used to supplement metallography to examine critical areas to help understand the mechanism of failure. Residual stresses were determined by means of X-ray diffraction to establish their magnitude in the areas in which cracking was observed.

Intergranular oxidation/cracking has been observed in Incoloy 800 tubular components operating at a temperature of 760°C in air. It has been shown that, in addition to a steady hoop stress, transient cyclic stresses exist due to instabilities in equipment operation. The combination of static stress and transient cyclic stress is believed to have caused localized intergranular cracking in Incoloy 800. This is supported by laboratory fatigue and creep/fatigue testing which shows a shift in fracture morphology from trans- to intergranular by the addition of a tensile hold time to a continuous fatigue cycle. An analogy is drawn to the oxide film rupture mechanism of stress corrosion cracking.

Introduction

This paper will deal with two cases in which intergranular cracking occurred in Fe-Cr-Ni-containing alloys. One of the alloys is nickel-based (Inconel 600) while the other is iron-based (Alloy 800). In both cases the cracking occurred as the result of the conjoint action of a tensile stress and a corrosive environment. An analogy to oxide film rupture mechanisms of stress corrosion cracking appears valid.

One type of attack described here is a localized (grain boundary) attack of an Inconel 600 alloy reactor shroud while in service. The shroud was exposed to a nitriding environment (predominantly NH_3) containing small amounts of H_2S and oxygen in the temperature range of 350° to about 510°C. The alloy shroud was fabricated from Inconel 600 hot-rolled sheet by die forming to produce the desired corrugations.

An investigation into the cause of intergranular cracking in Alloy 800 is also described. The components which suffered cracking were reformer furnace outlet connections which operated at a temperature close to 760°C. These tubular connections failed from the outside surface, which was exposed to air. Alloy 800 was originally chosen for the application to provide adequate resistance to oxidation. The components were subject to cyclic loading under conditions of strain control.

Experimental

Representative samples from the shroud and the outlet tube were examined metallographically and by electron probe X-ray microanalysis. Both energy dispersive and quantitative wavelength dispersive analysis techniques were used to determine elemental concentrations and to produce mappings of the distribution of elements of interest. Residual stress measurements on selected points of the shroud surface were carried out using the single exposure X-ray technique[1-5].

Examination of Inconel 600 Shroud

Metallographic examination of the surface of the shroud that failed in nitriding and sulfidizing environments revealed two different types of attack. In non-stressed areas (flat areas on the shroud surface shown in Figure 1) pitting was observed. Figure 2 shows a typical pit formed on the surface of the shroud in non-stressed areas. However, in highly stressed areas (such as shown in Figure 1) grain boundary cracking in the transverse

direction was quite common (Figure 3). Cracks like those shown
in Figure 3 were found along grain boundaries on a typical shroud
convolution shown in Figure 1.

The quantitative chemical compositions of the various
zones in the pit are shown in Table 1. Composition of the base
matrix (point A) was within the specification limits of the
Inconel alloy. The matrix near the pit (point A-1) was depleted
of chromium, with nearly normal amounts of nickel and iron. The
metal adjacent to the pit (point C) was enriched in chromium and
was found also to contain about 4% nitrogen. The corrosion
products within the pit (point D) are principally oxides with
about 50% chromium, 9% nickel, and 8% iron, and are spinels. The
corrosion products in a number of pits examined were found to
vary with position or, between pits, in composition. In a number
of instances corrosion products containing much higher amounts of
sulfur and nickel were observed.

A qualitative map of oxygen and sulfur distribution in
the crack area (Figure 3) is shown in Figure 4. The qualitative
chemical compositions of the various zones in the areas around
the cracks (Figure 3) are shown in Table 2. These include the
base matrix (point M, Figure 3a), the light grain boundaries of a
crack (point A, Figure 3a), the dark grain boundaries of a crack
(point B, Figure 3c), the light phase in crack (point C, Figure
3b), and the gray phase in the two phase mixture (point D, Figure
3a).

The composition of the matrix is within specification,
while the light colored grain boundaries (point A) show substantial chromium depletion (about 33%) with a high sulfur content
(13%). In contrast, the dark grain boundaries (point B) show
substantial chromium enrichment, some nickel depletion, some
sulfur and a more substantial amount of oxygen, suggesting the
formation of mixed oxide/sulfide or oxysulfides. The light phase
in the crack shows pronounced chromium and iron depletion and
substantial sulfur enrichment indicating it to be primarily
nickel sulfide or an eutectic mixture of nickel and nickel
sulfide. The gray phase at point D is rich in chromium oxide and
contains smaller amounts of nickel sulfide and iron oxides and
sulfides.

Metallographic examination of representative sections
of the shroud that failed in service indicated the cracking of
the shroud along grain boundaries occurred only in areas of
shroud convolutions (Figure 1). Tensile residual stresses
induced in the shroud convolutions during forming may have
assisted corrosion cracking along grain boundaries in these
areas. To determine the magnitude and sign of residual stresses
in these areas, the X-ray diffraction method was used. A piece

of the shroud not exposed to the environment was used for the
residual stress measurements and the stresses were determined on
several points located on the convex surfaces of a typical shroud
convolution shown in Figure 5 (points 1 to 7). Residual stress
measurements on the concave surfaces could not be carried out
because of interference by the shroud with the reflected X-ray
beam. However, it was possible to estimate* that the tensile
residual stresses on the concave surfaces were of about the same
magnitude as those in the convex surfaces.

Examination of
<u>Alloy 800 Furnace Connection</u>

Examples of cracked Alloy 800 connectors (pigtails)
were sectioned longitudinally to permit examination of a transverse section. Oxide-filled cracks were found along many of the
grain boundaries of the tube (Figure 6). In some cases there was
also evidence of creep damage in the form of cavity growth at
triple points. Creep failure has occurred in some instances and
has, in itself, led to a substantial reevaluation of the properties anticipated in Alloy 800. Nonetheless, it appears that the
intergranular oxidation is a separate mode of failure since it
sometimes occurs in the absence of creep damage.

All of the components showed some degree of grain
boundary carbide precipitation as a result of the service
exposure at 1450°F (Figure 7). One particularly interesting
observation was the presence of thin metallic layers within the
oxide layer in the cracks (Figure 8). As is shown in the figure,
short, 2-3 grains deep, intergranular cracks were found along the
entire length of the tube. However, growth of these cracks
beyond that level appeared to occur only at some localized area
along the pigtail. Grain boundary carbide precipitates were
evident throughout the material.

The grain boundary precipitates observed in the aged
(service) material are primarily chromium carbides (Figure 9).
Identification of the precipitates by Energy Dispersive X-ray
analysis showed that in some instances a Ni-Si rich phase was
also present. No depletion of chromium in the grain boundary

* This was done by using the measured residual gradient near the
  intersection of the convex and concave surfaces and analyti-
  cally predicting the nominal residual stresses in both the
  convex and concave surfaces and comparing with the measured
  values.[5]

region was detectable by the techniques available. The films of metal within the oxide were analyzed by wavelength dispersive X-ray analysis and found to be similar in chromium, nickel and iron content to the matrix (Figure 10).

In laboratory tests, the application of a sinusoidal cyclic tensile stress did not produce a failure having the intergranular characteristics of the field failures (Figure 11b). Furthermore, the fatigue endurance limit was far above the observed stress to cause failure in plant components. However, when a hold time under creep conditions (Figure 12) was super-imposed on the fatigue test at the peak of the tensile stress cycle, a drastic reduction in time to failure was obtained. The resultant crack path was intergranular, approximating in appearance the field failures (Figures 11b and 13).

Discussion

In the following paragraphs a mechanism is proposed to explain the role of the environment and stress in the two fail-ures. The mechanism is discussed in terms of two distinct steps: initiation of environmental attack and propagation of attack.

Initiation of attack occurs at grain boundaries at the surface exposed to the environment. Grain boundaries represent areas having lower thermodynamic stability, and enhanced rates of diffusion. While it might be expected that attack would be more pronounced along grain boundaries than in the grains, these alloys are almost exclusively attacked at the grain boundaries. Indications of a possible reason for grain boundary susceptibil-ity in these alloys can be seen in Figures 3 and 8, showing chromium carbides in the grain boundaries of the Inconel alloy in the as-received condition, and the Alloy 800 after service aging. Precipitation of the chromium carbides along the grain boundaries may cause a chromium depletion in the surrounding areas, thus increasing the sensitivity of the grain boundaries to attack by the environment. Susceptibility of the Alloy 800 to intergranu-lar attack is believed to be caused by this mechanism. Oxidation of chromium carbides in itself provides a possible mechanism for local grain boundary penetration.[6]

In the case of the Inconel, grain boundary susceptibil-ity to the sulfidizing and oxidizing environment is probably a result of preferential nitriding along grain boundaries. Local-ized nitriding of this alloy, along grain boundaries, could lead to preferential formation of chromium nitride, resulting in further solid solution chromium depletion in the grain boundary zones. Local nitriding and chromium depletion in solid solution

in the vicinity of the nitriding band, at the bottom of the pit
(Figure 2), was quantitatively demonstrated (Table 1). The pit
shown in Figure 2 was produced on flat areas of the shroud
surface where the material was in the hot-rolled condition.   In
these flat areas the net surface stresses were negligible.
Attempts to measure chromium depletion in the vicinity of crack
tips were not successful, probably because the chromium and
nitrogen gradients set up as a result of inward nitrogen
diffusion were much smaller. This is not surprising when one
considers that diffusion rates are known[7] to be much higher in
cold worked metals than in annealed ones.

Once the localized chromium concentration gradients
are set up (by local nitriding and/or chromium carbide precipitates), the alloy grain boundaries become susceptible to the
mixed sulfiding and oxidizing environment. Sulfur preferentially attacks the nickel-rich regions, while oxygen preferentially
reacts with the chromium-rich regions. Thus, the narrow zone
along the grain boundaries where chromium depletion occurs
becomes preferentially attacked by sulfur resulting in the formation of a mixed sulfide phase rich in nickel sulfide or perhaps[8]
low melting point eutectoid (mixture of $-Ni$ and phase Ni_6S_5
or NiS[9] [light phase in crack ΄C΄ in Figure 3b]). The area
immediately adjacent to this zone becomes preferentially attacked
by oxygen resulting in the formation of a two-phase mixture rich
in chromium oxide (gray phase in crack ΄D΄ in Figure 3a).
Whether or not the corrosion products produced by the environmental attack at a particular site would be primarily oxides (as
in the case of the pit shown in Figure 2), or mixtures of
sulfides and oxides (as in the case of the cracks shown in
Figure 3) would depend on the availability of oxygen and sulfur
in that particular site where the attack occurs. The above
mechanism is in accord with published work[10-15] relating to
the beneficial effect of chromium in mitigating corrosion of
nickel-containing alloys in a sulfur environment.

In both cases, it is postulated that under application
of a tensile stress, the protective films formed on the metal
surface are ruptured. Local penetrations of oxide at grain
boundaries provide stress concentration sites which, together
with grain boundary compatibility, cause oxide film rupture to
occur preferentially at grain boundaries. The local compositional differences at the grain boundaries cause oxidation
to continue at grain boundaries. Continued (repeated) rupture
of the oxide films in the grain boundaries permits penetration
of oxygen to the crack tip. A mechanism of penetration of
oxygen through cracks in the oxide film is postulated rather

than one involving accelerated diffusion of oxygen in the grain
boundary regions. Were accelerated diffusion the sole cause of
oxygen penetration, it seems unlikely that the very long narrow
cracks, observed in practice, would occur. It should be noted,
however, that McMahon, in work on nickel-base super alloys,
reported evidence of oxide penetration ahead of the crack tip
which may indicate oxygen diffusion along the grain boundaries.[16]
Repair of the ruptured oxide film at the crack tip in the grain
boundaries is slower than on the matrix because of the depletion
of alloying elements (chromium).

Initiation of the environmental attack along grain
boundaries in the form of a minute crack is very critical. Once
the environmental attack along grain boundaries is initiated,
the minute cracks are further "opened up" by the tensile stresses
exerted in these areas, and this facilitates the propagation of
attack by exposing new areas of the grain boundary zones to
the environment. Thus, more corrodents reach the tip of these
cracks to continue the attack. The availability of sulfur (in
the Inconel 600 case) and oxygen (in both cases) at the reaction
sites may be further enhanced by the higher rates of diffusion
expected in areas of high strain, as at the crack tip.

Oxide films which form at the crack tip are believed
to rupture in a brittle manner when plastic deformation occurs
in the metal substrate. Although, in non-stressed areas, attack
appears to initiate at grain boundaries (in the case of the
shroud), it fails to propagate because the oxide film remains
intact and prevents further ingress of the environment.

Under a steady tensile stress there may be sufficiently
little grain boundary movement that repeated oxide film rupture
either does not occur, or heals successfully. When a steady
stress was applied to the Alloy 800 pigtail, it failed by creep.
Although a nominally steady stress was applied to the shroud,
operating cycles are known to have caused stress fluctuations.
With the thicker pigtail, it was shown that with application
of a continuously cycling stress intergranular cracking did not
occur, it is believed, because the time at peak stress was
insufficient to permit substantial oxidation at the grain
boundaries. However the super-position of a creep hold time at
the peak of each fatigue stress cycle caused stress relaxation
to take place. Under the conditions of stress relaxation, oxide
film rupture at grain boundaries can take place and lead to the
observed intergranular attack.

It is postulated that oxidation occurs at the crack
tip when oxide film rupture exposes bare metal in that region to
the environment. While there is no concrete evidence to show
this is more likely than a mechanism including enhanced diffusion

of oxygen in the zone of stress concentration at the crack tip,
it is felt that the latter process would tend to lead to broader
cracks in view of the typical shape of the zone of plastic
deformation at the crack tip.

Future Work

The mechanisms proposed as the cause of failure in
each of the instances reported here are similar. A local
microstructural variation provides a susceptible path at the
grain boundaries; subsequent environmental attack is localized
in the region; under the application of a tensile stress
the corrosion product film is ruptured at the tip of the intergranular penetration; then the cycle repeats. The conjoint
action of oxidation and an applied tensile stress is necessary
to the occurrence of this form of cracking.

It is felt that the similarity to stress corrosion
cracking is worthy of further exploration. To this end, a study
of the behavior of these alloy/environment combinations under
conditions of constant applied strain rate or low frequency
strain cycling, coupled with an investigation of their oxidation
(or sulfidation) characteristics would be worthwhile. If
the analogy is correct, prevention of local attack ought to be
possible, both by increasing the corrosion potential of the
environment (general attack) and by reducing it (analogous to
passivation), and by changing the strain rate.

Acknowledgements

The authors wish to acknowledge the assistance of
their colleagues, particularly J.E. McLaughlin and D.S. Williams,
who provided much valuable input.

REFERENCES

1. AMR portable X-ray residual stress analyzer Model No. 4-301 Instruction Manual. Advanced Metals Research Company, Burlington, Mass.

2. C.S. Barrett. Structure of Metals, Chapter XIV. McGraw-Hill Book Co., New York, 1952.

3. M.J. Donachie, Jr., and J.T. Norton. X-ray Studies of Lattice Strains Under Elastic Loading, Trans. ASM 55, 51, 1962.

4. D.A. Bolstad and W.E. Quist. The Use of a Portable X-ray Unit for Measuring Residual Stresses in Aluminum, Titanium and Steel Alloys. W.M. Mueller, G.R. Mallett, and M.J. Fay, Editors, Advances in X-ray Analysis, Vol. 8. Plenum Press, New York, 1964.

5. H.R. Woehrle, F.P. Reilly, III, W.J. Barkley, III, L.A. Jackman and W.R. Clough. Experimental X-ray Stress Analysis Procedures for Ultrahigh Strength Materials. W.M. Mueller, G.R. Mallett, and M.J. Fay, Editors, Advances in X-ray Analysis, Vol. 8, Plenum Press, New York, 1964.

6. C.J. McMahon, Mat.Sci. & Eng., 13, 1974, 295.

7. D. Turnbull, in "Atom Movements" p. 129, American Society for Metals, 1951.

8. A.H. Cottrell, in "Theoretical Structural Metallurgy" p. 188, St. Martin's Press Inc., 1957.

9. L. Darken in "The Physical Chemistry of Metallic Solutions and Intermetallic Compounds" Vol II 4G, Nat. Phys. Lab. Symp. No. 9 (London) 1959.

10. Metals Handbook. 1948 Edition, American Society of Metals.

11. W. Betteridge, et al, J. Inst. Petrol., 41, 377, 1965.

12. J. Rosenqvist, J. Iron Steel Inst., 176, 37, 1954.

13. H.J.T. Ellingham, J. Soc., Chem. Ind., 63, 125, 1944.

14. M.B. Panish, et al, J. Phys. Chem., 62, 980, 1958.

15. Y. Matsunaga, Japan Nickel Review 1, 347, 1933.

16. C.J. McMahon and L.F. Coffin, Jr., Met. Trans., 1, 1970, 3443.

TABLE 1

ELECTRON MICROPROBE ANALYSIS OF THE
VARIOUS ZONES IN THE PIT SHOWN IN FIGURE 2

Area/Phase	Composition in Weight Percent						
	Ni	Cr	Fe	Ti	N	O	S
A (Matrix)	76.0	14.9	6.7	0.1	<0.5	–	<0.05
A-1 (Matrix Near Pit)	76.2	13.9	6.7	0.1	<0.5	–	<0.05
B (Yellow Precipitate)	–	–	–	>70	~20	–	<0.05
C (Phase Near Pit)	75.8	16.5*	7.3	0.1	~4	–	<0.05
D (Corrosion Products, Pit)	~9	~50	~8	2.5	<0.5	>25	1.0

* Possible Excitation of Cr-Rich Oxide in Pit

TABLE 2

ELECTRON MICROPROBE ANALYSIS OF THE VARIOUS ZONES IN
THE AREAS AROUND THE CRACKS SHOWN IN FIGURES 3 (a, b, c)

Area/Figure	Composition in Weight Percent					
	Ni	Cr	Fe	N	O	S
Matrix/Fig. 3a	75.0	15.0	6.5	<0.5	0	0
A*-(Lt.G.B.)/Fig.3a	65.0	10.0	4.0	<0.5	**	13
B*-(Dark G.B.)/Fig.3c	50.0	29.0	6.5	<0.5	~10	9
C-(Lt.Phase in Crack)/Fig.3b	68.0	4.0	2.0	<0.5	~0.5	24
D-(Gray Phase,Two Phase Mixture) Fig. 3a	30.0	35	7	<0.5	~15	7

  * Probable Matrix Excitation
** Oxygen Present Along G.B. Linings

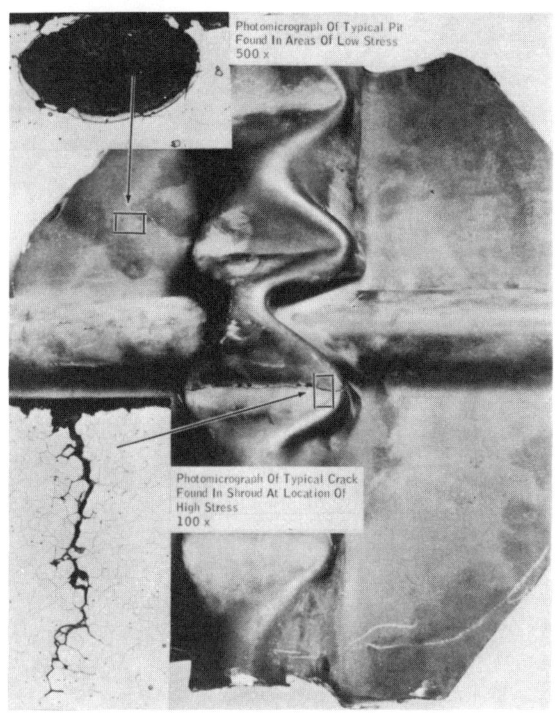

Figure 1

Section of the shroud that failed in service.
(Reduced to 50% in reproduction.)

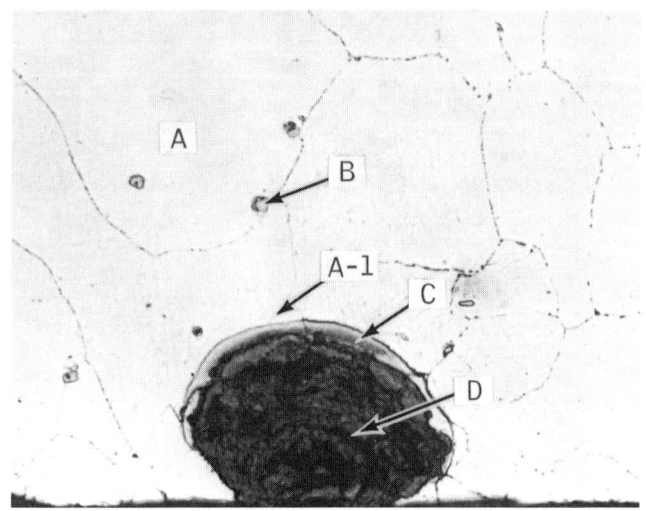

Figure 2

Typical pit on flat area of shroud.   500X

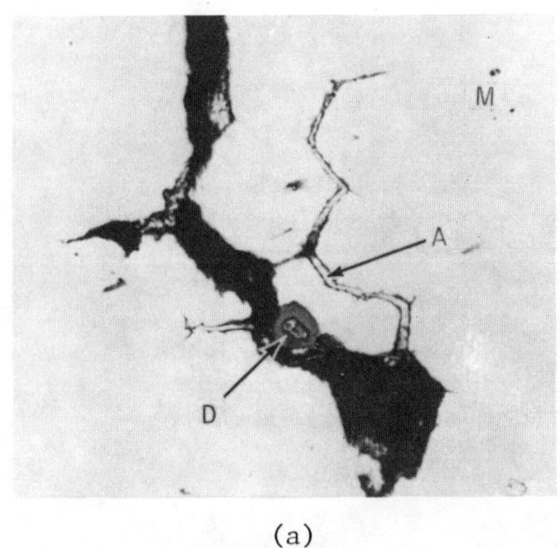

(a)

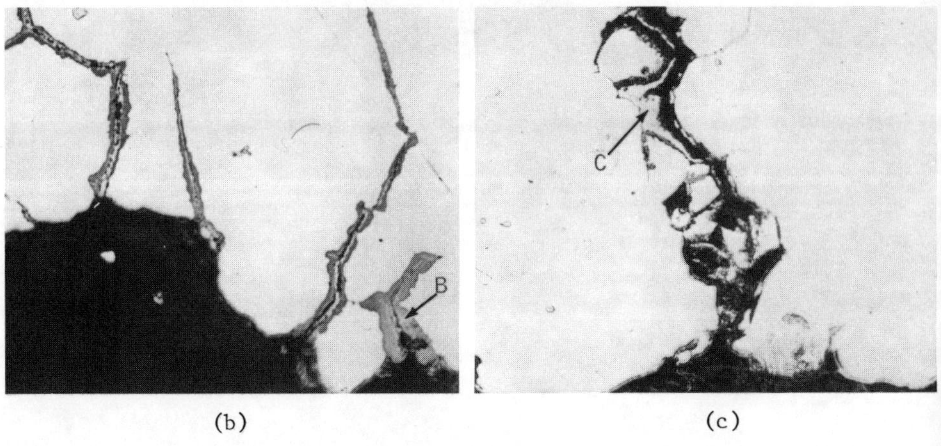

(b)   (c)

<u>Figure 3</u>

Microphotograph of cracks formed on highly stressed areas on the
surface of Inconel shroud exposed in a nitriding environment containing H_2S and oxygen.   500X

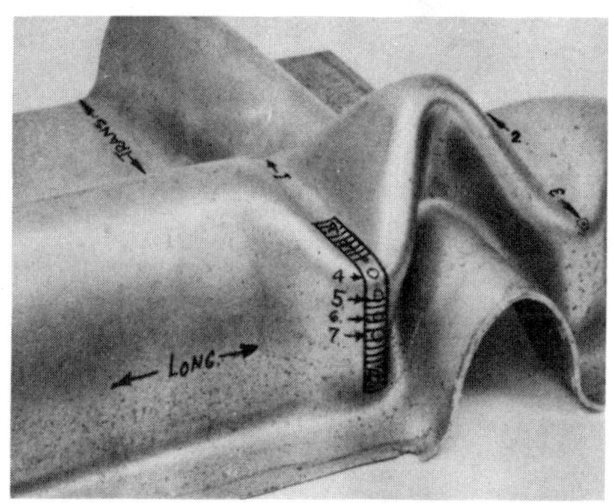

Figure 4

Typical shroud convolution showing locations of residual stress
measurements.

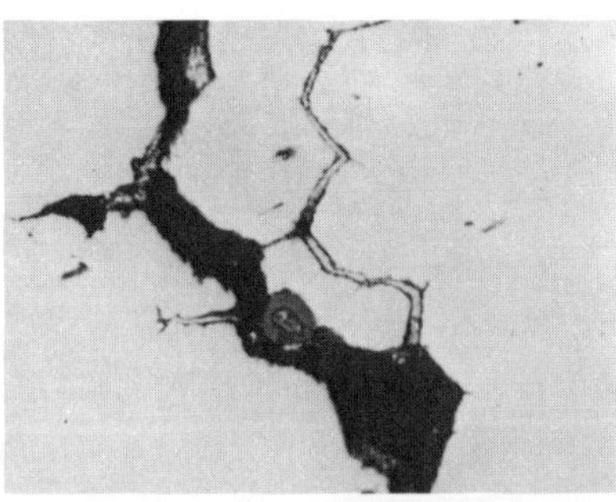

(a) Microphotograph of crack, 300X.

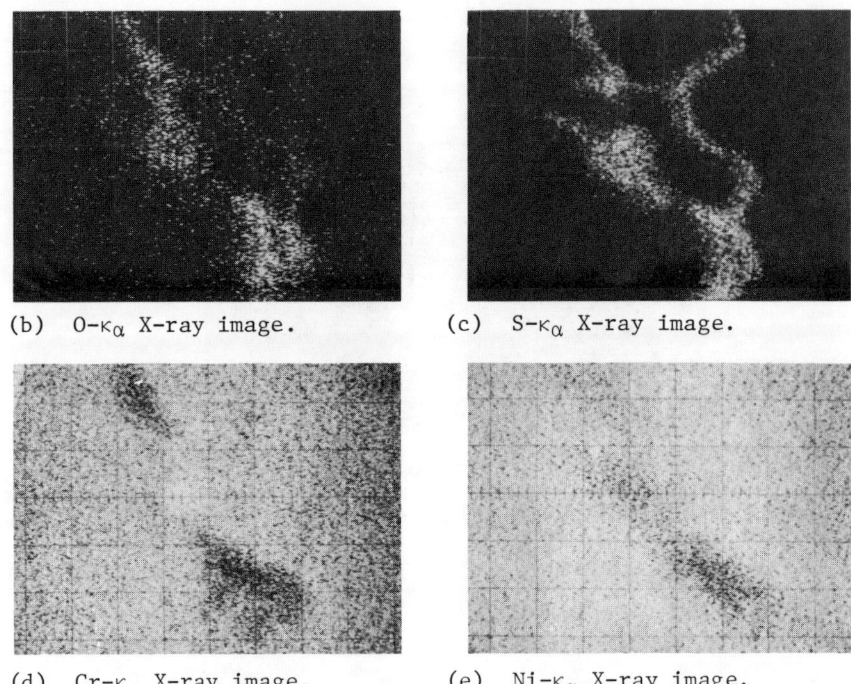

(b) O-κ_α X-ray image.

(c) S-κ_α X-ray image.

(d) Cr-κ_α X-ray image.

(e) Ni-κ_α X-ray image.

Figure 5

X-ray scanning images showing distribution of oxygen, sulfur, chromium and nickel in the area of the crack shown in Figure 1.

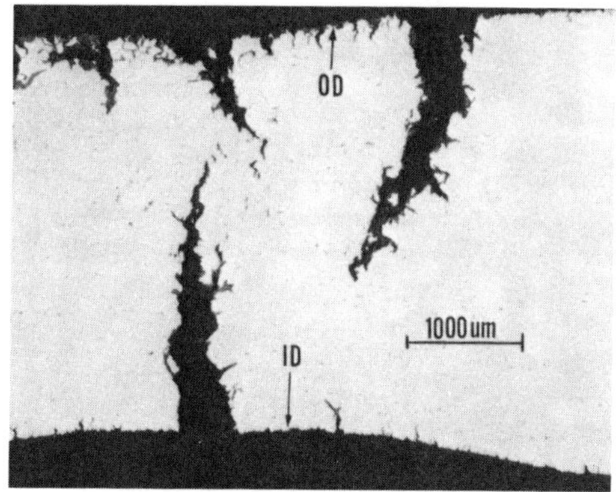

Figure 6

Cracks in Alloy 800 tubular furnace connection (pigtail) after
service.

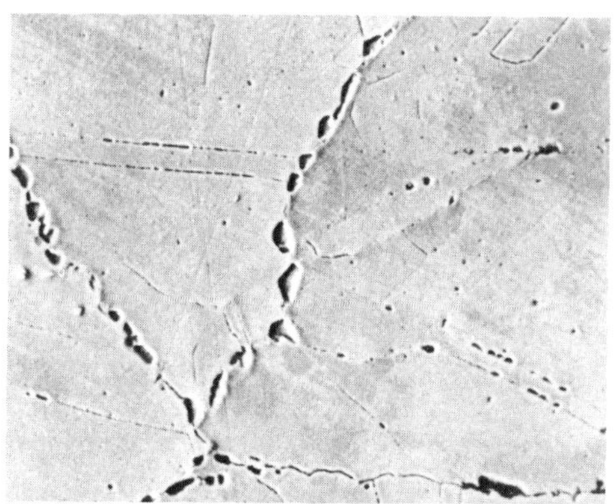

Figure 7

Carbide precipitation at grain boundaries of Alloy 800 after service at about 750°C.

                    P.A. COLLINS et al.

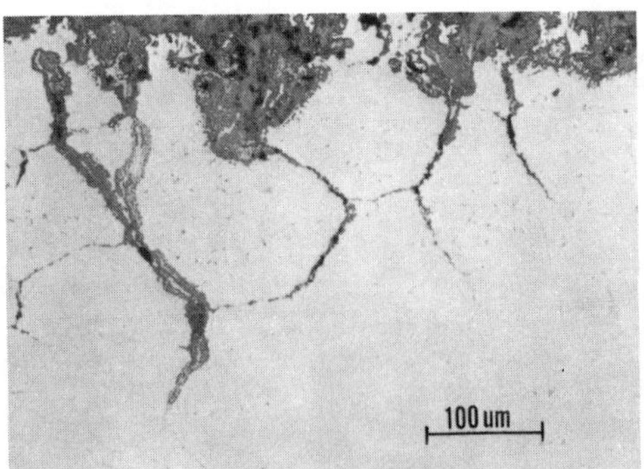

<u>Figure 8</u>

Metallic films observed within oxide-filled cracks in Alloy 800
pigtail.

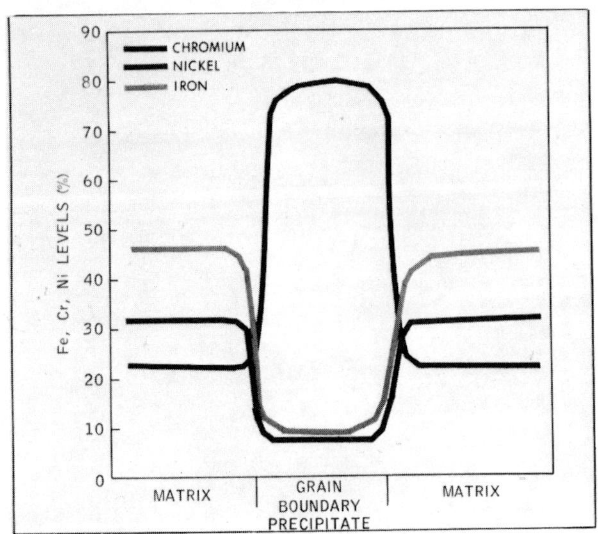

<u>Figure 9</u>

Elemental analysis of grain boundary region.

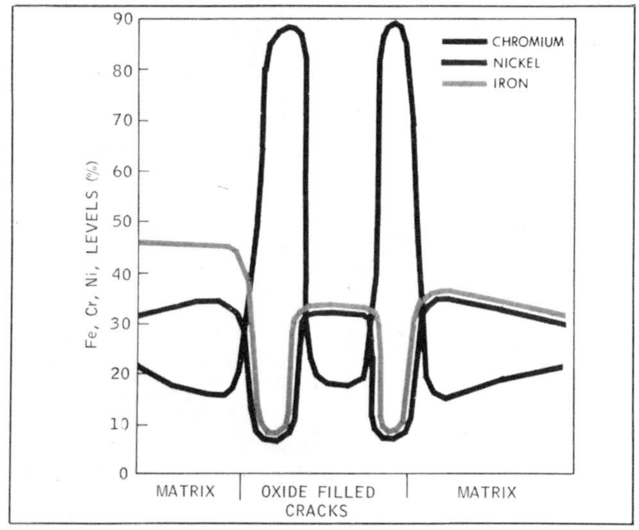

Figure 10

Chemical analysis across crack at grain boundary.

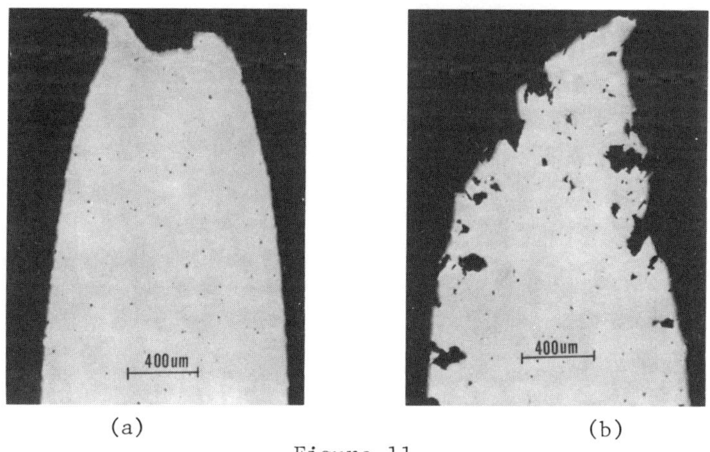

(a)   (b)

Figure 11

(a) Fatigue failure at high cycle under constant cycling.
(b) Creep-fatigue failure at low cycle with hold time at peak stress of each cycle.

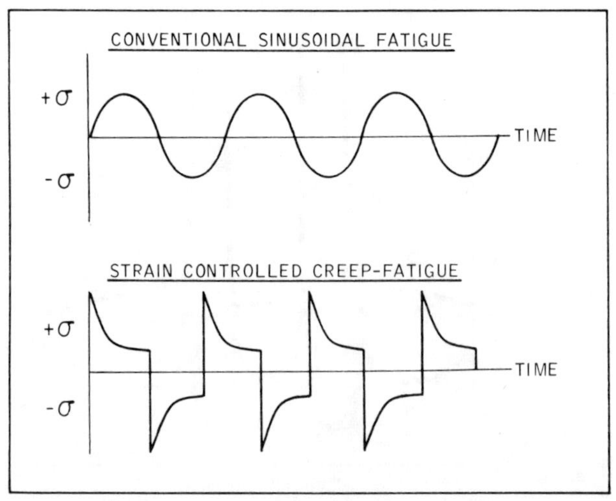

Figure 12

Stress cycle imposed in creep-fatigue test with hold time.

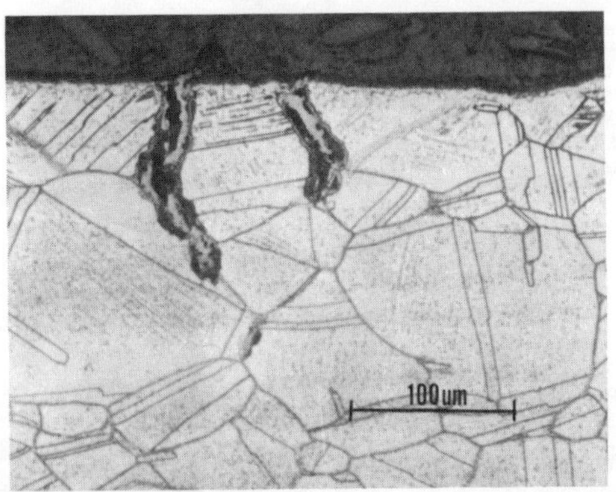

Figure 13

Intergranular cracks produced by laboratory creep-fatigue test.

DISCUSSION

R. Rolls: Is there a possibility that an 'oxide wedge' effect may contribute to crack growth in the Alloy 800 tubes?

Is there any likelihood that the metallic layer at the centres of cracks may have been assisted in its formation by a vapour transport process across the width of the crack from the walls?

M.J. Humphries: One cannot, on the basis of our studies, rule out the possibility that oxide wedging may contribute to the crack growth in the Alloy 800 tubes, however, the presence of open fissures down the center of oxide filled cracks does not support an oxide wedging mechanism.

It is unlikely that vapor transport of metal is the mechanism by which the films form. If one assumes that the mechanism involves transport of, say, nickel from the cracked regions to the crack tip, and that the rate limiting step is metal evaporation, one can show that the rates of deposition would be small in comparison to the observed growth rates.

T. Boniszewski: Did you find the metallic film sandwiched between oxide in the crack in the laboratory simulation studies which were supposed to reproduce plant failures?

M.J. Humphries: We did not find any evidence of the metallic film within the oxide in the cracks in the laboratory test specimens. The crack formed in these tests were short, in comparison to the service cracks, although they did follow a similar path. Additionally, the test duration was very short in comparison to the observed service life. Both of these factors may have been contributing to the absence of the metallic films.

D.B. Meadowcroft: My colleagues of CERL have previously seen metallic areas within the oxide in cracks, which they interpreted to mean that the particular crack had stopped growing. Is this interpretation compatible with your experience?

M.J. Humphries: The interpretation you offer is very interesting and perhaps ties back to other suggestions that the films are really nickel, formed from the nickel oxides in a plugged crack. However, we have found evidence of the metallic films in the oxide layer on the surface of the principal cracks (which were still propagating). Some of the films appear to be directly connected to the surface which also tends to refute the argument that they form only under conditions of low oxygen partial pressure. We believe that the metal film is a result of the way that the grain boundary precipitates are reduced to metal -- though definitive evidence for this is also absent.

<u>C. Coddet</u> :   Is there any evidence of the presence of a grain boundary inside the metallic remaining film.

<u>M.J. Humphries</u>: We have seen no evidence of grain boundaries remaining within the metal. In some cases the distribution of films is not along the center of the crack but instead, one or two parallel films are present.

<u>M. Victoria</u>: We have observed a similar type of cracking in failures in components made out from 304H stainless steel and 800 H Alloy:-
Conditions necessary for this type of cracking to appear are:

- High carbon content with a large amount of carbides precipitated at grain boundaries,
- High stresses.

<u>J. Bressers</u>:   As the temperature of interest, specifically in case of Alloy 800, one would expect creep damage to play an important role in failure.   An important parameter in that respect is the plastic strain-range $\Delta \varepsilon_p$.   What was, in your experiment, its magnitude and how does this compare with the plant failure situation.

<u>M.J. Humphries</u>: In our strain controlled laboratory experiments the strain range employed was from 0.5% to 0.2%. With the 0.2% strain, the plastic component was 0.045% and the stress cycle was $\pm$ 48 ksi. Our field measurements indicate that relatively high frequency strain cycles of a small magnitude (estimated that stress cycles were $\pm$ 15 ksi) occur, together with a superimposed macro deformation of the pigtail due to relative movement of the header and tube. This latter movement may result in several percent of plastic deformation. We believe that our laboratory tests show reasonable agreement with the work of Soo and Chow.

THE EFFECT OF MECHANICAL STRESS ON THE KINETICS AND MECHANISM OF SULPHIDATION OF A FeCrAl ALLOY AT HIGH TEMPERATURE

S. Toesca, Y. Pourantru and J.C. Colson

Laboratoire de Recherches sur la Réactivité des Solides associé au C.N.R.S., Faculté des Sciences Mirande, B.P. 138, 21004 Dijon Cedex, France

A study of dry stress corrosion of wire-drawn samples of a BULTEN-KANTHAL Fe-22Cr-5Al steel has been carried out with an apparatus specially designed for this task, in which the growth of thick sulphide layers is followed by conductivity measurements.

In the range 650-780°C, 20-400 torr for pressures of hydrogen sulfide and 0-150 daNmm^{-2} for G, tensile stress, the kinetic curves are parabolic until a conversion ratio of about 0.4 . The rate constant obtained after linear transformation have a variation given by :

$$K_v(T,P,G) - K_v(T,P,G = 0) = bG^{(0.85 \pm 0.10)}$$

The morphological and micropobe studies show that the increase of the corrosion rate is induced by macro and micro-cracking but the layer exhibits a protective behaviour because of the plasticity of the FeS phase and the increase of the dispersion of Cr_2S_3 in the external part of the corrosion layer.

Introduction.
The influence of mechanical stress on the mechanisms and the kinetics of the growth of thick corrosion layers on metals and alloys has been little investigated and chiefly dry corrosion. This type of study, however, is of growing interest due to the development of high-temperature technologies related to the shortage of raw materials. We approached this question by designing a specific apparatus (1) and by studying the behaviour of some pure metals (2) submitted to a uniaxial stress in an agressive medium where the reacting gas was dry hydrogen sulfide.

Here are given the characteristics of the behaviour of a Fe-22Cr-5Al steel, at rest and under stress, in presence of dry

hydrogen sulfide. This study is part of a research on metallic materials with a satisfactory behaviour in a sulfiding medium at
high temperature (3).

1. <u>Experimental device ; characteristics of initial samples and
   experimental conditions.</u>
               1.1. Initial samples.
               The investigated alloy was a wire-drawn Fe-22Cr-5Al
steel from BULTEN-KANTHAL, 0.20 mm in diameter. The hydrogen sulfide, from Matheson, of theoretical purity 99.9 % was distilled
several times before being used.

               1.2. Apparatus.
               The kinetic curves were obtained by the apparatus we de-
signed (1). The progress of corrosion was followed by measuring
the variation in electrical conductivity of the residual part of
the initial sample. Corrosion may be studied with or without
stress ; figure 1 illustrates this device. Comparatively, thermogravimetry was also used for rest corrosion.
               The morphological study and the location of sulfided
phases were made by scanning electron microscope and wavelenght
dispersion X microprobe. The sulfided phases were identified by
radiocrystallography.

               1.3. Experimental conditions.
               The experimental conditions were selected from a prior
study (4) carried out with the same alloy from standard massive
samples ; these were (10x10x2) mm sheets. This prior study showed
that, on the whole, alternate corrosion layers of various composition occur. The lower the temperature and pressure, the more numerous they are. In addition to obtain a single duplex corrosion
layer from wire samples, experiments show that we must operate at
temperatures between 650 and 850°C for pressures ranging between
20 and 400 torr of pure and dry hydrogen sulfide. Wire samples of
minimum diameter must also be used.

2. <u>Rest corrosion.</u>

               2.1. Morphology of sulfided samples ; nature and distri-
                    bution of the sulfided phases formed.
               The observation of samples sectionned at different con-
version extents shows the existence of a compact corrosion duplex
scale. At the outer interface, rather big crystal stuck together,
of low covering power, occur. The inner part of the corrosion
layer has a diameter almost equal to that of the initial wire
whatever the reaction progress. Microprobe analysis first shows
that the distribution of sulfur is homogeneous throughout the
double layer and allows to delineate accurately the alloy residual
core. Then, the iron is noticed present in the outer part of the
corrosion scale whereas chromium prevails in the inner part,

without a negligible concentration in the outer part.

As for aluminum it seems to be mainly present in the inner part of the corrosion scale with an interface maximum concentration between the iron-rich and chromium-rich part of the corrosion scale.

Phases $Fe_{1-y}S$ and Cr_2S_3 were identified accurately; alternatively the Al_2S_3 phase could not be put in evidence.

2.2. Kinetic curves and their linear transforms.

In the range 700 - 850°C under a pressure of 100 torr of hydrogen sulfide, the kinetic curves $x = f(t)$ are parabolic (fig. 2). This suggests the existence of a diffusion mechanism which would be the determining step of the growth of the thick corrosion layer. In this case, the characteristic transform of these conditions (5) i.e. :

$$F_D(x) = \left(\frac{1}{\Delta-1} + x \right) \log \left[1 + (\Delta-1)x \right] + (1-x) \log (1-x) \quad (E)$$

where x is the conversion extent of the reaction and Δ the Pilling and Bedworth coefficient should be linear.

If it is easy to calculate Δ in the case of a pure metal when the corrosion scale contains only one phase, it is not the same for an alloy when the corrosion scale is not always homogeneous even if we assume that the different phases are not miscible. Indeed the dispersion state of minority phases within the main phase cannot be well known.

In these conditions we have chosen to select for which values of Δ our curves would admit linear transforms according to equation (E). The suitable Δ values are comprised between 1.60 and 1.80 and hence we adopted an experimental Pilling and Bedworth coefficient equal to 1.70. This experimental value of Δ will also be assumed as suitable to transform the stress kinetic curves.

The transforms thus obtained are linear up to conversion extents of about 0.8 and lead to an activation energy of (32 ± 3) Kcal.mole^{-1}.

2.3. Interpretation.

It is most probable that the determining step of corrosion scale growth is the diffusion of present ionic species through that scale. Account taken of the nature of cations and point defects identified in iron sulfide (7) and in chromium sulfide (8) the prevailing role of cationic diffusion should be kept in mind. This is confirmed by the increasing number of voids at the alloy-sulfide interface as the reaction proceeds. However, it is probable that most of them are filled in, while the inner part of the corrosion scale develops and where Cr_2S_3 is majority, along with a dispersion of aluminum sulfide and iron sulfide. It is reasonable to think that this sublayer is partly built up by a process of inner corrosion and that the stress extent in the form of cracks, scaling off and a juxtaposition of crystals at the periphery of the outer part of the corrosion scale is limited by the plasticity of the set {alloy, sulfides}.

3. Growth of a thick corrosion layer under uniaxial stress.

3.1. Kinetic results.

The reaction curves x = f(t) deduced from the conductivity curves of the alloy considered, obtained at 760°C under 150 torr of H_2S for G values, stress force per section unit, comprised between 0 and 150 daNmm^{-2} are given in figure 3. They are parabolic until a conversion extent of about 0.35 .

The $F_D(x)$ transforms, specific to diffusion in cylindrical symmetry and calculated using the experimental Δ determined in the study without stress are linear (fig. 4) until a conversion extent of 0.3 account taken of measurement errors. Near rupture the sample conductibility is always far smaller than that which would predict the conversion extent measured at the end of the experiment.

3.2. Variation in the reaction rate, at constant temperature and pressure, v.s. stress force per surface unit.

Let $K_V(T_O, P_O, G = 0)$, be the rate constant derived from the linear transform corresponding to rest corrosion (G = 0) of a sample at temperature T_O and pressure P_O ; let $K_V(T_O, P_O, G)$ be the rate constant of the same corroded sample at T_O and P_O and submitted to traction per surface unit G. Experiments show that the difference $\Delta K_V(G) = K_V(T_O, P_O, G) - K_V(T_O, P_O, G = 0)$ increases as a function of G, but not linearly. If the curve Log $\Delta K_V(G) = f(\log G)$ is plotted, the points are in line account taken of errors and we obtain a law of type :

$$\Delta K_V(G) = bG^{(0.85 \pm 0.10)}$$

3.3. Morphological study, identification and distribution of sulfided phases.

Refering to samples sulfided at rest in the same temperature and pressure conditions, observation under scanning electron microscope allows to point out the following facts :

The corrosion layer at the outer interface has rather the same aspect under stress (fig. 6) as at rest (fig. 5) . It always consists of small crystals ; their size decreases as stress increases. Some rare large cracks also appear for conversion extents above 0.30 (fig. 6).

Observation of sections at T and P constant for various G values reveals the occurrence of microcracks in the corrosion scale as well as within the residual material ; the greater the stress, the more numerous those microcracks ; but the general aspect of the corrosion scale is the same as at rest.

Crystallographic analysis results in the identification of Cr_2S_3 and $Fe_{1-y}S$; as at rest it is not possible to characterize either Al_2S_3 following experiment or hydrated alumina after leaving the samples in wet air.

X microprobe analysis of samples that have been submitted to maximum stress (150 daNmm^{-2}) shows that the distribution of iron is practically the same throughout the corrosion scale. In addition the inner part of the corrosion layer is relatively poor in chromium and rich in aluminum whereas the outer part shows a layer chromium concentration bigger than that observed during reaction at rest (fig. 7 and 8). The observation of sections of samples submitted to increasing G shows that this concentration increases with G and that the relative volume of the inner part of the corrosion scale decreases correlatively.

3.4. Interpretation.

The parabolic profile of the curves when a stress is applied and their linear transforms during the first third of the corrosion reaction show that the determining step of the kinetics of the chemical transformation is always the diffusion of the present ionic species through a layer which remains protective. However, the nature of the corrosion scale changes all the more as the stress is greater ; the relative volume of the inner part decreases at the same time as the dispersion of the Cr_2S_3 phase increases in the outer part in concentration, as well as the crystallite size which decrease as shown by the surface state of the samples. Therefore, the number of grain joints and areas rich in defects increases with stress, which undoubtedly causes the diffusion rate of ionic species to increase at constant temperature and pressure, hence a rate law written as :

$$\Delta K_v(T_o, P_o, G) = bG^\alpha \text{ where } \alpha \text{ is positive.}$$

In the present case, α is less than unit whereas in the same rate range at a lower temperature this exponent was around unit in the case of iron (2) allowing for measurements. Thus, another factor is added to the favorable contribution of plasticity of the iron sulfide matrix in the resorption process of microcracks induced by stress. It is probably the expansion characteristic of the formation of Cr_2S_3 ; when this phase forms alone, the calculated value of Δ is 3.92. In our case this order of value is probably conserved. In these conditions the crystallites of chromium sulfide would partially restore the protective aspect of the scale.

When the reaction extent reaches 35 %, microcracks occur and then, the deterioration of the material is inevitable. Finally the fall in conductibility observed in very slow working conditions and at more rapid conditions when the conversion extent is high may be reasonably attributed to cracking and inter-grain corrosion observed.

At the limit, for very slow chemical reactions our method might only account for cracking in dry corrosion conditions (9) as in the case of wet corrosion.

Conclusion.
        The study of dry corrosion under stress with growth of
thick corrosion layer in a Fe22Cr5Al steel enabled us to put in
evidence the part played by chromium sulfide in the process of
healing of the scale, whose covering power tends to be reduced by
stress. But, in this case, it is obvious that the main role is
played by the iron sulfide matrix, main constituant of the scale
whose plasticity is determining in the experimental conditions.

References.

(1) Y. Pourantru, S. Toesca, J.C. Colson, Rev. Sci. Instrum.,
    1979, 50, 7, 916-917.
(2) Y. Pourantru, S. Toesca, J.C. Colson, J. Chem. Research (S),
    1979, 7, 235 ; J. Chem. Research (M), 1979, 2, 2739-2750.
(3) Contrat D.G.R.S.T., n° 76-7-1108 et n° 78-7-1047.
(4) P. Mari, Y. Pourantru, J.P. Larpin, S. Toesca, to be published.
(5) M. Billy, G. Valensi, J. Chim. Phys., 1956, 53, 832.
(6) M.B. Pilling, R.E. Bedworth, J. Inst. Metals, 1923, 29, 529.
(7) C. Mathiron, J.C. Colson, P. Barret, Bull. Soc. Chim., 1969,
    2, 427.
(8) C. Mathiron, J.C. Colson, Bull. Soc. Chim., 1970, 1, 79.
(9) K. Moerbe, Chem. Tech. (Berlin), 1969, 21, 3, 155-159.

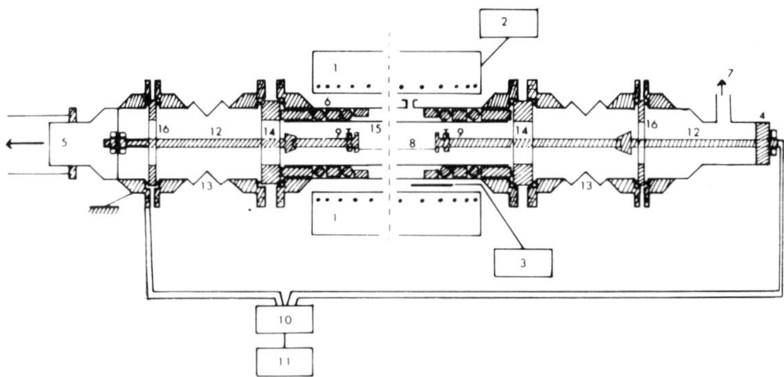

FIG. 1 Experimental apparatus.

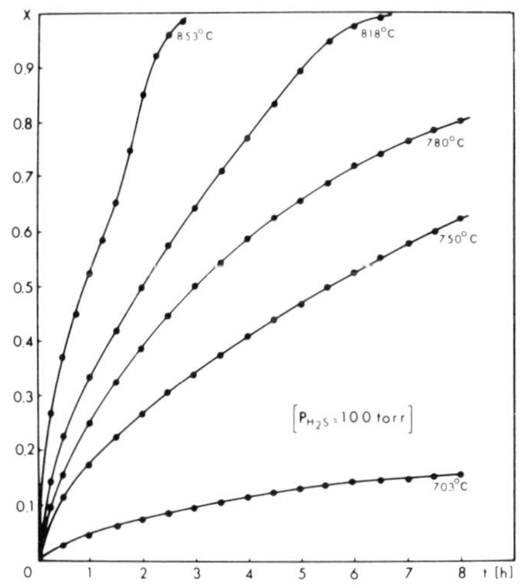

FIG. 2 Curves x = f(t) at rest.

S. TOESCA et al.

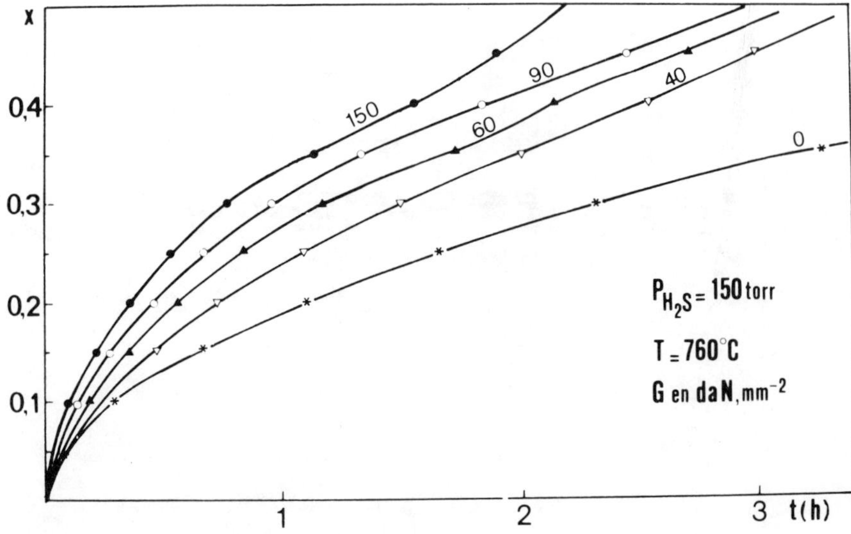

FIG. 3 Curves x = f(t) under stress.

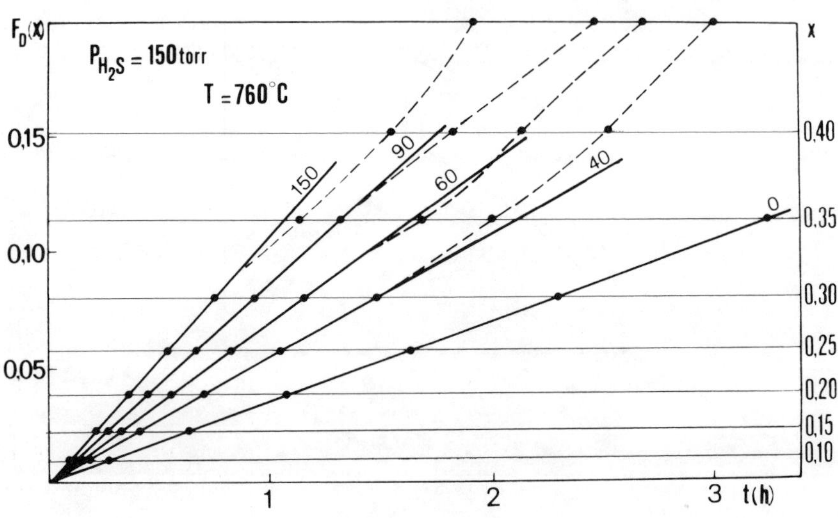

FIG. 4 Linear transforms F(x) = g(t) under stress.

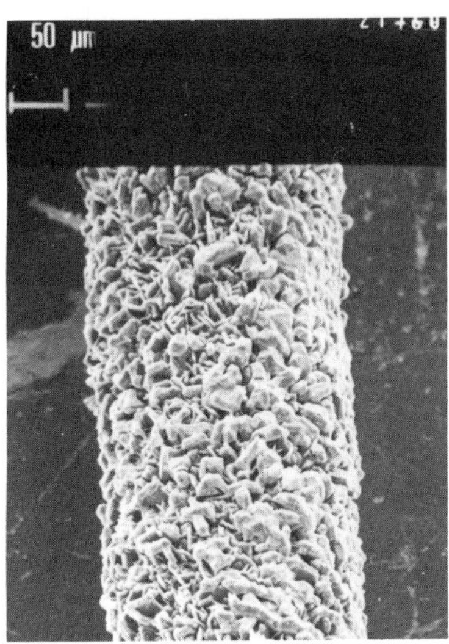

FIG. 5   Corrosion at rest
x = 0.35

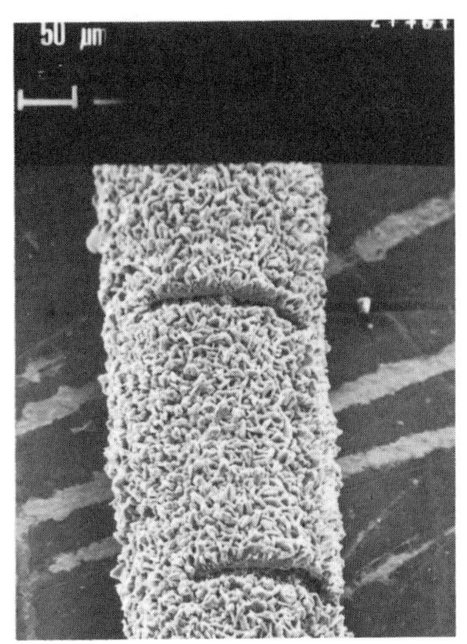

FIG. 6   Stress corrosion, x = 0.35
G = 150 daN.mm^{-2}

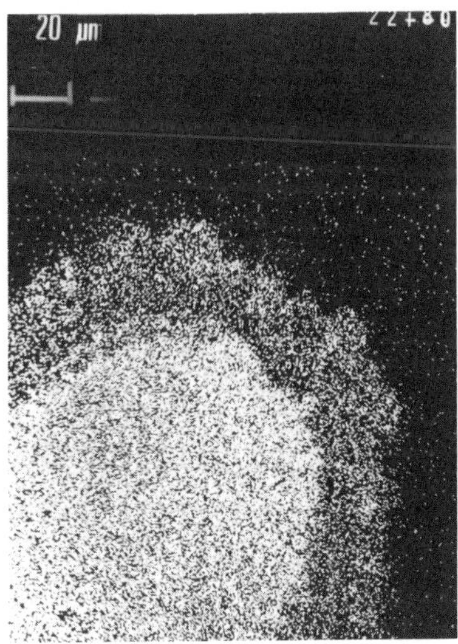

FIG. 7   Corrosion at rest
x = 0.30
chromium spectrum

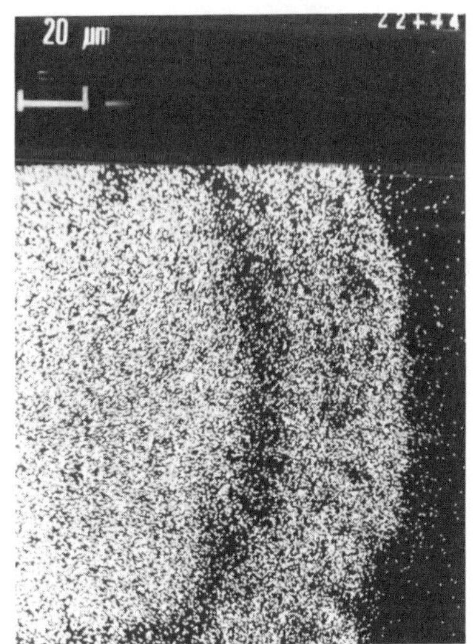

FIG. 8   Stress corrosion, x = 0.30
chromium spectrum
G = 150 daN.mm^{-2}

DISCUSSION

R. Rolls: In the macro-cracking of the scale on the sulphidated
FeCrAl wire specimens, did you observe a large number of small
cracks or a small number of large cracks? It would seem that if
the small cracks were formed then these would assist in a self-
healing after stress relaxation which might favour a more
protective behaviour in addition to the plasticity of the FeS
phase as Toesca suspected.

S. Toesca: We have observed two sorts of cracks: microcracks,
which always appear in large number, during the whole duration
of the corrosion process; and large cracks which appear just
before the rupture of the samples. This last type of cracks is
very difficult to observe. Indeed, we think too that the microcracks assist in a self-healing of the scale and favour its more
protective behaviour.

H.J. Grabke: Toesca has presented a very nice method to measure
corrosion and creep simultaneously. However, I doubt if in the
case of sulfidation he can deduce reasonable imformation on the
corrosion kinetics from the measurement of electrical conductivity of the sulfidized specimen. The electrical conductivity
of the formed sulfides is very high, and should be considered.
Could you do this? The sulfides have metallic conductivity.
The method would be better applicable to studies of corrosion
and creep under formation of insulating or semi-conducting
oxides.

S. Toesca: Of course, this method is not convenient to measure
corrosion rate and creep in any case. It gives good results only
when the global conductivity of the formed products is high, compared with the one of the remaining alloy or metal. This conductivity has always been considered. Generally, temperature and reaction extent must not be too high, many oxides and sulphides being
semiconductors. The ideal case, difficult to encounter, is that
of insulating products.

THE INTERACION BETWEEN MECHANICAL STRESS AND GASEOUS

CARBURISATION AT HIGH TEMPERATURES

H.J. Penkalla[*], V. Guttmann, J. Timm

CEC, JRC, Petten Establishment, The Netherlands

[*]Institut für Reaktorwerkstoffe, KFA Jülich,F.R.Germany

INTRODUCTION

The technical application of high temperature alloys is mostly
characterised by a simultaneous mechanical and chemical exposure.
The chemical attack by reactive environments not only leads to
material wastage but it can also cause remarkable structural
changes due to various reactions such as oxidation, carburisation
etc. These processes of course are accompanied by changes in
mechanical properties. A special problem making the quantitative
determination of the material behaviour under the influence of an
aggressive environment difficult is the fact that not only the
mechanical properties depend on the corrosive attack but in turn
the chemical reactions can be influenced by the applied stress.
Therefore results gained in pure corrosion tests carried out on
unstressed samples are not necessarily fully transferable to the
state of exposure in a stressed condition, especially if the long
term behaviour is considered. Sometimes the influence of deform
ation on the corrosive attack is investigated by exposing
predeformed material. This method of course can give only an
indication about the influence of a predeformed structure and it
is certainly not typical of the situation where corrosion occurs
during deformation itself.

This work presents results gained in an investigation of the
carburisation behaviour of different high temperature steels in an
unstressed condition compared to their behaviour under creep
deformation.

EXPERIMENTAL PROCEDURE

The experiments have been carried out on two high temperature steels. Three further steels have been used for carburisation tests in the unstressed condition only. The compositions of all five alloys are given in Tab. 1. The two HK 40 type steels and the 40/28/4 Alloy were tested in the as received centrifugally cast condition. Alloy 800 H and Alloy 802 were solution annealed at 1150°C/1h and 1220°C/1h, respectively, with subsequent water-quenching.

All tests have been performed in a purely carburising environment of about 1% CH_4 and 99% H_2 with a carbon activity of a_c = 0.8 at the test temperature of 1000°C.

Carburisation tests were carried out on all alloys using unstressed specimens of 12 x 12 x 12 mm. In order to determine the influence of the specimen size on carburisation, rod shaped specimens of different diameters have been used.

For experiments carried out to compare the behaviour of stressed and unstressed material, respectively, the low carbon version of HK 40 (HK 40 (A)) and the Alloy 800 H steel have been chosen. Creep tests were performed on specimens with a gauge length of 50 mm and a diameter of 9 mm. Reference samples in the unstressed condition were of the same diameter. For all specimens the same surface treatment has been applied using a final grinding by a 800 grit SiC paper.

Carburisation progress has been determined by different methods. For samples which were tested only in the unstressed condition both the weight gain and the depth of the carburised zone have been determined. Carburisation of samples in the stressed and unstressed reference conditions has been compared by using depth measurements only. In some cases the carbon uptake has been determined also by chemical analysis of thin layers which have been machined from the sample surface in steps of 0.1 mm.

In the case of depth measurement the relation:

$$\Delta d_c = \frac{d_c(\sigma) - d_c(0)}{d_c(0)} \cdot 100\%$$

has been used, where $d_c(\sigma)$ is the depth of the carburised layer under stress and $d_c(0)$ that for the unstressed reference condition.

In order to separate the M_7C_3 outside layer and the $M_{23}C_6$ inside layer a two step etching process has been applied. Fig. 1 shows an example of the etching of HK 40 (A) and Alloy 800 H. After short etching times only the $M_{23}C_6$ carbides can be detected whereas both types of carbides become visible after longer etching times. Whereas the boundary between the M_7C_3 and the $M_{23}C_6$ containing layer is a very sharp one for HK 40 (A) it is less exactly pronounced for Alloy 800 H which still shows $M_{23}C_6$ islands in the M_7C_3 zone. For both alloys, but in particular for Alloy 800 H, an exact determination of the total depth of the carburised layer became difficult and therefore most of the measurements have been restricted to the M_7C_3 zone determination.

RESULTS AND DISCUSSION

CARBURISATION OF UNSTRESSED SAMPLES

The results gained on the carbon uptake of unstressed samples of cubic shape are shown in Fig. 2. The weight gain per unit area w can be described by

$$w^n = k_p . t \qquad (1)$$

where k_p is the rate constant and t the exposure time. The exponent n has been found to be between 1.5 and 1.8. Thus it deviates slightly from 2. A value of 2 would confirm the parabolic rate law which has been reported for similar investigations (1,2) and which is normally observed for the condition where strongly protective reaction products cover the surface. No correction has been made in the present case for the reduction of the effective exposure area which decreases with exposure time. Such a correction would lead to slightly smaller n-values.

From Fig. 2 the further conclusion can be drawn that the rate constant becomes smaller when the Ni-content of the alloy increases. Comparing the Alloy 800 group steels and the Alloy 40/28/4 the higher Si-content of the latter has also to be taken into account. Both elements, Ni as well as Si are reported to decrease the carburisation rate (2,3,4,5).

The present investigations have not shown the incubation periods which have been observed for similar carburisation tests (1) although this feature has not been confirmed in other investigations (2).

The linear relation between weight gain and time shown in a log/log scale (Fig. 2) already suggests in the first approximation that there is no incubation period. This has been verified by some short term experiments, Fig. 3. After the shortest exposure time of 4 h a weight gain is observed. Moreover the extrapolation through

the origin shows that also below 4 h of exposure no incubation
period is to be expected within the experimental limits.

A further method to describe the corrosion rate during carburisation is given by the determination of the depth of the carburised
layer. An example for HK 40 (A) and Alloy 800 H is given in <u>Fig. 4.</u>
For the latter only the depth of the M_7C_3 zone is presented.
Similarly to the results gained on the weight increase, the growth
rate of the carburised zone can be expressed by an equation
similar to equation (1), i.e.

$$d^n = k_p \cdot t \qquad (2)$$

where d is the zone depth. For HK 40 (A) the value n is 1,4 for
both the $M_{23}C_6$ and the M_7C_3 zones. For the M_7C_3 containing layer
of Alloy 800 H n = 1,2. Thus the values are comparable to those
observed for the kinetic weight gain (n for HK 40 (A) = 1,8 and
n for Alloy 800 H = 1,5) especially if it is taken into account
that the weight gain data should be somewhat smaller if the
reduction of the net area is taken into account.

The exponent deviates from the value of 2 which has been found for
internal oxidation (6) and which has been proposed also for
carburisation based on the similarity of both processes (1).

The use of specimens with a variation in geometrical shape can lead
to remarkable differences in the carbon uptake due to differences
in the change of the net area of exposure during testing. This has
to be taken into account for instance when mechanical testing is
performed on specimens with different cross sections. In this case
the comparison of test results can become extremely difficult. An
example of the influence of the diameter of rod shaped specimens of
HK 40 (A) on the carburisation behaviour is given in <u>Fig. 5.</u> As is
to be expected carburisation proceeds faster with decreasing sample
radius (R), being directly proportional to 1/R. Such a dependence
can be expected from the diffusion equation which in the case of a
diffusion field of radial symmetry about a cylindrical axis becomes

$$\frac{dc}{dt} = D \left(\frac{d^2 c}{dR^2} + \frac{1}{R} \cdot \frac{dc}{dR} \right) \qquad (3)$$

Moreover, <u>Fig. 5</u> shows that the ratio between the thickness of the
M_7C_3 containing layer and the total depth of the carburised layer
(i.e. including the inside $M_{23}C_6$ zone) remains nearly constant
being about 0.5.

<u>CARBURISATION OF STRESSED SAMPLES</u>

The results showing the influence of a creep deformation on the carburisation behaviour are given in Fig. 6. As already pointed out the most reliable results are gained by measuring the depth of the M_7C_3 containing layer. Initial tests (Fig. 4) had shown that both the M_7C_3 and the $M_{23}C_6$ layer exhibit the same growth tendency and therefore the indication has been made mainly by means of the M_7C_3 depth determination.

Fig. 6 shows that in the stress range employed carburisation resistance of HK 40 (A) is found to be higher when tested under stress. For one stress level ($\sigma = 16$ Nmm^{-2}) the effect has been determined for three different exposure times. No significant time dependence has been observed. Probably this will hold for other stresses as well, as would be expected from the fact that there is no break between the 200 h-exposure curve and the 50 h-curve, gained at low and high stresses, respectively.

For Alloy 800 H results similar to those for HK 40 (A) have been observed in the low stress region, i.e. the stressed samples show the higher carburisation resistance. At higher stresses the difference in carbon uptake becomes smaller and finally it disappears.

A chemical analyses of cylindrical layers has been carried out for HK 40 (A) samples after 100 h and 200 h exposure time under a stress of $\sigma = 16$ Nmm^{-2} and for the associated stress free samples, respectively. The plot of carbon content against penetration depth, Fig. 7, exhibits parallel curves in the radial direction for both material conditions with a strong slope change at the M_7C_3 - $M_{23}C_7$ boundary. The carbon content of the stressed sample is always lower than that of the unstressed one, measured at the same radial distance from the surface. It will be seen that the scatter of individual points about each curve is small. Additionally, the difference between the lines remains almost constant along the radius and it does not change with time.

No simple explanation can be given for the differences in carburisation rate between the deformed and the undeformed material condition. From the theoretical view point a plastic deformation can lead either to an increase or a decrease of diffusion rate. There are two mechanisms thought to be responsible for an increase of diffusibility; the dislocation core diffusion process and the dislocation sweeping effect. The latter has been worked out especially for hydrogen in iron and steels (7). It describes the transport of foreign atoms by mobile dislocations which enables the diffusion rate to exceed that in the dislocation free lattice. On the other hand the diffusion of impurities in a deformed material can be smaller compared to that in undeformed material as has been reported for instance for hydrogen diffusion in iron and steels (8,9) at ambient temperature and for carbon in austenitic

steel at temperatures round about 1000°C (10). This effect can be
explained by a trapping mechanism (11,12) which is due to material
defects as dislocations and microvoids formed during deformation
and which are able to attract the diffusing impurities.

Although it would seem that the trapping theory could be applicable
to explain the present investigations (at least when the high
stress region for Alloy 800 H is excluded) the mechanism is not
consistent with all results. The main point is that it is not in
agreementwwith the carbon profiles shown in Fig. 7 because
trapping should not only reduce the progress of the carburisation
front but consequently it should lead to an increase of the local
carbon content. This however has not been observed. In contrast,
as documented in Fig. 7 , the decrease of carbon diffusion rate
has been found to be combined with a decrease in carbon content
regarding the same distance from the sample surface.

From the fact that the difference in the carbon content between the
stressed and the unstressed condition remains nearly the same all
over the carburised zone, it can be concluded that the carburisation rate does not differ during the whole exposure time. The
results rather indicate that a difference in carbon uptake is
restricted to the starting phase, i.e. carbon uptake for the
stressed material is blocked or at least slowed down at the beginning of the test, but with progressing time the diffusibility in
both materials becomes the same.

There is no doubt that dislocations formed during deformation will
have an influence on carbide distribution. Dislocations act strongly as nucleation sites as has been observed for the generation of
primary carbides in the uncarburised material, Fig. 8 (13).
In which way the dislocations and the associated carbides can
reduce carbon uptake at the onset of exposure is still unclear.
Two mechanisms which seem to offer a suitable basis of explanation
are built up on the basic approach that at the beginning the
deformed material is characterised by finer carbide particles and
eventually by a higher carbon uptake. The latter assumption need
not be in contrast to the results presented in Fig. 7 if only a
very thin surface layer is considered. In this case the chemical
analyses would not separate this layer from deeper parts included
in the first 0.1 mm cut.

Considering, firstly, the situation where both features apply, one
would expect some barrier effect due to the higher carbide content
because the cross section remaining for carbon penetration is
reduced (in comparison to the matrix the diffusibility in the carbides can be considered as negligible).

A second mechanism to be proposed can be based on a finer carbide distribution, thus preventing any discrepancy with the chemical analyses. It is based on the general thermodynamical effect that for smaller carbide particles the carbon concentration in the matrix will be higher than for coarse particles. This of course will lead to a slower carbon uptake.

With increasing exposure time the mentioned mechanisms should become progressively less important. This can be explained by significant structural changes due to ripening processes. The finer carbides will be transformed into coarser ones and the diffusion rate will come to be governed only by the relatively thick carburised zone of coagulated carbides. Moreover the difference in dislocation density between the unstressed and the stressed sample becomes smaller during exposure. This is due to the fact that carbide formation leads to tensile stresses in the "unstressed" sample because the carbides have a higher specific volume than the matrix. Thus the differences in carbide distribution at the carburisation front must become smaller.

The ripening process will also allow an explanation for the reduction in difference of carburisation rate at higher stresses, as shown in Fig. 8 for Alloy 800 H. At higher stresses i.e. at high dislocation densities coagulation becomes a relatively fast process (13). Probably it can be fast enough to avoid any finer carbide distribution at the beginning of exposure. Hence the differences between the unstressed and the stressed condition will become less relevant and finally they should disappear.

In addition to the foregoing proposed mechanisms which are based on the carbide nucleation at dislocations it seems worthwhile to consider a third alternative which is based purely on dislocation dynamics and which is directly related to the sweeping mechanism.

Samples of HK 40 (A) and Alloy 800 H both in predeformed and undeformed conditions were exposed without external stress. Hardly any difference in carburisation behaviour has been observed in this case. This of course leads to the conclusion that moving dislocations instead of sessile ones are responsible for the observed differences between unstressed and creep stressed material. The relevance of gliding dislocations could be explained in the following way. During the onset of deformation,dislocations will leave the sample through the surface. This egress effect can be accompanied by a certain force opposing the inward diffusion process of carbon, depending on the capability of the dislocations to attract the C atoms. Such an outgassing effect has been confirmed for instance for hydrogen in steel (14,15). This effect of course will not be very strong at the high temperature but nevertheless it should exist. The egress of dislocations and hence

the outward force should become reduced at high stresses because
of the formation of networks, cell walls and subgrain boundaries
which all act as barriers for long path dislocation movement. Thus
the results for Alloy 800 H in Fig. 6 are in accordance with the
TEM findings that no cells or subgrains are formed in HK 40 (A)
whereas they exist in Alloy 800 H (13).

FINAL REMARKS

Different theories have been proposed to explain the observation
that carburisation in the high temperature steels HK 40 and
Alloy 800 H is different in the unstressed and the stressed conditions. All foregoing discussed mechanisms must be considered at
the moment as theoretically possible but no direct confirmation
can be presented now. From the results available it seems that
short term tests of only a few hours are necessary which would
enable clearer information to be obtained about the processes
which are active during the onset of carburisation.

REFERENCES

1)   A. Schnaas und H.J. Grabke, Werkstoffe und Korrosion,
     1978 (29) p. 635

2)   J.M. Harrison, J.F. Norton, R.T. Derricott and J.B. Marriott,
     Werkstoffe und Korrosion, 1979 (30) p. 785

3)   W. Steinkusch, Werkstoffe und Korrosion, 1977 (28) p. 1

4)   W. Steinkusch, Werkstoffe und Korrosion, 1979, p. 837

5)   H.J. Grabke, U. Gravenhorst und W. Steinkusch, Werkstoffe
     und Korrosion, 1976 (27) p. 291

6)   J. Kapteijn, Z. Metallkunde, 1974 (65) p. 157

7)   J.K. Tien, N.F. Panayotou, R.J. Richards, Second Inter-
     national Congress on Hydrogen in Metal , 6-11.VI, 1977
     Paris - France

8)   J.H. Andrew, K.V. Bhat and H.K. Lloyd, J. of the Iron and
     Steel Inst., Aug. 1950, p. 382

9)   G.M. Evans and E.C. Rollason, J. of the Iron and Steel Inst.
     Nov. 1969, p. 1484

10)   I.N. Kidin, G.V. Shcherbedinskii, V.I. Andryushechkin and
      V.A. Volkov, Met. Sci. Heat Treat. 1971, vol. 13, p. 1021

11)   J.F. Newman and L.L. Shreir, J. of the Iron and Steel Inst.,
      Oct. 1969, p. 1369

12)   J.P. Hirth, Met. Trans. A, 1980, vol. 10A, p. 753

13)   V. Guttmann and R. Bürgel, Joint Research Centre, Petten
      (The Netherlands), unpublished research, 1980

14)   M.R. Louthan, S.R. Caskey, J.A. Donovan and D.E. Rawl,
      Mat. Sci. Eng., 1972 (10) p. 357

15)   J.A. Donovan, Met. Trans. A, 1976, vol. 7A, p.145

TABLE I

	C	Si	Mn	Cr	Ni	Al	Ti	N	Fe	N	P	S
HK 40 (A)	0.30	1.36	0.35	24.05	20.50	0.01		0.154	balance	0.154	0.017	<0.010
HK 40 (B)	0.46	1.46	0.26	24.70	21.00				balance			
Alloy 800 H	0.06	0.37	0.66	20.05	31.35	0.28	0.48	0.01	balance	0.01		
Alloy 802	0.33	0.37	0.85	20.45	31.20	0.27	0.66	<0.01	balance	<0.01	0.027	<0.01
Alloy 40/28/4	0.36	1.91	0.78	26.75	40.35				24.50			

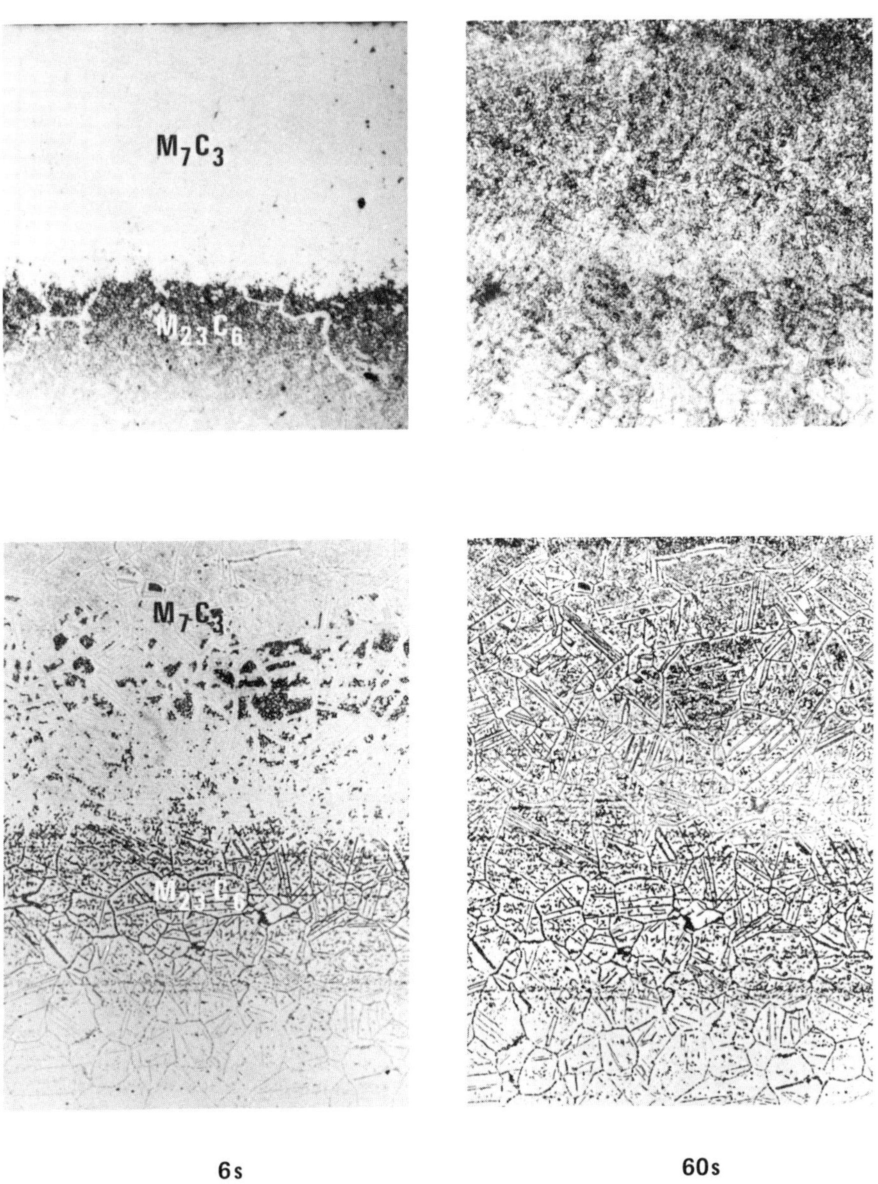

Fig. 1 Identification of carbides by MURAKAMI-etching.
(top: HK40 (A); bottom: Alloy 800H).

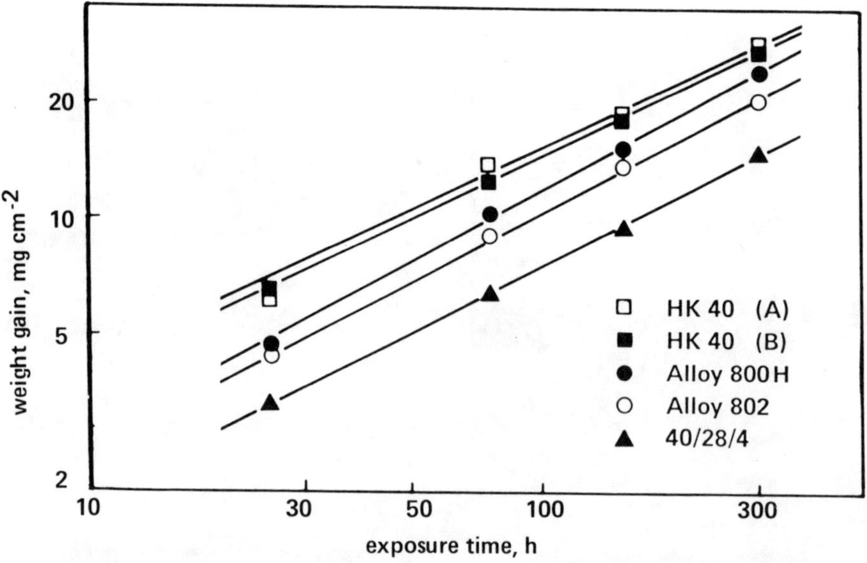

Fig. 2   Kinetic weight gain on different alloys exposed to carburising environment.

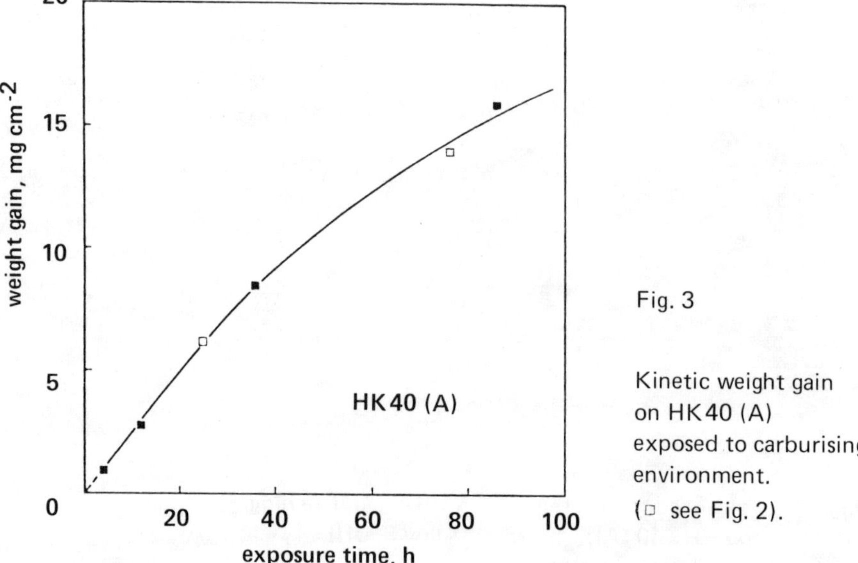

Fig. 3

Kinetic weight gain on HK 40 (A) exposed to carburising environment. (□ see Fig. 2).

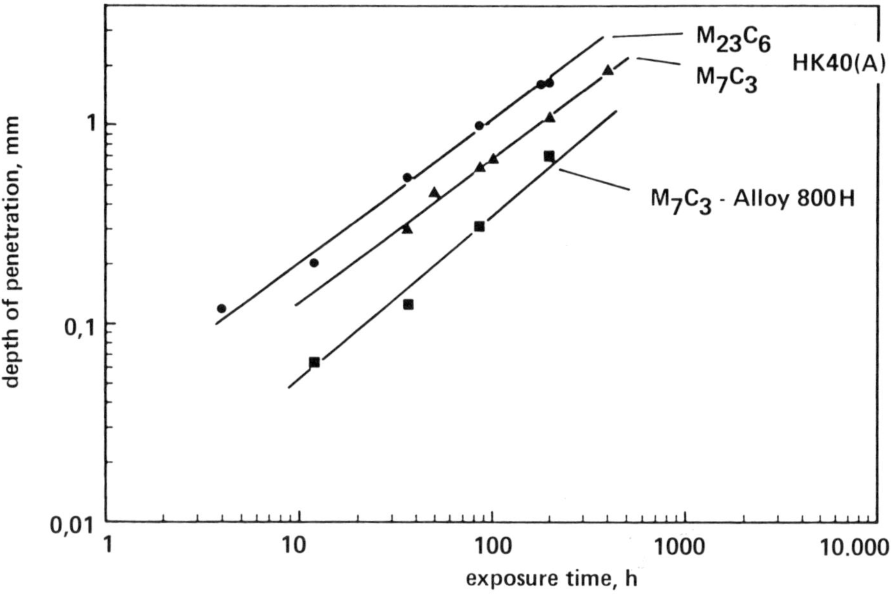

Fig. 4   Effect of exposure time on carburisation depth.

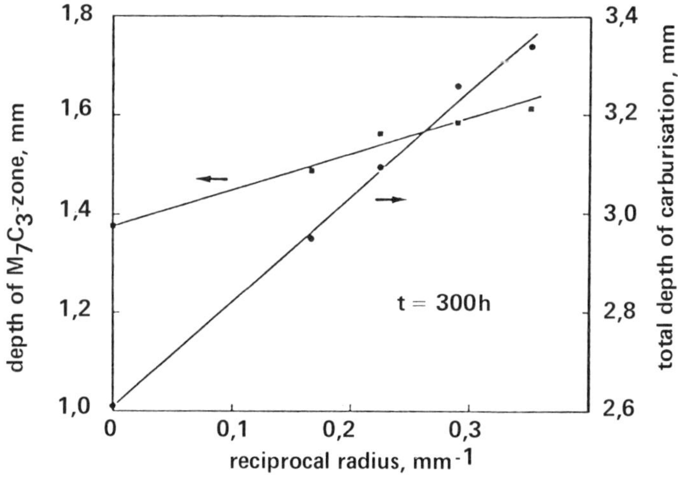

Fig. 5   Effect of specimen geometry on carburisation depth.

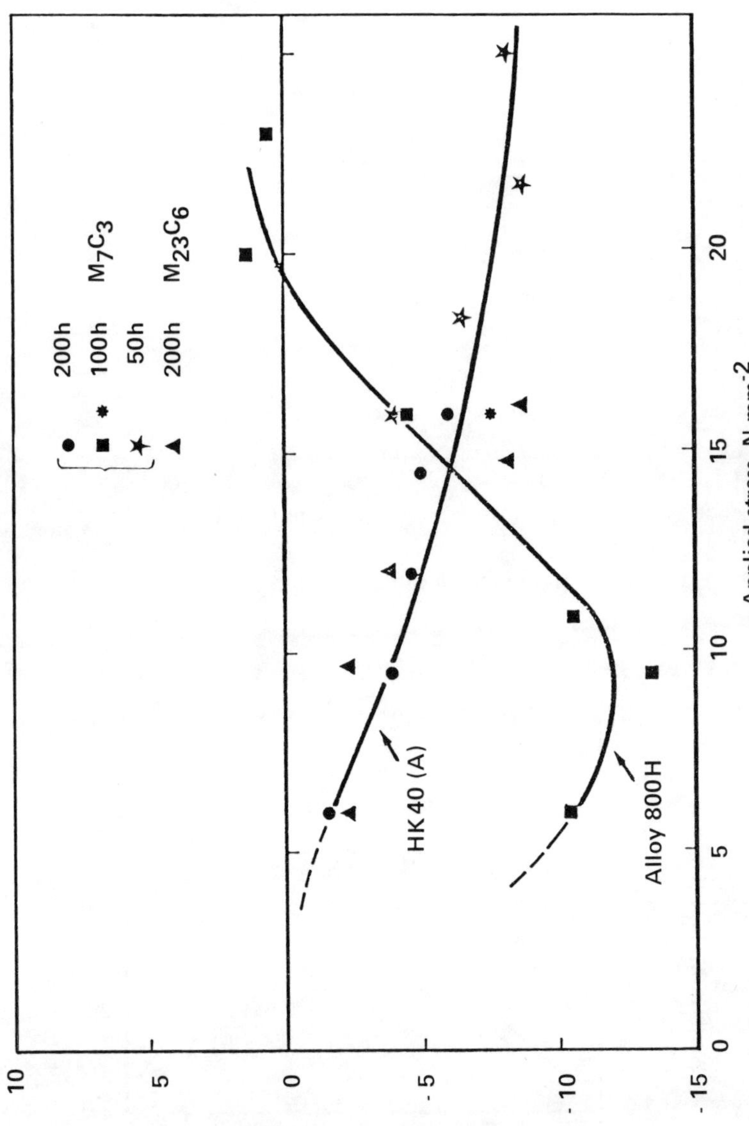

Fig. 6   Effect of creep stress on related difference of carburisation depth.

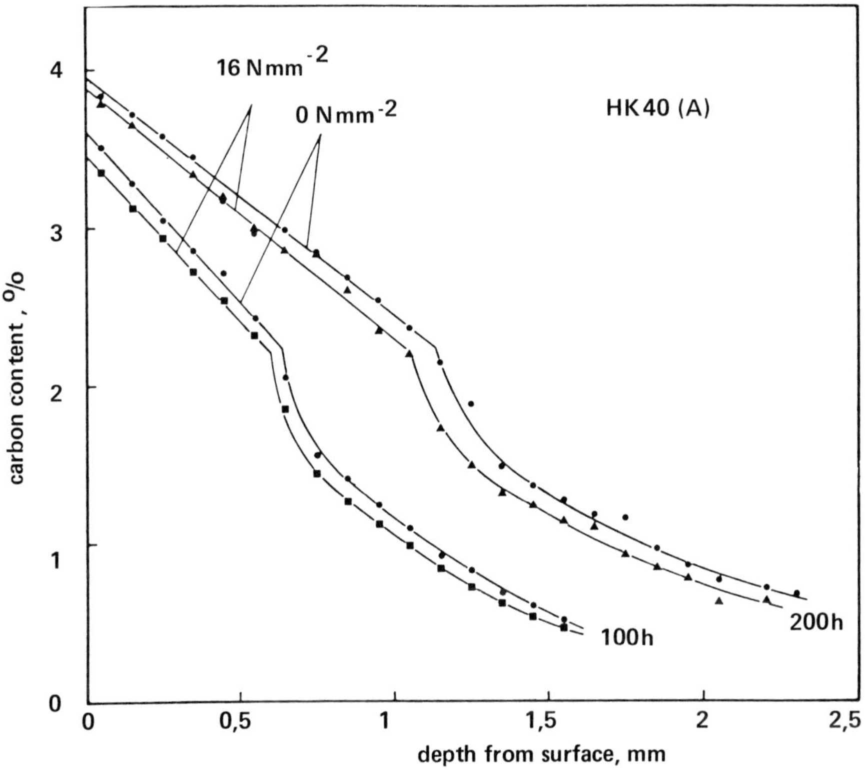

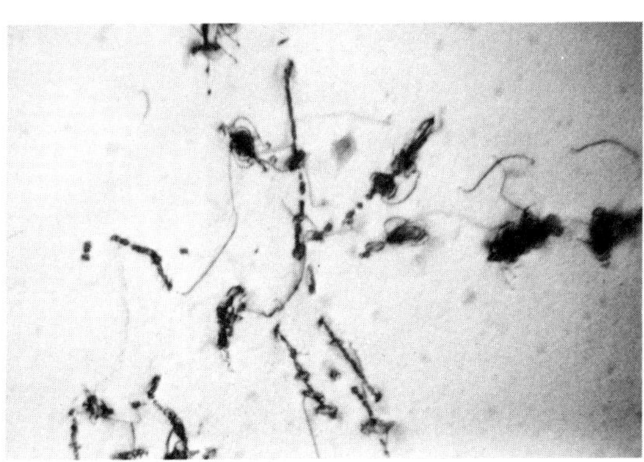

Fig. 7   Carbon concentration profiles after different testing conditions.

Fig. 8

Carbide formation
at dislocations
(Alloy 800 H).

DISCUSSION

H.J. Grabke: In the paper presented by Penkalla, I miss any attempt to explain the effect of creep on carburisation rate. The retardation of carburisation by the creep is against any expect-- ation. The diffusion of carbon into the samples should be accelerated by the mechanisms of creep.

H. Penkalla: I cannot fully agree that a creep deformation process should always increase the diffusion rate of carbon instead of de- creasing it. Possible exeptions related to the similar case of hydrogen diffusion at lower temperatures are mentioned in the paper, although these mechanims do not seem to be applicable to our results. A satisfactory atomistic interpretation of our investigations cannot be given for the moment and more results will be necessary to establish a complete theory.

EMBRITTLEMENT OF STAINLESS STEEL BY CONTAMINATION

WITH LIQUID Zn

R.de Backer[+], A.Dhooge[++], P.Moyaert[+], M.Victoria[X], A.Vinckier[XX]

Laboratory for Strength of Materials and Welding Technology

St.Pietersnieuwstraat 41, 9000 Gent, Belgium.

ABSTRACT

To study the effect of the chemical composition on the liquid
zinc embrittlement of stainless steels, high temperature tensile
tests were performed on 304 and 321 type stainless steels in
contact with zinc. The effect of temperature, stress and strain
on the intergranular attack have been determined. In particular
the minimum levels of stress and plastic strain and the lowest
temperature to be attained for the attack to start has been
established.

An effort has been made to evaluate tho effects of grain size,
in seperating those due purely to a change in the susceptibility
to embrittlement from those due to a change in mechanical
properties.

The results are discussed in terms of a model for grain
boundary penetration of liquid Zn.

+   Graduate students, University of Ghent.
++   Research Engineer, Research Center of the Belgian Welding
     Institute.
X   Research Assistant, University of Ghent.
XX   Associate Professor, University of Ghent.

1. INTRODUCTION

The embrittling effect caused by exposing stainless steel into
contact with liquid zinc has been studied amongst other
authors, by Rädeker [1] and Hebsleb and Schwenk [2]. Under
certain conditions of stress and temperature both the strain
to fracture and the tensile strength are found to be sharply
reduced.

The practical importance of this type of behaviour has been
highlighted in recent years by reports of a number of
failures in plant components and particularly by that of the
Flixborough disaster in England in 1974 [3].

This paper describes the preliminary results of a program on
the embrittlement of stainless steel by various liquid
metals at present being conducted at the University of Ghent.

2. EXPERIMENTAL PROCEDURE

Tensile specimens of 6 mm and 12 mm gage diameter were cut
from two types of stainless steels : 321 H and 304 L. Their
respective chemical analysis are given in Table I. Specimens
with various grain sizes were obtained by annealing treatments
for periods up to 1 1/2 hour at 1200°C under pure argon.

All testing was performed in a "Hot ductility" machine. The
specimens are heated under controlled and programmable
conditions by running an electrical current through them.
Temperature is controlled through a thermocouple welded to
the center of the gage length. This procedure provides a
constant temperature, within $\pm$ 5°C, at the center of a gage
length approximately 15 mm long. In order to obtain reasonably
clean surfaces for further observation all testing was done
under a pure argon atmosphere.

To obtain a good contact between the liquid Zn and the
specimen, the surface was cleaned with abrasive paper and
Zn was applied by one of the following methods :

- by applying a coat of epoxy charged with 60% in weight of
  Zn filler powder
- by wrapping a pure Zn wire around the center of the
  specimen.

During testing, the specimen is heated to the desired
temperature in a few seconds, then a constant load is
applied and the time to fracture and the reduction in area
is measured. If no fracture occurs after one hour of loading,
then the test is interrupted and the metallographical cross
sections are examined for cracking under the optical microscope.

3. EXPERIMENTAL RESULTS
3.1. Stainless steel type 321 H

Figures 1a and 1b show schematically the test results for two grain sizes. The numbers at the measuring points indicate reduction in area. In figure 1a the grain size is ASTM 8 to 9, while in figures 1b grain size is ASTM 2 to 3. The reductions in area at fracture observed in the material with larger grain size are generally greater.

The test results are plotted in figure 2, showing that, at same temperatures the time to failure under constant load increases with decreasing load. In figure 3, values of the time to failure as a function of the applied stress at a constant temperature are shown. No cracks could be detected metallographically close to the surface of specimens tested at 600°C, even after prolongued testing (1 hr.) at 300 N/mm^2. An equivalent threshold in stress can be extrapolated from the data shown in figure 3 which should be below 85 N/mm^2. At this stress, the specimen did not fail after 1 hr of testing at 700°C, but showed surface cracks.

In figure 4, the test values corresponding to the two grain sizes have been superimposed. At the same temperature and stress, the smaller grain size specimens require longer times to fracture. The curves tend to converge for times longer than 100 sec.

3.2. Stainless steel type 304 L

An overall view of the test results is shown in figure 5. It can be noticed that the reductions at fracture are in general lower than those of the 321 steel. This higher degree of embrittlement is illustrated more clearly in figure 6, where the time to fracture is plotted as a function of temperature for various stress levels. For each stress level, there is a temperature region for which the specimen fails almost immediately with only minor plastic deformation. If the test temperature is decreased, a behaviour similar to that seen in the 321 steel is observed, in that a certain amount of plastic strain occurs before the specimen fails. In order to confirm this behaviour, specimens with a notch in the center of the gage section were tested and the results are shown in figure 7. For the same level of stress, the notched specimens cracked immediately at temperatures where unnotched specimens,showed a reduction in area of 19% before failure. One of the notched specimens did not crack at 675°C, but its behaviour could be traced to a lack of wetting by the liquid Zn at the notchtip.

Metallographic observation of both types of steel after
fracture showed the cracks to be intergranular. This is
shown by optical microscopy in figure 8 for the case of the
304 L steel and by scanning microscopy in figure 9 for the
case of the 321 H steel.

4. <u>DISCUSSION OF RESULTS</u>

In the results obtained so far certain characteristics of the
embrittlement of stainless steel by liquid Zn are apparent :

- no embrittlement takes place below a certain temperature,
  which is in the vicinity of 650°C.
- A minimum tensile stress is necessary. In the case of the
  321 H steel this value is approximately 80 N/mm^2.
- Although the data are still incomplete, larger grain sizes
  seem to be more sensitive to embrittlement.
- In the lower temperature region (below 750°) a certain
  amount of plastic deformation takes place before cracking.

This characteristics are similar to those described in the
literature [2,4] for steels embrittled by liquid metals.

In the particular case of stainless steel, the data are in good
agreement with the mechanism proposed by Bastien et al [5,6]
These authors have shown that the temperature at which
embrittlement is observed coïncides with a change of the
diffusion mode. At 750°C only the interdiffusion of Zn and
Ni is important, Cr and Fe behave as more or less inert
solvents. The Zn diffuses intergranularly and the higher
diffusion rate of Ni in Zn depletes of Ni the immediate
vicinity of the grain boundary. In this region impoverished
in Ni, a $\gamma \rightarrow \alpha$ transformation takes place, which together with
the formation of 3 Zn Ni in the grain boundary is responsible
for the embrittlement.

Evidence for this model is shown in figure 10 obtained with the
electron microprobe. Figure 10a shows the image of a secondary
crack while figure 10b shows the Zn K$_\alpha$ image of the same region.
Figure 10c shows the results of various line scannings in this
region and the results are schematized in figure 10d. Although
no quantitative analysis is offered, the relative distributions
of Zn, Ni, Fe and Cr seem to correlate well with the
mechanism described above.

ACKNOWLEDGMENTS

The present work as part of the Program of Embrittlement of
stainless steels by liquid metals, has been funded by IWONL,
whose assistance is gratefully acknowledged.

REFERENCES

1. W. Rädeker : Stahl und Eisen (1953) 73, 654.
2. G. Herbsleb and W. Schwenk : Werkstoffe und Korrosion (1977)
   28, 145.
3. The Flixborough Disaster : Report of the Court of Inquiry,
   Her Majesty's Stationery Office London, 1975.
4. W. Rostoker, J.M. Mc Caughey and H. Markus : Embrittlement
   by Liquid Metals (1960) New York Reinhold Publishing Corp.
5. M. Andreani, P. Azon and P. Bastien : C.R. Acad. Sc. Paris
   (1966) 263, 1041.
6. M. Andreani, P. Azon and P. Bastien : Mémoires Sc. Rev.
   Métallurgie (1969), LXVI 21.

R. DE BACKER et al.

TABLE 1 : CHEMICAL ANALYSIS OF STAINLESS STEEL

Steel Type	C	Si	S	Mn	Cr	Ni	Mo	Ti	Fe
304L	0,022	0,33	0,017	1,79	18,2	9,5	0,26	-	Bal.
321H	0,075	0,3	0,02	1,6	18,3	10,1	-	0,45	Bal.

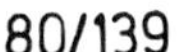

80/139

Fig:1a Stainless steel 321H (grain size ASTM 8-9).

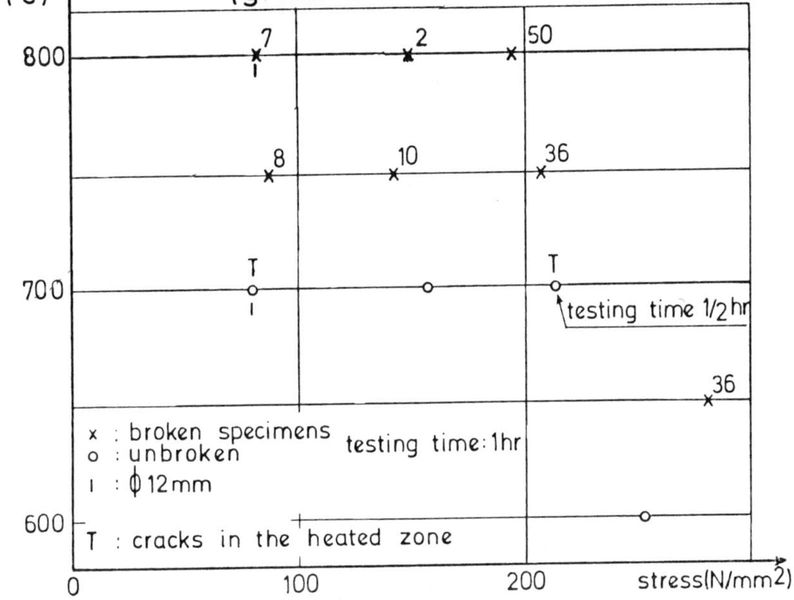

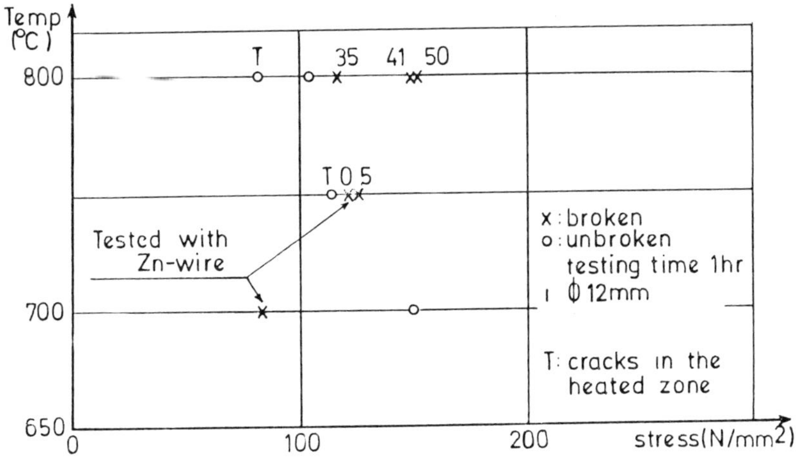

Fig: 1b Stainless steel 321H annealed (grain size ASTM 2-3).

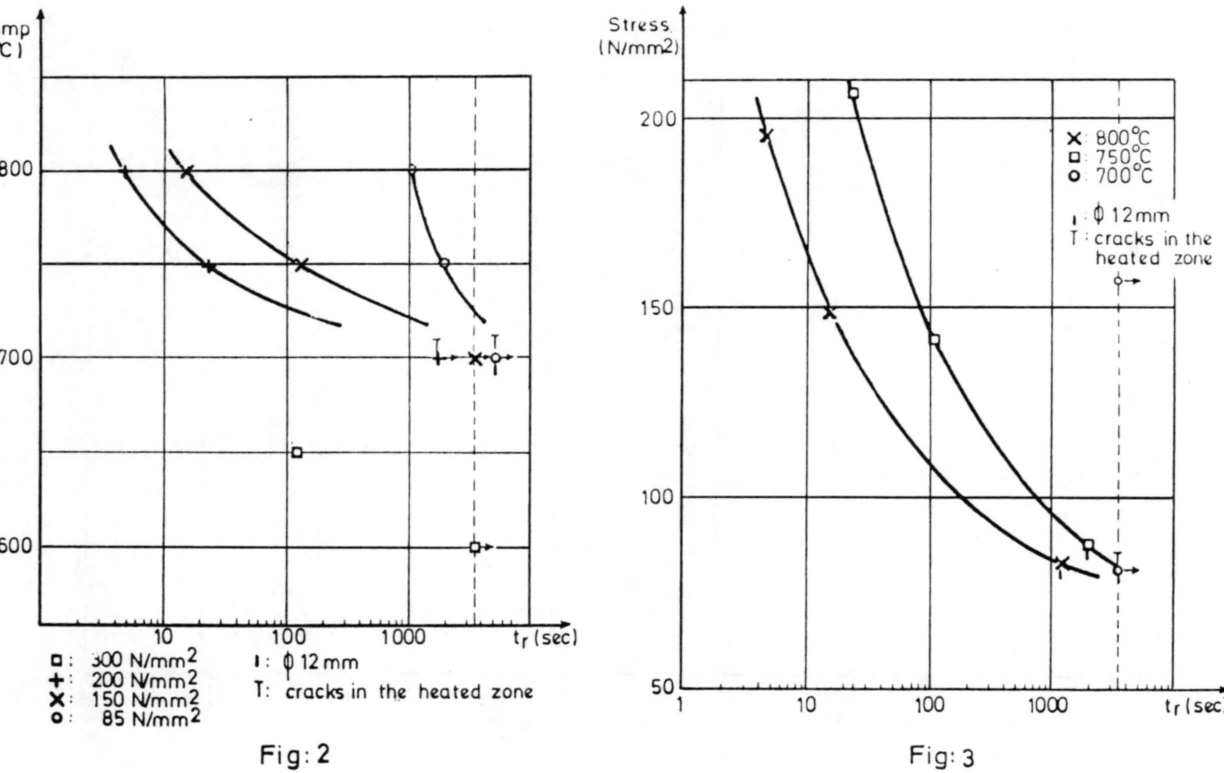

Temp (°C)
800
700
600
10
100
1000
tr (sec)
□ : 300 N/mm2
+ : 200 N/mm2
X : 150 N/mm2
o : 85 N/mm2
I : φ 12mm
T : cracks in the heated zone
Fig: 2
Stress (N/mm2)
200
150
100
50
1
10
100
1000
tr (sec)
X : 800°C
□ : 750°C
o : 700°C
I : φ 12mm
T : cracks in the heated zone
Fig: 3

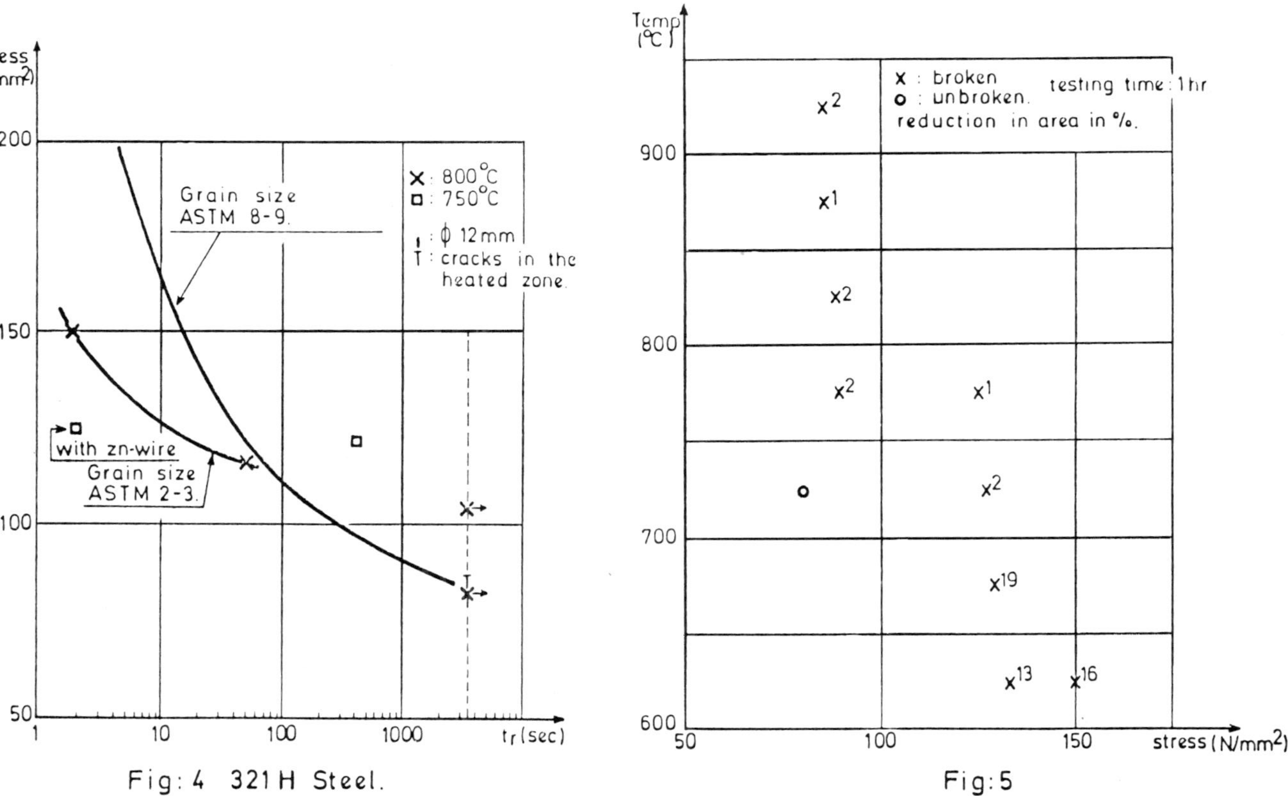

Fig:4 321H Steel.

Fig:5

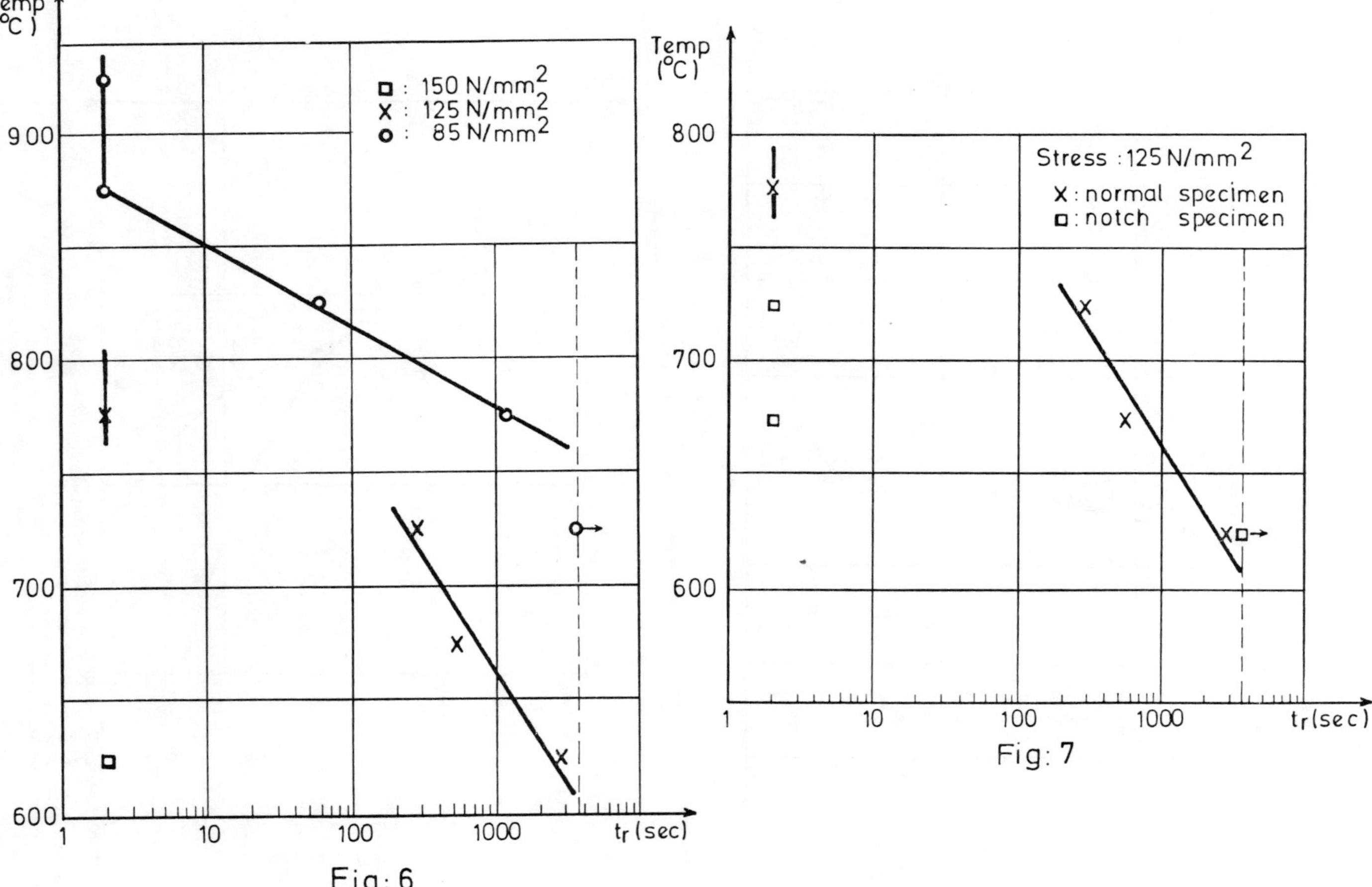
Temp
(°C)
900
800
700
600
□ : 150 N/mm²
X : 125 N/mm²
O : 85 N/mm²
1
10
100
1000
tr (sec)
Fig: 6
Temp
(°C)
800
700
600
Stress :125 N/mm²
X : normal specimen
□ : notch specimen
1
10
100
1000
tr(sec)
Fig: 7

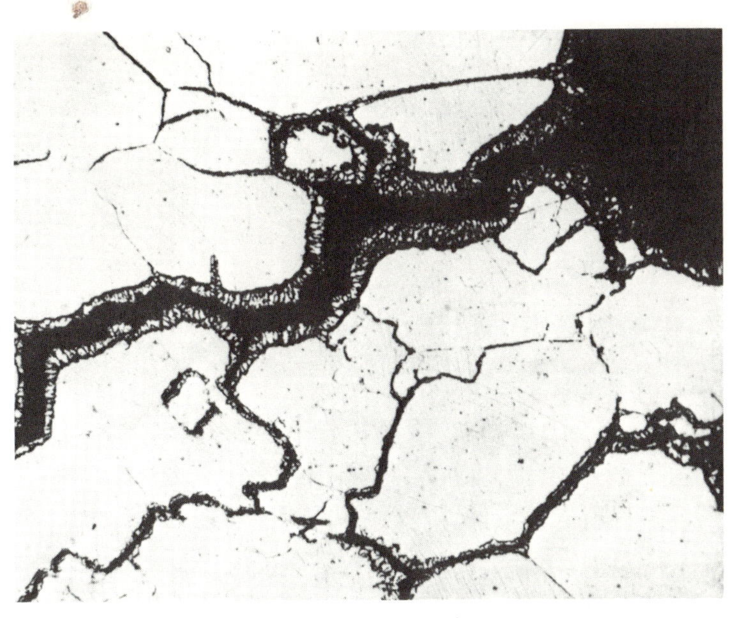

Fig: 8

Fig: 9   100 x

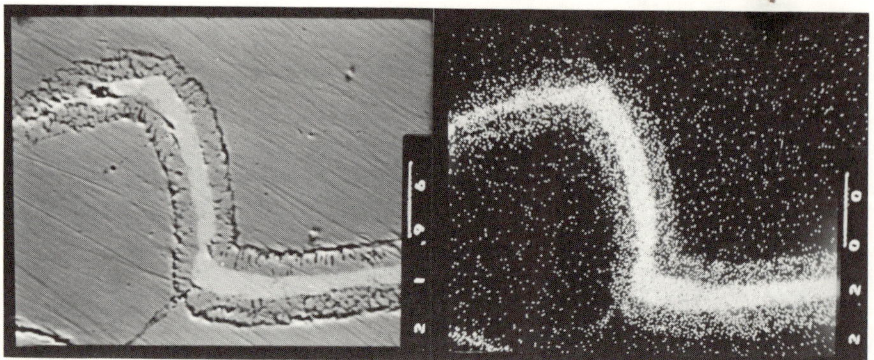

Fig: 10 a   Fig: 10 b

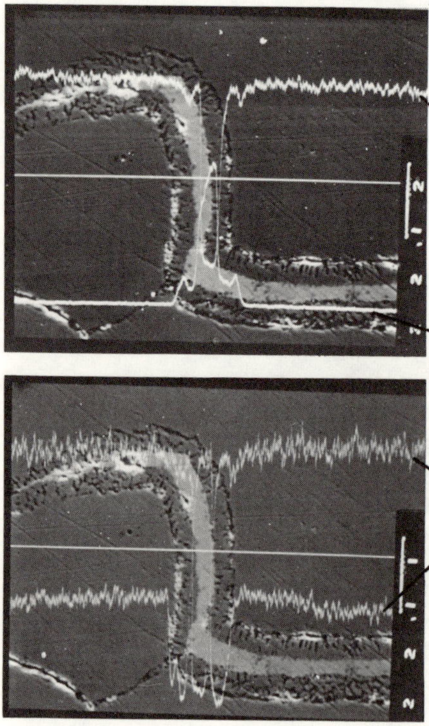

Fig: 10 c

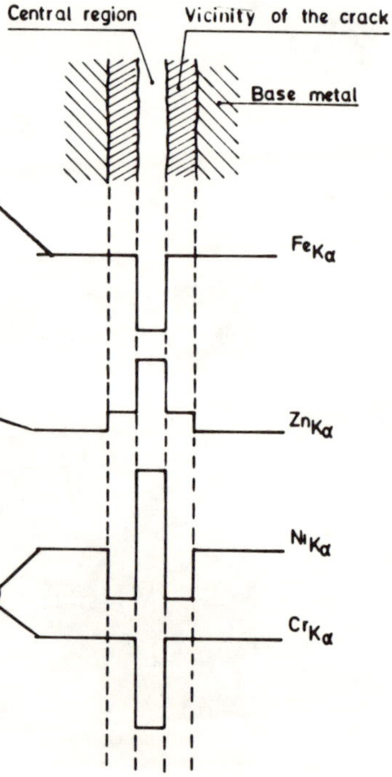

Fig: 10 d

INFLUENCE OF THERMAL CYCLING BETWEEN 20 AND 400°C

ON THE OXIDATION OF COPPER

C. CODDET, G. DE BARROS and G. BERANGER

Division Matériaux, Département de Génie Mécanique
Université de Compiègne

B.P. 233 - 60206 COMPIEGNE CEDEX - FRANCE

ABSTRACT

During periods of overheating the copper windings of high power
electric machines undergo important deteriorations due to an oxidation under undetermined conditions.

Several experimental approaches were followed in order to specify
the most unfavourable conditions for which the oxidation of copper
occurs at very high rates.

The experimental results here described were obtained by the monitoring of the acoustic emission produced during the oxidation of
copper samples both under isothermal and thermal cycling conditions.

INTRODUCTION

In case of overload, electric machines are submitted to overheating and in some circumstances, the copper windings undergo very
important deteriorations due to oxidation. The final aim of this
study is to determine the most unfavourable conditions, for which
the oxidation of copper occurs at very high rates ; however, the
work presented here is only concerned with the study of stresses
developped during the oxidation of copper and especially during
the cooling periods. The stresses generated during isothermal
oxidation or by the mismatch of thermal expansion coefficients
during temperature evolutions can be relieved by many different
processes ; some of these processes such as cracking generate
elastic waves which propagate into the material. The monitoring

C. CODDET et al.

of these elastic waves (acoustic emission) with the help of pie-
zoelectric transducers gives the means to follow the irreversible
processes occuring into the sample . [1-4]

EXPERIMENTAL TECHNIQUE

The principle of the acoustic emission technique has already been
described elsewhere [1-4], also we shall only specify here the main
parameters used in this study ; the copper samples (electrolytic
99,9 %) in the form of discs (40 mm diameter, 0.5 mm thickness)
were orthogonaly fixed to a stainless steel (Fe- 18 Ni- 10 C r) wave
guide by a screw. The signal amplification was 96 dB and we used
a high pass filter of 100 KHz cutting frequency. The oxidation
was performed in pure oxygen (N 45) in a resistor furnace allowing full control of the atmosphere and temperature programming.
The thermal cycles were the following : lower temperature : 60°C ;
upper temperature : 400°C ; linear increase and decrease of 4
hours each.

EXPERIMENTAL RESULTS

The first experiments consisted in the monitoring of the acoustic
emission produced during the cooling of a sample after an isothermal oxidation. The oxidation temperatures (550 and 750°C) were
chosen in order to enclose the temperature of 650°C which is believed to be, in the case of copper oxides, the transition temperature between elastic and plastic behavior. The oxidation durations (30 hours at 550°C and 30 mn at 750°C) were selected so as
to obtain the same oxide thickness of about 30 μm. Figure 1 shows
the acoustic emission (cumulated intensities) recorded in each case. It demonstrates clearly that if oxidation is performed at a
temperature above that of the elastic-plastic transition, the
growth stresses are relieved by plastic deformation and thus, during the cooling, only the mismatch between the thermal expansion
coefficients leads to some deteriorations of the oxide scale while, when oxidation is performed at a temperature under that of the
elastic-plastic transition, the addition of the accumulated growth
stresses and thermal stresses leads to a more important deformation of the oxide scale during the cooling.

The results obtained under thermal cycling are reported on figures 2 to 4. One can observe that after an induction period which
lasts for about 6 cycles, the acoustic emission during each cycle
increases rapidly to reach a maximum around the tenth cycle, then
it decreases in a quasi parabolic way (fig. 2 and 3). An evolution
is also observed inside each cycle (fig. 4) : as the number of cycles increases, the emission begins earlier and earlier (i.e. for
a lower and lower ΔT) but the total intensity becomes smaller and
smaller. These observations seem to indicate that at the beginning
of the oxidation, the stress level in the scale increases as the

oxidation proceeds and this up to a critical thickness and then it decreases. The same kind of behavior was already observed during stress measurements by the bending method in the case of isothermal oxidation of copper under 400°C.[5]   One can notice that in our case, the intensity of the thermal stresses was quite low (the temperature evolution was smaller than 100°C per hour) ; under those conditions, the oxide scales remain compact and adherent to the metal and moreover they present numerous CuO whyskers at the oxide-gas interface (fig. 5). Consequently, it seems that the recorded acoustic emission can be attributed to microcracks and probably to grain boundary slidings and that the thermal stresses developed during the cooling help in this case to characterize the level of the growth stresses into the scale. As already seen before in the case of the oxidations at 550 and 750°C, the acoustic emission lasts a very long time at room temperature after the last cooling (fig. 6), thus showing that the stress relaxation processes involved are time-dependant.

CONCLUSION

The use of the acoustic emission method to follow the stress relaxation processes during the oxidation of metals and alloys seem to be very appropriate : the experimental system is very simple and the samples are not disturbed by the technique. However, more results are still necessary to allow a good understanding of the phenomenons involved in order to distinguish between different mechanisms. But actually, one can hope to be able to characterize the effect of superimposed stresses on the oxidation mechanisms, and especially to specify the highest thermal evolution which can be withstanded by different materials if the protectiveness of oxide layers has to be preserved.

REFERENCES

1.   C. CODDET, J.F. CHRETIEN and G. BERANGER
     C.R. Acad. Sci., Paris (1976), t 282, série C, 815.

2.   C. CODDET, G. BERANGER and J.F. CHRETIEN
     Materials and Coatings to resist high temperature oxidation
     and corrosion, p 175 - 1978 - LONDON. Appl. Sci. Pub.

3.   J.F. CHRETIEN and N. CHRETIEN
     Non destruct. test., (1972), 8, 220.

5.   W. JAENICKE, S. LEISTIKOW, A. STADLER and L. ALBERT
     Mem. Scient. Rev. Metallurg., (1965), LXII, 5, 231.

4.   C. CODDET, J.F. CHRETIEN and G. BERANGER
     Titanium and Titanium alloys - Scientific and Technological
     aspects, p. 1097 - 1980 - NEW-YORK Plenum Pub.corp.

             C. CODDET et al.

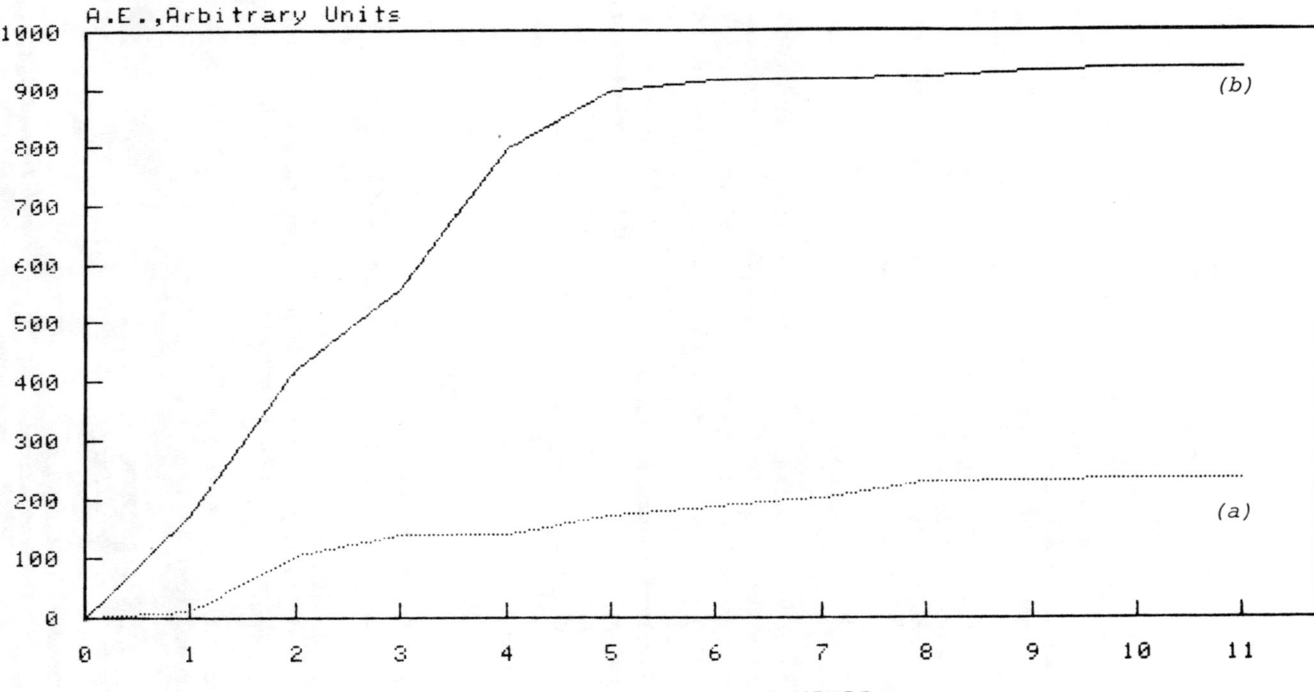

Fig. 1 - Acoustic Emission recorded during the cooling of two samples oxidized :
(a) 30 mn. at 750°C ; (b) 30 hr. at 550°C (same oxide thickness of about 30 μm).

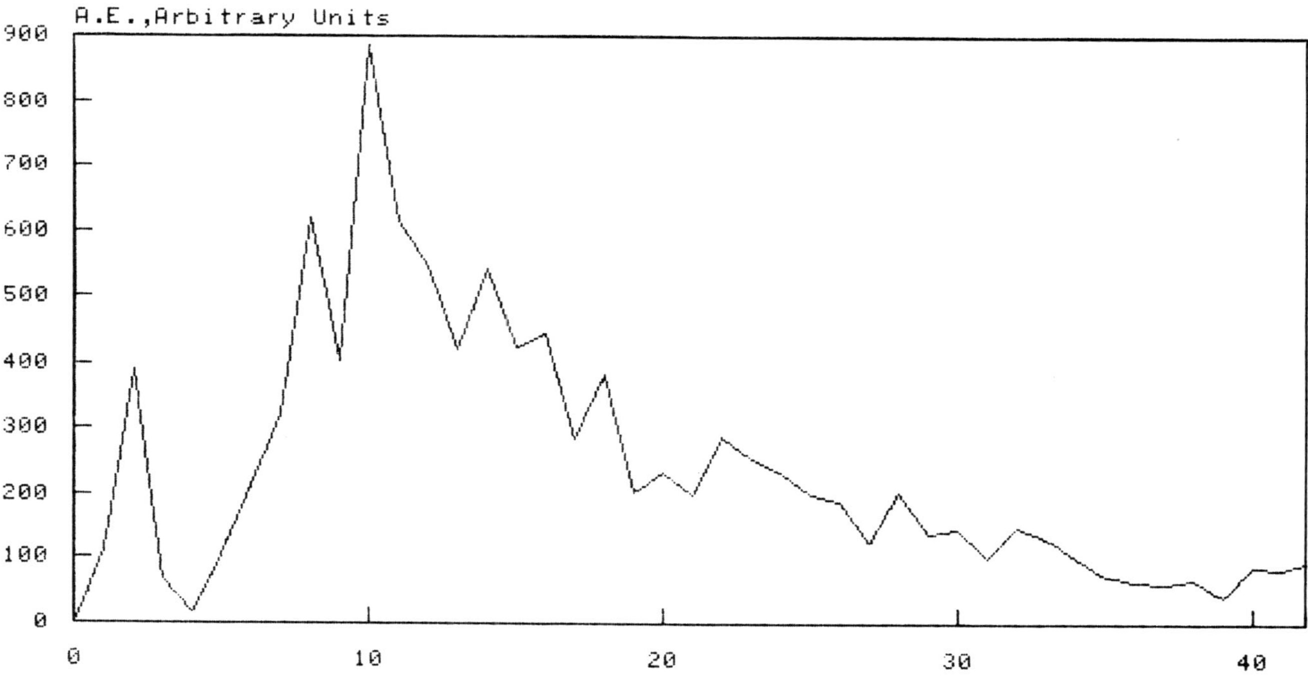

Fig. 2 - Acoustic Emission versus number of thermal cycles.

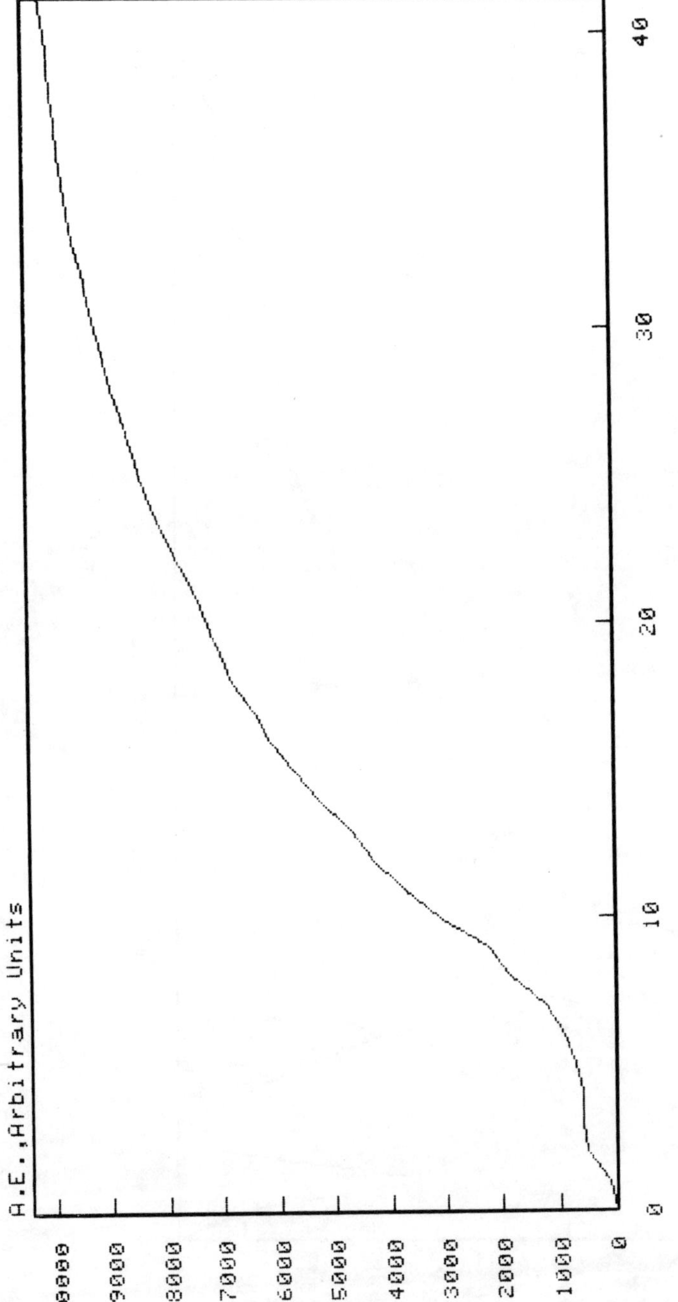

Fig. 3 - Acoustic Emission versus number of thermal cycles. Cumulated curve corresponding to fig. 2.

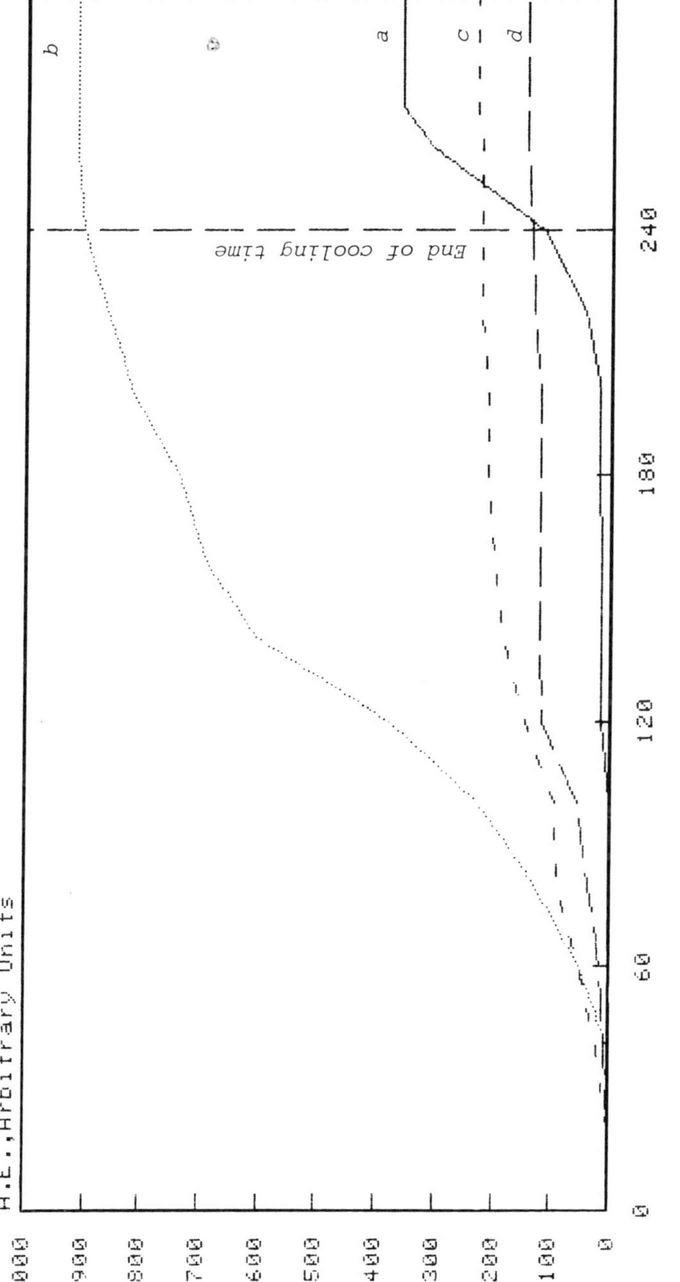

Fig. 4 - Acoustic Emission versus time during one cycle
a : 2 nd cycle ; b : 10th cycle ; c : 20th cycle ; d : 30th cycle.

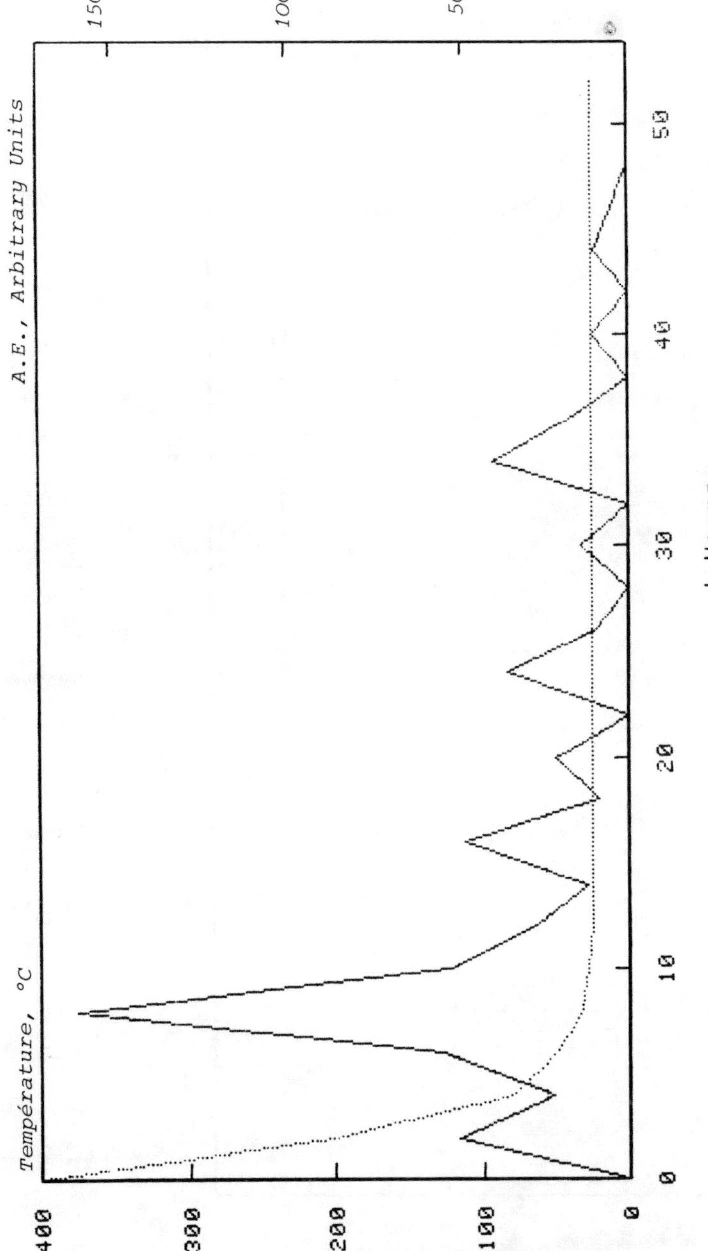

Fig. 6 - Acoustic Emission versus time during the last cooling after 42 cycles.

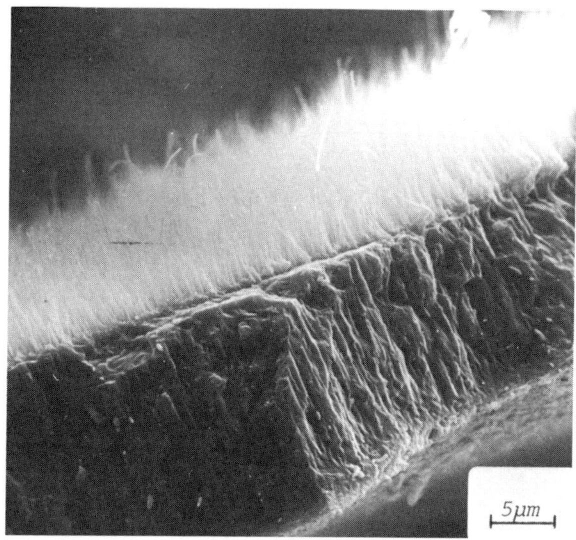

*Fig. 5 - Cross section of the oxide scale after 38 cycles
in oxygen.*

DISCUSSION

<u>M.I. Manning</u>: Have you used spectral analysis of acoustic emission
signals to determine the nature of the processes contributing to
the signals? This seems worthwhile since scales sometimes remain
adherent whilst emitting large A.E. signals and sometimes detach
completely with very little signal emission.

Do you feel that transformations between the two copper oxide
phases could give rise to any emission signals in your thermal
cycling tests, since at CERL we have detected emissions due to
Fe_2O_3/Fe_3O_4 conversions in scales growing on steels during isothermal cycling of gas compositions?

<u>C. CODDET</u> : Spectral analysis of acoustic emission signals
is actually difficult to achieve especially in our experimental
conditions. Due to the high frequencies involved, it is
necessary to use very high rates of sampling and in those
conditions the actual data storage capabilities are small
(around 15 mm) while the duration of the experiments is quite

large. Moreover, the signals are very complex and analytic
methods which are actually studied by other people in our
University require large amounts of computer time.
Nevertheless, this technique will be probably applied in the
near future to our oxidation problems.
As to the Cu_2O/CuO conversion, it is a possible source of
acoustic emission but this source has not been isolated from
others up to now.

STRAIN GENERATION DURING THE OXIDATION OF ZIRCALOY 2

S. Brooks, A.E.T. Nye, J.E. Forrest

Central Electricity Research Laboratories,
Leatherhead, Surrey, U.K.

SYNOPSIS

A technique is described which enables the strain generated during
the oxidation of a metal to be measured as a function of time.
The convex surfaces of a thin walled tube are oxidised whilst the
inner surfaces are protected by an inert or mildly reducing atmosphere. By measuring the change in capacitance between the tube
and a ring held concentrically within it, the variation of the
internal diameter with time can be continuously monitored. Hence
it is capable of measuring the strain induced in the metal during
oxidation of the outer tube surface. Results are reported for
Zircaloy 2 oxidised between 550 and 800°C in high purity argon containing 2% water vapour. During oxidation the tube diameter was
found to increase at a constant rate which cannot be correlated
with the transition from near parabolic to linear oxidation kinetics
observed on this alloy during gravimetric monitoring of its oxidation
behaviour in a thermobalance. A theory is proposed whereby oxygen
dissolution in the metal causes the tube to dilate with the transition from parabolic to linear rate being attributed to oxide fracture due to the distention of the metal substrate.

INTRODUCTION

Since the pioneering work of Pilling and Bedworth[1], numerous experiments have been designed to demonstrate the existence of growth
stresses in oxides. The theories of the origins of these stresses
have been reviewed on numerous occasions (for example ref. 2 and 3)
but always their corroboration has been hampered by a lack of
quantitative data on real systems.

Most experiments designed to measure stresses generated during oxidation can be placed in one of three categories:

1. Strip Bending Experiments

The bending of a metal foil clamped at one end is observed either at temperature, with one side of the strip protected from oxidation, or it is allowed to oxidise on both sides, cooled to room temperature and the oxide on one side subsequently removed. In the latter variant the effects of differential thermal expansion are an added complication whilst in the former experiment difficulty is always encountered in protecting one side of the specimen. There is also the difficulty of accurately measuring the foil deflection and the effects of the specimen edges make it hard to analyse the results.

2. Expansion and Contraction of Helical Specimens

This is an adaptation of the first method, in which one or both sides of a helical metal strip are oxidised and its subsequent movement observed. This movement arises because the difference in curvature between the inner and outer surfaces provides a means of generating differential stresses. Hence using this technique the requirement for protecting one side of the specimen is eliminated, although the other criticisms given above still apply. Both techniques are not readily amenable to continuous monitoring of oxide stress generation at high temperature.

3. Deformation of Unconstrained Metal Discs

Appleby and Tylecote[4] observed the doming of a metal disc at temperature whilst one side was oxidised and the other protected by blowing an inert or slightly reducing gas across its surface. By having free edges the maximum stress was at the centre and oxide fracture stresses could be calculated. In common with bending strip experiments the protection of the underside of the disc poses problems since a delicate balance is required between sufficient gas flow to prevent oxidation whilst the pO_2 at the upper surface must be sufficient for oxidation to occur. The movement of the disc is also difficult to measure either continuously or accurately. Appleby and Tylecote used a discontinuous cathetometer technique.

For any new technique to be a significant advance in the study of stress and strain generation in oxide growth the change of a physical parameter must be measured and four major problems apparent in the above experiments need to be resolved. Complete protection of one side of the sample is required; the kinetics of oxide formation must remain unchanged; the sample's stress system must be easily analysed, and the effects of thermal mismatch be eliminated by continuous monitoring at temperature.

A technique has been developed which overcomes these deficiencies.
A thin walled tube is oxidised on its outer surface whilst the inner
surface is protected by an inert or reducing gas. As the oxide grows,
any growth stresses change the internal diameter of the tube, this
being monitored continuously at temperature as a change in capacitance
between the tube and a concentric inner ring. End effects are
eliminated because the width of the ring is small ($\sim$3mm) compared
with the uniformly heated length of the tube ($\sim$100mm).

This method has already been used to study strain generation during
the oxidation of nickel and AISI 316 in air[5] and the work has now
been extended to Zircaloy 2 (Zr,1.5Sn,0.1Fe,0.1Cr,0.05Ni), a nuclear
fuel can material. It is well known that this material undergoes
a transition from parabolic to paralinear kinetics and many papers
have sought to explain this phenomenon[6,7]. Two basic theories have
been proposed, the first in which oxide stress generation causes
oxide cracking, and the second where recrystallisation generates a
network of porosity in the oxide. Using the capacitance technique
it was hoped to provide evidence in support of one or other of these
theories.

APPARATUS

A schematic diagram of the apparatus is shown in Figure 1. The
capacitance transducer assembly is held in place by silica balls
positioned between the tube and transducer support assembly. The
sample tube, $\simeq$14mm internal diameter, is enclosed in an outer tube
so that a controlled oxidising atmosphere can be used. "O" ring
seals connect the tubes at each end so that the sample tube is free
to move relative to both the transducer and the containment tube,
ensuring that any stresses close to the measuring point are hoop
stresses.

The active area of the transducer is shown in more detail in Figure
2. Using a three terminal capacitance technique in conjunction
with a low impedance monitoring circuit a capacitance of 1pf can be
measured to better than 1% in the presence of a cable shunt capacitance of 1000pf. By measuring the capacitance between the tube
and the 3mm wide ring radial changes of less than 1μm could be resolved. The strain was continuously monitored by feeding the
deviation from balance signal of the capacitance measuring bridge
onto an x-t recorder.

EXPERIMENTAL

In the previous work on Ni a slightly reducing atmosphere of 5%H$_2$
in Ar was passed through the inside of the tube to protect the
transducer and the concave surface of the tube from oxidation.
However as zirconium oxide has a very low dissociation pressure

P_{O_2} $\sim 10^{-55}$atm at 600°C) it is not practicably possible to provide a
reducing environment. Therefore the interior was first flushed
with high purity argon and then sealed so that when heated the zirconium tube itself gettered any residual oxidising species.

The apparatus was then heated to temperature with a high purity
argon atmosphere on the outside of the sample tube. The capacitance change was monitored during heating enabling a check to be
made of the calibration of the transducer from a knowledge of the
thermal expansion coefficients of the transducer material and
Zircaloy 2. Once at temperature the tube's outer surface was exposed to Ar-2%H_2O and the capacitance change as a function of exposure monitored.

After exposure the tube was sectioned and examined using optical
and electron optical microscopy.

A thermobalance accurate to ±1µg was used to obtain weight gain
data for similar material in an identical atmosphere.

RESULTS

The radial increase as a function of time for Zircaloy 2 tubes
oxidised in Ar-2%H_2O for periods of up to 200h between 550 and
800°C is shown in Figure 3. These curves were reproducible to
±2% and the error in the radial change is ±0.5µm.

It is evident that the tube expanded in a linear manner from the
time at which Ar-2%H_2O was introduced into the apparatus (approximately ½h after starting to heat). Superimposed on Figure 3 are
the positions at which the transition from parabolic to linear kinetics occurred, these being obtained from the thermogravimetry experiments.

The degree of protection afforded to the inner surfaces of the tube
is shown in Figure 4. Two types of oxide are evident, white "breakaway" oxide on the outer surface and black protective oxide formed
inside the tube.

DISCUSSION

Previous investigations have shown that initially a non-stoichiometric oxide grows at a near parabolic rate by inward oxygen transport via short circuit paths. Two main mechanisms have been proposed to explain the transition from parabolic to linear kinetics
and the formation of a white stoichiometric oxide instead of the
initial anion deficient black oxide. Bradhurst and Heuer[6] proposed a mechanical failure hypothesis whereby the stresses generated in the oxide as a result of film growth lead to mechanical
cracking and a transition to non-protective kinetics. Cox[7] pro-

pounded a recrystallisation hypothesis whereby a pore network is developed during initial oxidation, which, when it penetrates to the metal substrate causes a transition to linear kinetics.

It is difficult to reconcile either mechanism with the present findings that a linear tube dilation is observed from time zero and, as shown in Figure 3, is unchanged with the transition in oxidation kinetics or morphology. Hammad et al[8] measured the strains generated in Zr-2.5%Nb during air oxidation and found that strips elongated at an approximately constant rate such that after 20h at 600°C extensions of 1.8 to 2% were observed. These values were dependent upon the orientation of the strip to the rolling direction. In the present work on Zircaloy 2 a strain of 0.12% was measured after 20h at 600°C. Two possible explanations of the dilation observations can be proposed:

1. Reaction of the metal and oxidant at the metal surface could yield large compressive stresses at the oxide-metal interface, which could be relieved both perpendicular and parallel to the metal surface. This stress relief would cause metal distortion which would be dependent upon the rate of oxidation and hence, for the initial parabolic kinetics, time. However no time dependence of distortion rate was observed either in the present work or that of Hammad et al. Hence this hypothesis is not feasible.

2. Solution of oxygen within the metal would produce a lattice expansion and hence dilate the tube. X-ray diffraction of exposed and unexposed tubes showed that such an expansion did in fact occur and it is well known that oxygen has an appreciable solubility in zirconium. For example Hussey and Smeltzer[9] found that 23% of the reacted oxygen was absorbed into the metal at 600°C and 16% at 550°C. Electron probe microanalysis has shown that the oxygen concentration in the Zircaloy 2 tubes used in these tests increased with exposure time and temperature. For example after 20h at 750°C the oxygen content of the metal increased by ∿12 atomic %. Hence experimental evidence indicates that this oxygen solution mechanism explains the tube dilations observed. As the oxygen was uniformly distributed within the alloy, in contrast to the profile expected if volume diffusion in the alloy provided the rate control, then control must either be via oxygen absorption at the metal surface or transport of oxygen through the scale. As a constant tube dilation was observed which was independent of the oxide type, linear oxygen uptake by the metal must have occurred and it is suggested that this is rate controlling.

This satisfactory interpretation of the dilation results indicates that the theories put forward by Bradhurst et al[6] and Cox[7] to explain the transition from parabolic to linear kinetics may not be

valid. It is proposed that the transition from parabolic to linear
kinetics is caused by scale fracture, not primarily due to oxide
generation, but by oxygen dissolution in the metal substrate.
Wiedemann[10] has shown that compressive stresses applied to
Zircaloy 2 specimens corroding in high temperature steam reduce the
corrosion rate and delay the time to transition. This observation
is contrary to the predictions of the mechanical failure due to
oxide stress generation hypothesis. However such a compression
would oppose the dilation due to oxygen dissolution in the metal
and hence suppress the transition. Other support for the tube
dilation mechanical failure mechanism is provided by the results
shown in Figure 3. This shows that, within experimental error, the
linear oxidation rates initiated at approximately the same tube
dilation. A strain in the metal and the oxide, assuming no oxide
detachment, of 0.07% occurs at transition. This is approximately
equal to the fracture strain of monolithic zirconia.

Hence a refinement of the mechanical failure hypothesis is proposed
to explain the transition from parabolic to linear kinetics observed in the oxidation of Zircaloy 2.

<u>CONCLUSIONS</u>

1.   The capacitance measuring technique can be successfully used to
     measure strain generation at temperature in oxidising Zircaloy 2
     tubes.

2.   The diameter of Zircaloy 2 tubes oxidised in Ar-2%H_2O at 550 to
     800°C increases linearly with time.

3.   No correlation was noted between the tube dilation and the
     breakaway oxidation transition from parabolic to linear oxide
     growth kinetics.

4.   The tube dilation is attributed to oxygen dissolving in the
     alloy controlled by surface reaction at the oxide/metal inter-
     face.

5.   It is proposed that the transition from parabolic to linear
     oxidation kinetics is caused by oxide fracture due to substrate
     deformation by oxygen dissolution.

<u>ACKNOWLEDGEMENTS</u>

This work is published by permission of the Central Electricity
Generating Board.

REFERENCES

1. Pilling, N.B., Bedworth, R.E., J. Inst. Met., 29, 529, 1923.

2. Stringer, J., Corrosion Science, 10, 513, 1970.

3. Fromhold, A.T., Stress Effects and the Oxidation of Metals, Ed. J.V. Cathcart, AIME, 1975.

4. Appleby, W.K., Tylecote, R.F., Corrosion Science, 10, 325, 1970.

5. Brooks, S., Nye, A.E.T., Forrest, J.E., Corrosion Science Symposium, Nottingham, 1978.

6. Bradhurst, D.H., Heuer, P., J. Nucl. Mats., 37, 35, 1970.

7. Cox, B., J. Nucl. Mats., 41, 96, 1971.

8. Hammad, F.H., El-Raghy, S.M., Ibrahim, A.E.A., Unpublished Work. Nuclear Research Centre, Cairo, Egypt.

9. Hussey. R.S., Smeltzer, W.W., J. Electrochem. Soc., 111, 1221, 1964.

10. Wiedemann, K.H., Z. Metalikde, 65, 735, 1974.

S. BROOKS *et al.*

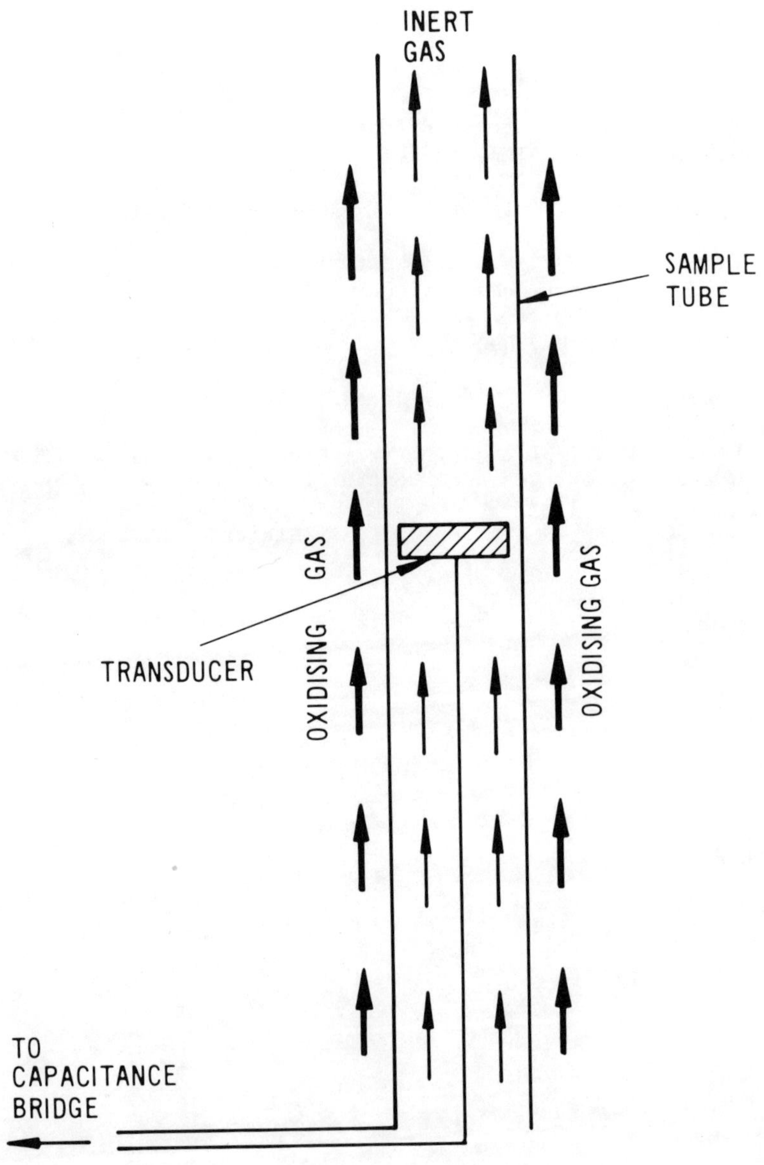

FIG. 1   SCHEMATIC OF APPARATUS

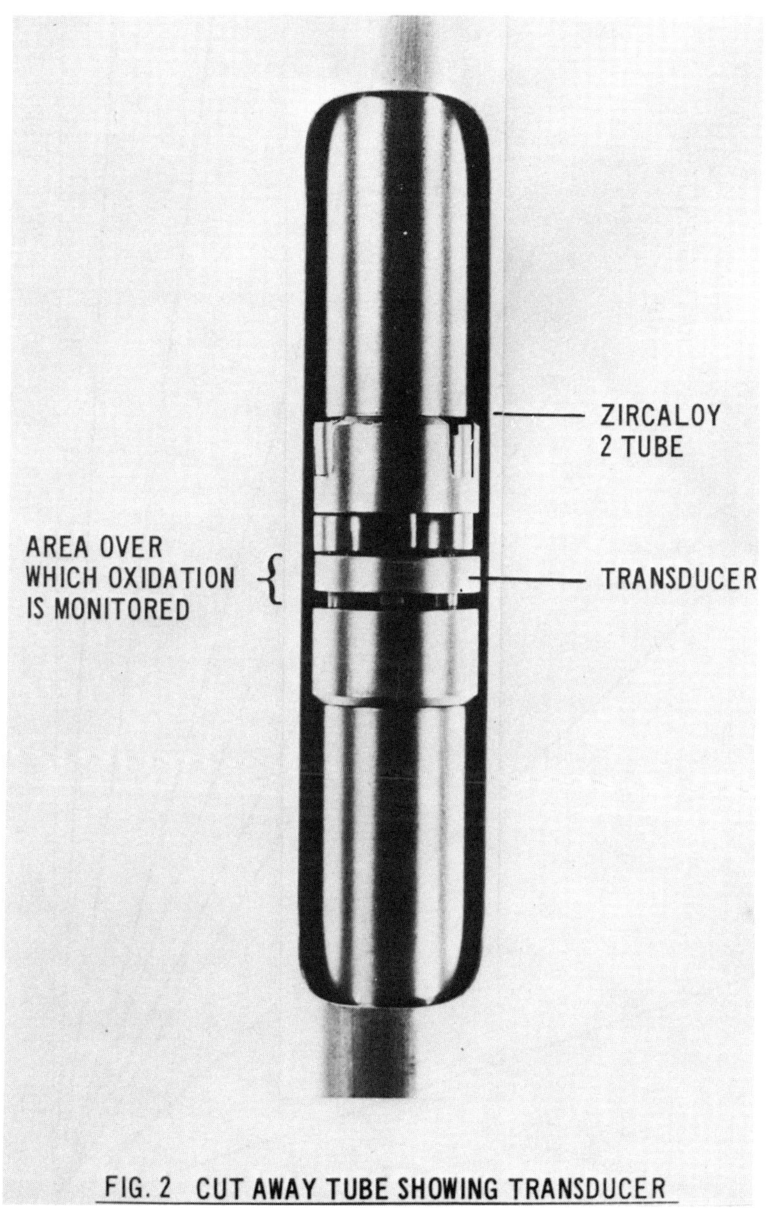

FIG. 2 CUT AWAY TUBE SHOWING TRANSDUCER

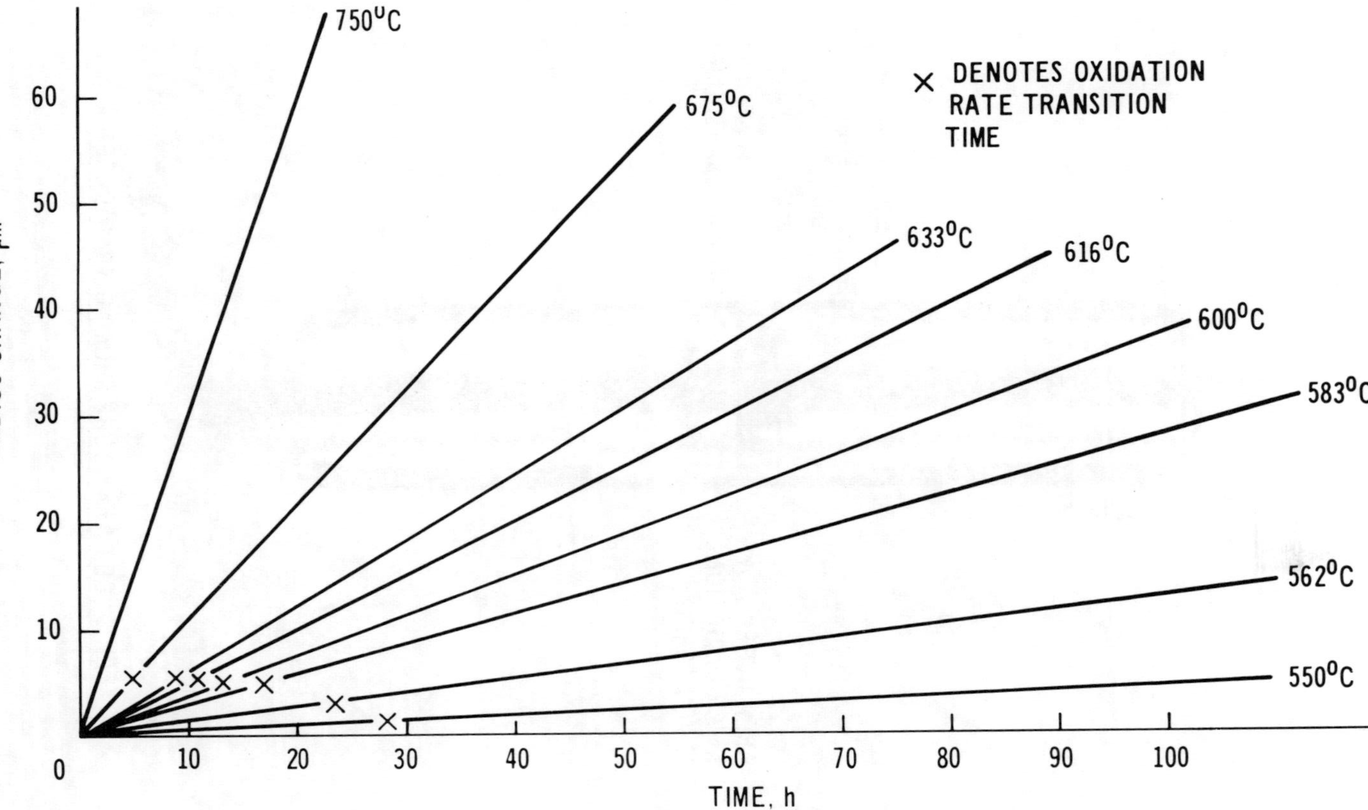

FIG. 3   RADIUS CHANGE v TIME FOR ZIRCALOY 2 TUBES

FIG. 4   LONGITUDINAL SECTION OF TUBE SHOWING BLACK OXIDE ON PROTECTED INSIDE AND WHITE OXIDE ON EXPOSED OUTSIDE. 200 h AT 640°C

DISCUSSION

B. Ilschner: Since I am not an expert in Zirconium alloy oxidation, my question may have a very simple answer which I overlooked: since you work with a thin-walled tube, I would expect that saturation is achieved after some time, thus ending any expansion of the tube due to oxidation uptake. From then on further diameter changes should be due only to stresses generated by the oxide; by pre-saturation of the Zirconium alloy tube specimen prior to the experiment proper, this point might be shifted to shorter time.

D.B. Meadowcroft[x]:   Our use of thin walled tube would obviously make saturation occur much more quickly than if thick walled tubes were used, but as the solubility of oxygen in zirconium is $\sim$ 30% this would have occurred only after prolonged exposure. Presaturation would certainly be a good test.

B. Pieraggi: I would like to ask if the observation of linear variation of the measured deformation against time that you have observed during your work, cannot be related to the small thickness of the specimens. Indeed, we have made some observations during a study of the oxidation of hafnium and hafnium-copper alloys, which show that the occurrence of a linear variation of deformation against time can be related to the relative magnitude of oxide formation and oxygen dissolution rates. We have observed that the deformation remains low (less than 0.1%) when the oxygen dissolution rate is lower than the oxide formation rate, and that the deformation rate becomes linear when the oxygen dissolution rate is faster than the oxide formation rate.

D.B. Meadowcroft[x]: This is a most interesting observation in view of our finding that the deformation rate of Zircaloy 2 was constant. If your analysis is applicable to this system in the temperature range we indicated, it would indicate that the oxygen dissolution rate into Zircaloy 2 is faster than the oxide formation rate.

P. Kofstad: May I ask a general question on the mechanism of oxidation of zirconium. Both measurements of electrical conductivity and of diffusion across growing ZrO_2-scales clearly indicate that ZrO_2 is an ionic conductor – at least over large oxygen pressure ranges. This must mean that the growth of ZrO_2 scales is at least partially governed by electron transport through the scale. Is this taken into account in considering the oxidation and breakaway behaviour of Zirconium.

D.B. Meadowcroft[x]:   I think a distinction must be drawn between the black protective oxide and the white breakaway oxide. Whilst the rate of formation of the white oxide could be limited by electron diffusion, I doubt if this is so for the protective oxide. The black colour indicates absorption by free electrons, and measure-

---

[x]: The paper: Strain Generation during the Oxidation of Zircaloy 2 by S. Brooks, A.E.T. Nye, J.E. Forrest, C.E.R.L. Leatherhead, Surrey, U.K. was presented by D.B. Meadowcroft.

ments on bulk black zirconia (as is produced if a high current density is forced through zirconia to make it sub stoichiometric) show that it has a high electronic conductivity. Certainly it has been shown that protective oxidation rates in water at lower temperatures are uninfluenced by change in the electronic conductance of the oxide film. When dilation of the substrate causes fracture of the protective black oxide, oxidant ingress may then cause the oxide to become white and stoichiometric. Breakaway could possibly be a repeated breakdown of the protective layer, and indeed a thin black oxide is occasionally seen at the base of a breakaway scale.

Two different test rigs for creep experiments in aggressive environments

K.-H. Döhle, A. Rahmel, M. Schmidt, M. Schütze

Dechema-Institut

Postfach 970 146, D-6000 Frankfurt/M.

Abstract

In high-temperature technology the materials are simultaneously subject to high temperature, mechanical stress and corrosion. Therefore the knowledge of the behaviour of such materials in aggressive environments requires test methods which combine these three kinds of attack.

At Dechema-Institut developed testing facilities allow creep tests in corrosive atmospheres.

The first kind of test rig was built for long term creep tests under constant load in a coal gasifier atmosphere at temperatures between 800°C and 950°C. During the test the creep strain of the specimens is recorded continuously. The tests are part of the PNP-program which is carried out in cooperation with Rheinische Braunkohlenwerke AG, Köln and Kernforschungsanlage Jülich.

The second kind of test rig was built for the performance of constant strain rate creep tests in corrosive gas atmospheres. The obtainable strain rates are between $9 \cdot 10^{-7}$ s^{-1} and $3 \cdot 10^{-11}$ s^{-1} when using a specimen with a gauge length of 40 mm. Experiments can be performed at temperatures up to 1100°C. Stress and strain are recorded continuously.

The design of the two test rigs will be discussed in details in this paper.

1. Introduction

In most cases high temperature materials are subject
to a combined attack by high temperature, mechanical
stress and corrosion. Until now the behaviour of such
alloys was usually studied in either mechanical tests
or corrosion tests separately which means that either
creep data or corrosion data can be obtained but no
data of a combined attack. The influence of corrosion
on mechanical strength at high temperatures should
not be neglected, as shown in [1], thus the necessity
of creep tests in corrosive environments is evident.

Data of the mechanical strength at high temperatures
can be obtained by different tests. The most often
used stress rupture test usually supplies data of
rupture life only. As pointed out in [1] corrosive
attack does not only influence the time to rupture
but also may change the entire course of the creep
process. Therefore continuous recording of strain(ε)
- time(t) data is necessary for information about the
influence of corrosion on creep behaviour. Very often
the ε-t diagrams obtained by this way are not really
qualified to reveal clear differences in creep behaviour, but these differences are evident in the $\dot{\varepsilon}$-t
diagram.

The mechanical strength at high temperatures can also
be investigated by constant strain rate creep tests.
The analogy to the constant stress (or load) creep
test appears in the basic creep equation

$$\dot{\varepsilon} = B \sigma^n \qquad \text{for constant stress (or load) creep tests}$$

and

$$\sigma = C \dot{\varepsilon}^m \qquad \text{for constant strain rate tests}$$

at a constant temperature and at identical structure
parameter S [2]. In case that a material shows second-
ary creep the $\dot{\varepsilon}$-t diagram of the constant stress creep
test, as well as the σ-ε diagram of the constant strain
rate creep test will show the 3 regions of primary,
secondary and ternary creep.

Corrosion can affect the creep behaviour in many re-
spects. An increase of the strength may occur e. g. by
the precipitation of corrosion products in the material affecting either reduction of grain boundary
sliding or precipitation hardening within the grain.
Materials with surface scales formed by corrosion may

act as a composite material and thus the creep strength
may increase if the strength of the scale is higher
than that of the metal.

A decrease of the creep strength will happen by reduction of the supporting cross section, by increasing
grain boundary sliding in the case of ductile corrosion products on grain boundaries, by depletion of
solution and/or precipitation hardening elements, etc.
A summary of the effects of corrosion on creep behaviour and vice versa is given in [1]. All these effects
will lead to a change in the course of the creep diagrams as shown schematically in Fig. 1 and Fig. 2.

2. Test rigs for creep tests in corrosive environments

At Dechema-Institut creep tests in corrosive environments are performed under different aspects. Long term
creep tests, i. e. stress rupture tests with continuous
recording of the strain, provide engineering data for
a planned coal gasifier plant. Constant strain rate
creep tests are performed to reveal basic interactions
between corrosion and creep.

2.1 Test rig for long term creep test in a synthetic coal gasifier atmosphere

The German PNP-project sponsored by "Der Bundesminister für Forschung und Technologie" and "Das Land
Nordrhein-Westfalen" includes investigations of the
high temperature strength of materials in a simulated
coal gasifier atmosphere at temperatures between 800°C
and 950°C. Part of this program is performed at Dechema-
Institut in close cooperation with Rheinische Braunkohlenwerke AG, Köln, and Kernforschungsanlage Jülich.
Therefore a set of 8 test rigs has been built for the
performance of long term creep tests in a 50% H_2O
vapour, 35% H_2, 5% CO, 5% CO_2, 5% CH_4 gas atmosphere.

2.1.1 Mechanical equipment

Each test rig consists of a loading bar system connected to a pull-rod, a furnace and a gas tight container in which two specimens and the extensometers
are arranged (Fig. 3, Fig. 4 and Fig. 5). The shape
of the loading bar keeps the effective load at the

specimens constant during elongation. The pull rod
(IN 713 LC) leads through a gas tight metal bellow
out of the container (IN 600). The furnace with 3
separately controlled zones can be heated up to 1000°C
and the constancy of temperature is better than $\pm$ 2°C
in the gauge length during the measurements. The creep
specimens with gauge collars are according to DIN
50125. At each of the two specimens two extensometers
are installed to control whether the creep strain occurs in a uniform way (Fig. 4). There are two different ways of measuring the creep strain. Four test rigs
are supplied with extensometers leading through gas
tight metal bellows out of the container to mechanical
dials. In the other four test rigs the creep strain is
measured by inductive electric transducers fixed at
the end of the extensometers. These transducers are in
a second container below the main container in the
same gas atmosphere as the specimens (see Fig. 4).
This second container is kept at a temperature of
80 ... 150°C to prevent condensation of the water vapour in the coal gasifier atmosphere. Using electric
transducers is more advantageous since the strain data
are recorded automatically by a data logger whereas
the mechanical dials have to be read in certain intervals. At the bottom of the container gas in- and outlet, thermocouple inlet and (in the case of electric
transducers) all electric wire inlets are installed.
This arrangement facilitates easy handling when replacing the specimens after each test.

2.1.2 Gas supply

The dry premixed process gas containing CH_4, H_2, CO_2,
CO is flowing from gas cylinders to a gas humidification station where it is enriched with 50% water
vapour by bubbling through water of 82°C. From this
humidification station the gas pipings are running to
each test rig. A second gas pipe system supplies argon
as flushing gas. At each test rig there are gas regu-
lators for argon and process gas and magnetic valves
which are remotely actuated from the data logger. The
gas enters the container at the bottom and leaves in
the upper part. After leaving the gas is cooled down
to about 3°C to remove the water vapour by condensation.
The flow rate of the dried gas can now be read from
subsequent flow meters. The whole gas piping system
has to be kept at a temperature of about 100°C by heating ribbons to avoid condensation of water vapour in
the pipes.

2.1.3 Control system and data recording

Dangers during the experiments may arise from process gas containing inflammable H_2 and poisonous CO. The CO-contents in the air are controlled by a CO-meter which is connected to the data logger. CO-alarm by the logger is already given at 50 ppm and the furnaces are switched off automatically at 150 ppm CO. Additionally the room ventilation moves with maximum speed in the case of gas alarm and the gas supply is switched from process gas to argon flushing. The same happens in the case of a broken pipe which leads to a decrease in pressure and thus activates the alarm system. Therefore no accident should happen by H_2 and CO. The control of the room temperature (which must not exceed 40°C) is also included into the alarm system of the data logger. Another control function supervises the temperature of the heating ribbons (which must not fall below 80°C) and the temperature of the furnaces. Apart from the control and alarm-functions, the data logger is used for the recording of experimental data. This data is evaluated by a desk top computer which provides lists of $\varepsilon, \dot{\varepsilon}$, t-data and plots the ε-t and $\dot{\varepsilon}$-t diagrams.

2.2 Test rigs for constant strain rate creep tests

For basic investigations of the interactions between corrosion and creep and especially of the behaviour of scales during creep, the constant strain rate technique is more qualified than constant load experiments. Therefore two constant strain rate creep test rigs were constructed at Dechema-Institut (test rigs of the required kind were not commercially available). The design of these test rigs is partially based on experiences gained at BBC Mannheim [3] and Allianz Zentrum für Technik, Ismaning/München [4].

2.2.1 Mechanical equipment

Each rig consists of a drive unit which moves a pull rod, a gas tight container in which the specimen is placed and a furnace (Fig. 6).

Drive unit (Fig. 7):

A synchronous electromotor drives a gearbox with 10 gears. The speed is further reduced by a harmonic drive and a worm gear. This worm gear is connected with a spindle drive which moves the pull rod (Nimonic 90) with a speed of about 0,1 mm/h to about $4 \cdot 10^{-6}$ mm/h. The resulting strain rates are between $9 \cdot 10^{-7}$ s^{-1} and $3 \cdot 10^{-11}$ s^{-1} when using a specimen with a gauge length of 40 mm. The drive unit can be separated between worm gear and harmonic drive and the pull rod can be moved by a winch so that it takes less time to replace the specimen. The whole drive unit was designed for a tensile force of 50 KN at the pull rod.

Reaction container:

Again the specimen is in a gas tight container (IN 600) from which the pull rod leads out through a gas tight metal bellow. Alumina extensometers (Fig. 8) lead from the specimen to an electric transducer which is situated above in a small second gas tight container. Gas supply and thermocouple come in from the top of the main container and thus an easy change of the specimen is possible. The furnace with 3 separately controlled zones can be heated up to 1100°C with a temperature derivation of $\pm$ 2°C over the gauge length.

2.2.2 Electrical control

The electronics for the registration of the experimental data, for temperature control of the furnaces and electric control of the drive units are in a 19" cabinet which is connected with the data logger. The data of ε, σ, T are registrated on a recorder as well as collected by the data logger and put into a desk top computer to obtain σ-ε plots. The alarm system of the data logger switches off the furnace and the drive in case of overheating, mechanical overload and when a programmed value of stress or strain is reached. The data logger also enables the drive unit to perform low cycle fatigue experiments in a programmed range of stress or strain values or to simulate constant load experiments.

A schematic description of the arrangement and the connections of all test rigs is given in the sketch of Fig. 9.

3. Preliminary experiences

The test rigs for long term creep tests with mechanical dials have been working since April 1979 and proved to be reliable. Experiments with electric transducers were started in November 1979 but had to be interrupted since the metal bellows in these test rigs proved to be too sensitive and broke due to corrosion. Now these experiments have been started again and from first experiences it may be concluded that this system will work well.

The test rigs for constant strain rate creep tests were recently finished and the experiences gained until now are limited to the drive unit. The drive unit was tested with aluminum specimens at room temperature and proved to fulfil the design data.

The experimental results from the long term creep tests are not yet complete and the experiments in the constant strain rate creep test rig have not yet begun. Therefore no results can be reported in this paper but perhaps in the not too far future.

Acknowledgements

The authors would like to thank Prof. Dr. H. Nickel, Mr. Düsterwald (Kernforschungsanlage Jülich GmbH) and Dr. Schuhmacher, Dr. Bruch, Mr. Küppers, Mr. Pütz (Rheinische Braunkohlenwerke AG, Köln) for their support at the construction of the long term creep test rigs, Dr. Grünling, Dr. Bauer, Mr. Hartnagel (Brown, Boveri and Cie AG, Mannheim) and Dr. Schüller, Dr. Hagn, Mr. Frank (Allianz Zentrum für Technik, Ismaning bei München) for design data and construction diagrams of their constant strain rate test rigs, and Der Minister für Wirtschaft, Mittelstand und Verkehr des Landes Nordrhein-Westfalen, Düsseldorf, via Rheinische Braunkohlenwerke AG, Köln (long term creep tests) and Deutsche Forschungsgemeinschaft (constant strain rate creep tests) for financial support.

References

1. H.W. Grünling, B. Ilschner, S. Leistikow, A. Rahmel, M. Schmidt: Proc. of International Conference on the Behaviour of High Temperature Alloys in Aggressive Environments, Petten, NL, Oct. 1979

2. B. Ilschner: Hochtemperaturplastizität, Springer-Verlag, Berlin - Heidelberg - New York, 1973

3. W. Hartnagel, R. Bauer, H.W. Grünling: Proc. of
   European Symposium on the Interaction Between
   Corrosion and Mechanical Stress at High Tempera-
   tures, Petten, NL, May 1980

4. R. Frank, L. Hagn, H.-J. Schüller: Der Maschinenschaden 50 (1977) Heft 6

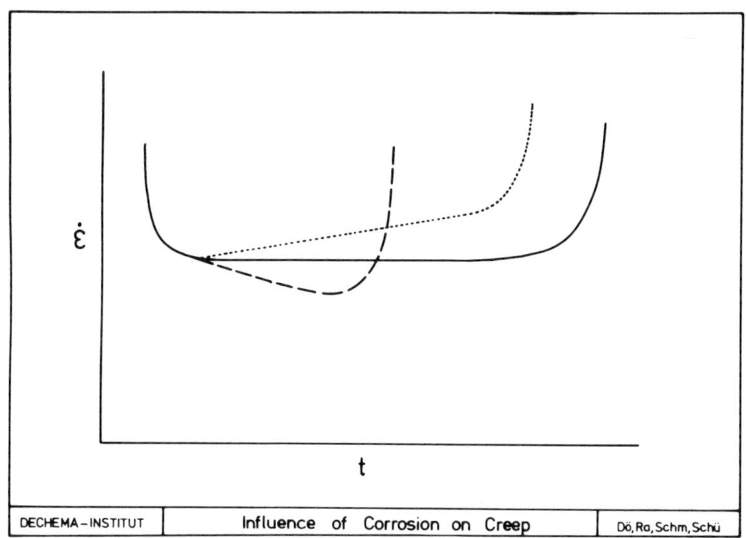

Fig. 1 - Schematic diagram of the influence of corrosion on the creep rate in constant stress tests
________ without corrosion
...... decrease of strength by corrosion
------ increase of strength by corrosion

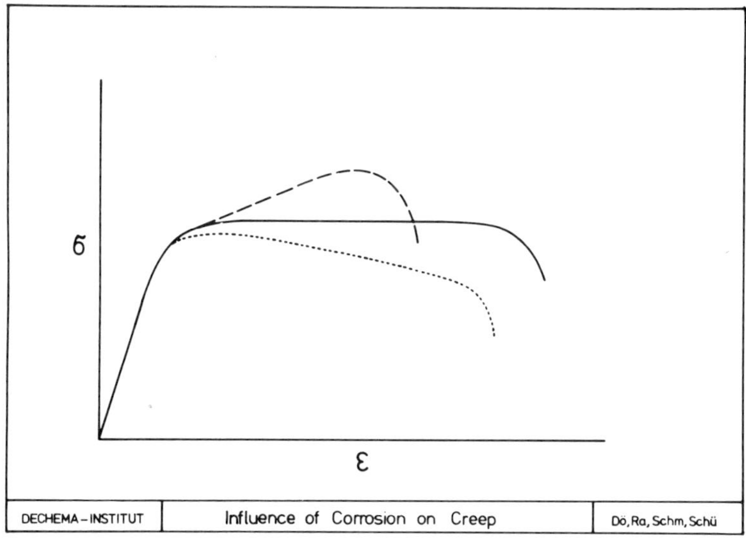

Fig. 2 - Schematic diagram of the influence of corrosion on the creep stress in constant strain rate
tests
________ without corrosion
...... decrease of strength by corrosion
------ increase of strength by corrosion

K.H. DÖHLE et al.

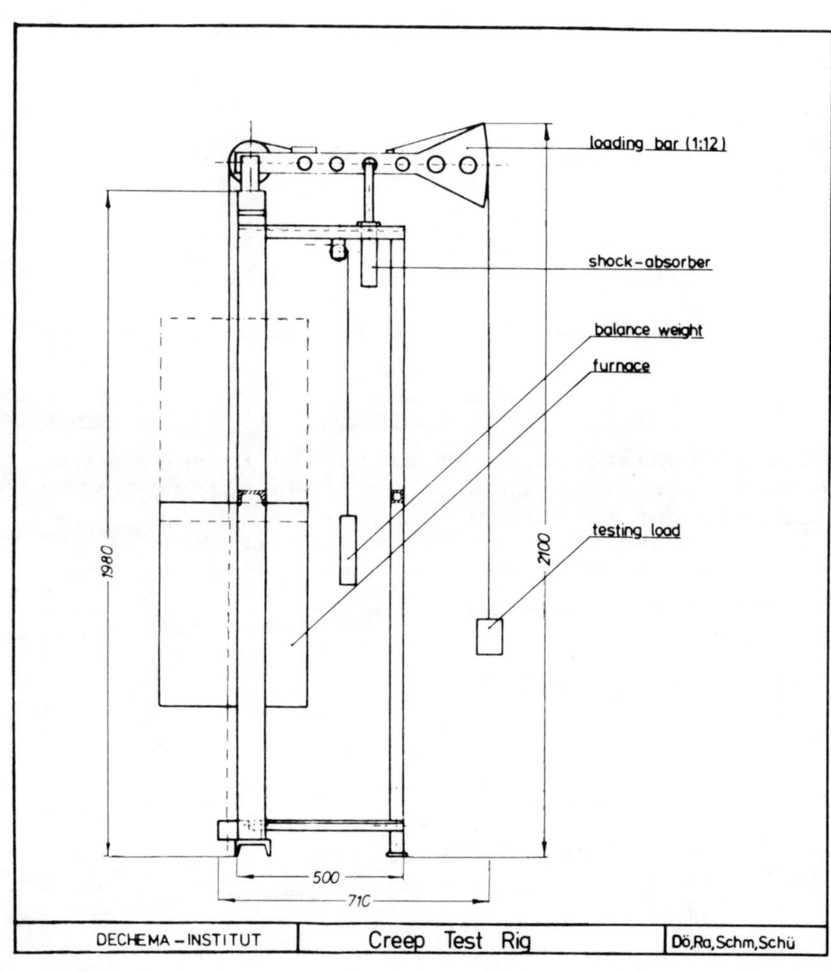

Fig. 3 - Schematic illustration of the constant load creep test rig

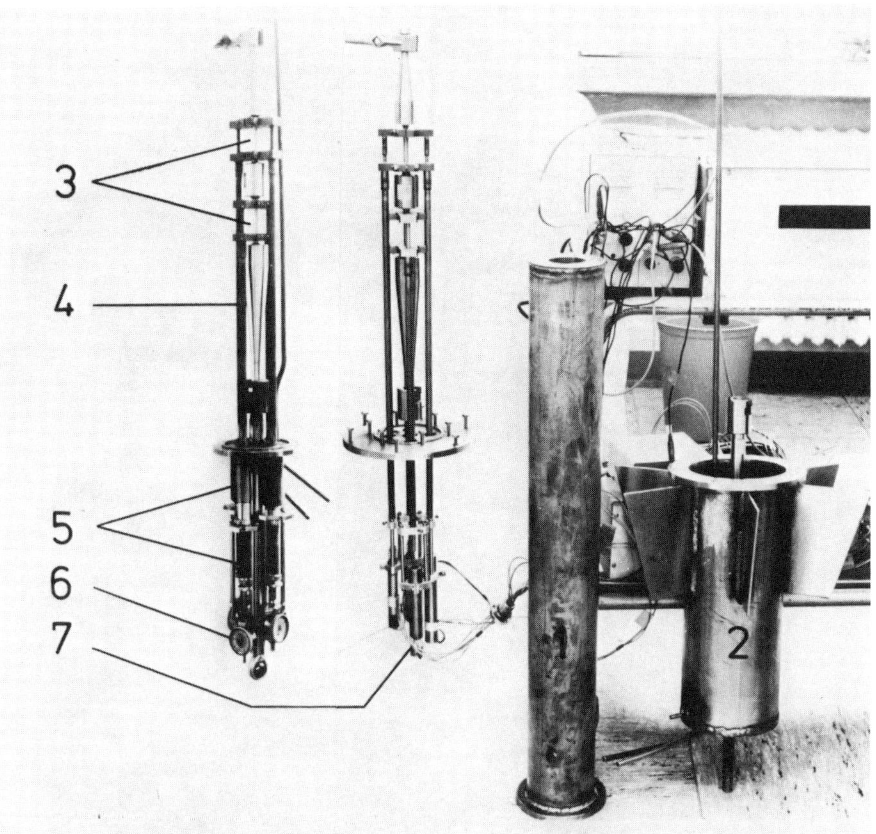

Fig. 4 – The two extensometer systems
Left : Extensometers with mechanical dials
Right: Extensometers with electric transducers
       (surrounded during the tests by container (2))

1 –   main container

2 –   container for the transducers

3 –   specimens

4 –   extensometers

5 –   metal bellows

6 –   mechanical dials

7 –   electric transducers

Fig. 5 – View of three constant load test rigs

Fig. 6 - View of one of the constant strain rate
test rigs and the 19"-electronics cabinet

Fig. 7 - Drive unit of the constant strain rate
test rig

Fig. 8 - Pull-rods, specimen and alumina extensometers of the constant strain rate test rig

K.H. DÖHLE et al.

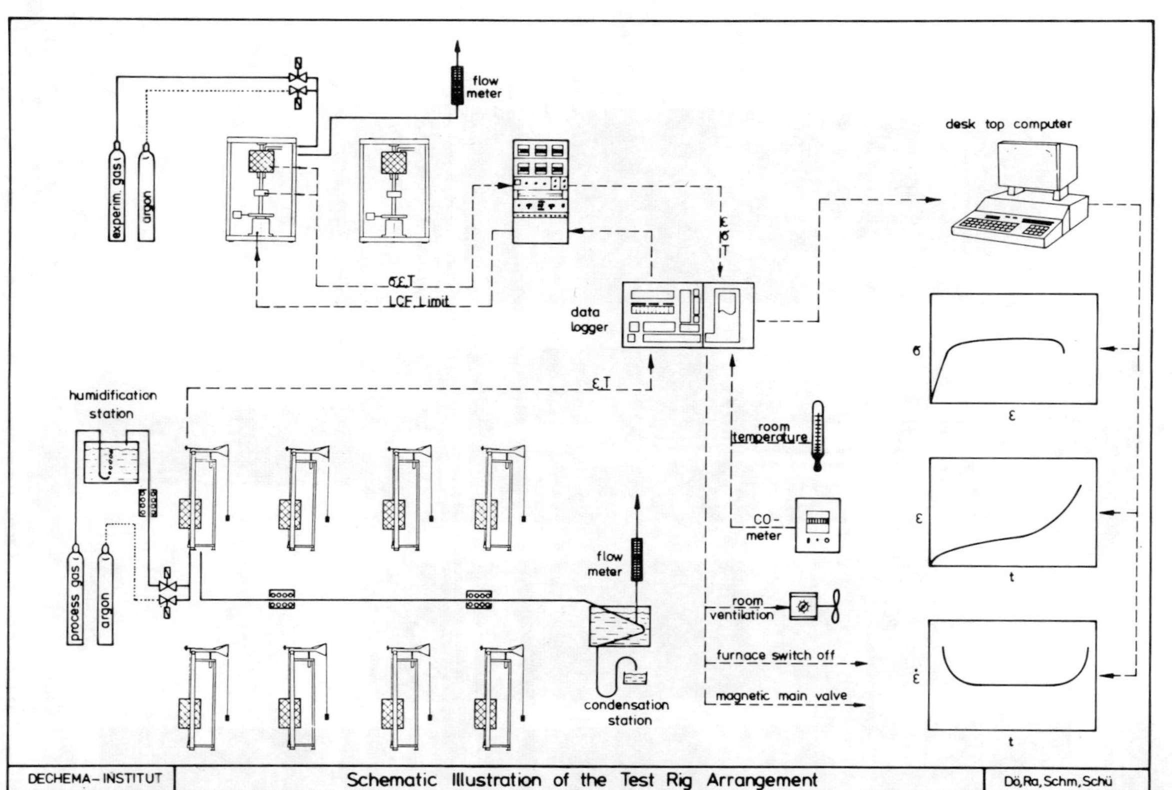

Fig. 9 – Schematic illustration of the arrangement of the constant strain rate test rigs (upper part) and the constant load test rigs (lower part), the gas flow system and the data recording and control system

DISCUSSION

J. Bressers: Does the LCF testing possibility of the constant
strain rate type machines described refer to push-pull fatigue.
If so, how is buckling in compression avoided given the long
loading bar system used?

M. Schütze: There was no need for push-pull experiments or compression tests in our programme. Therefore our machines were designed
for the performance of tensile tests and LCF-tests in the pull
region only.

B. Ilschner: In addition to mechanical data generation, it is
necessary to withdraw specimens quickly from the apparatus to
avoid considerable structural changes during cooling/unloading.
Does your equipment provide facilities for this purpose?

M. Schütze: The testing facilities have to be seperated into two
groups:
The main purpose of the constant load test rigs is to obtain data
of the effect of corrosive attack on the mechanical properties. To
avoid further corrosion after fracture of the specimens the furnace
is switched off automatically and the gas supply is changed from
process gas to argon flushing. Since investigations concerning the
dislocation structure are not planned in this programme no rapid
cooling of the specimens is needed to freeze the dislocation
structure.

The constant strain rate test rigs were constructed for the investigation of basic interactions between corrosion and creep which may in
include the influence of the dislocation structure on the corrosion
behaviour as well. For freezing the dislocation structure the furnace and the reaction container can be opened very quickly in order
to cool the specimen. To avoid any damage of the specimen (especially of the sensitive oxide scales) when taking it out of the
specimen holders in the furnace, special specimen holders were
designed which ensure quick and easy dismount of the specimen even
after severe corrosion.

H.W. Grünling: I wonder whether one can apply potential drop
method in monitoring defect formation and cross section changes
during constant strain rate experiments.

W. Hoffelner: Although we have no direct experience in using the
potential drop technique under these circumstances, from our own
and from other experiments with this technique for measurements
of crack growth I would expect mainly two problems. The one is the

required long time stability of the measuring system as pointed
out also by K. Krompholz, E.D. Grosser, K. Evert in Materials
Technology and Testing 11, 2, 60 - 67, 1980 and the other originates from resistivity changes due to plastic deformation of the
sample. I could imagine that these effects make accurate measurements rather difficult.

OPENING CONTRIBUTIONS FROM PANEL MEMBERS

A. Palazzi (chairman)

I think that the field of the subject of this symposium is one of
the most difficult ones of the metallurgical sector in which we
are always faced with scatter of results, which is partially due to
uncertainty in instrumentation and measurements. Most of the scatter
originates from the characteristics of the material and the deviation from nominal composition, because alloys must be obtained
by conventional methods, i.e. melted, cast into castings or ingots
for further deformation. These conventional procedures give rise
to two unavoidable phenomena, i.e. dendrites and segregation. Moreover, the existence of grain boundaries, i.e. the area where
every dramatic happening is met, is unavoidable with the conventional melting and casting procedure.
I will begin with some provocative topics saying that preparation
methods for alloys and components are now in germ, where attempt
is made to destroy dendrites, taking profit that superalloys and
many ternary alloys have a big distance between liquidus and so-
lidus surfaces during the descent of temperature; between these
two surfaces a mechanical action is possible in order to destroy
dendrites and to have alloys which in every point correspondend
to the average or nominal composition. Another method for making
future life easier in this field is to proceed through the so-
called atomization starting from the liquid state. With this method
powders or droplets of an alloy are prepared with diameters in
the range of five to ten microns. Micro-dendrites of this dimension
are not so disturbing as dendrites of five or ten centimetres length
as in conventional preparation.
This sophisticated future method of alloy preparation is surely
more expensive than conventional ways, but there is an advantage
to be considered as a conterbalance. The advantage of this phar-
maceutical preparation method is that nearly no material is lost,
whilst with conventional methods the croploss especially in superalloys is sometimes in the order of 50%.
The best way to present provocatice ideas in the course of the
panel presentation is to proceed with the order of magnitude of
the things dealt with, i.e. from atomic up to the industrial
plant scale. We think that Professor Kofstad will speak about some
theoritical needs and some scientific models, that Professor

Ilschner will enlighten the role of fracture mechanics in our field; that Dr. Guttmann will speak about the transposition of the results of sample testing into fullsize components behaviour, and that Dr. Schubert finally will speak about the technological implications of the tests and the scientific work presented here.

P. Kofstad

I think this is as difficult a task as I am ever been given: Four
minutes is to be filled with wise and golden words. But let us
turn to the theme of this meeting: the interaction between corrosion and mechanical stress at high temperatures. This title may
suggest that in addition to what we may call "normal" corrosion,
we have at this meeting been dealing with the added complicating
feature of mechanical stress. I think it is important to recall
that normal corrosion is in no way relaxed or stress-free. In
general there are always mechanical stresses associated with
"normal" corrosion. This is particuarly marked when the corrosion
takes place under cycling temperatures. So when we emphasize
mechanical stress in the title we are really not encountering
new problems. We are just superimposing an additional external
stress to already existing stresses. The basic problem is the same.
When considering protective corrosion behaviour at high temperatures
- with or without an external mechanical stress - we are concerned
 with various aspects of the reactions. One important aspect is
to achieve dense and continuous scales with low rates of transport
of the reactants. But - as has been demonstrated in so many of the
papers at this meeting - we must be equally concerned with the
ability of scales of reaction products to deform along with the
deformation of the substrate metal, i.e. we must consider creep,
plastic deformation and stress relief in the scales. Cracking of
scales often give rise to accelerated corrosion rates.
If we are going to understand such behaviour and to alleviate the
problems of scale cracking, it is important to have better knowledge of the creep behaviour of oxides. As regards creep of solids,
there is considerable amount of data on metals and alloys. Data
on oxides are, however, very limited; such studies have only been
done on a few selected oxide systems. In future work it would be
wise to put greater emphasis on creep in oxides and how this varies
with the nonstoichiometry or the partial pressure of oxygen. Only
with such data available can we get a real understanding of the
basic processes during hightemperature corrosion and to have a
better control of the ability of scales to deform in a desired
manner.
This lack of basic data not only applies to oxide creep, but to

many other aspects of high-temperature corrosion. In alloy oxidation,
for example, we often lack simple basic data on the relevant oxide
systems, what phases are formed in the reactions and what are
their properties, etc. All in all, a main point of my remarks is a
plea for a greater emphasis on relevant basic studies.
Let me continue in the same vein. Prof. van de Voorde said in his
opening remarks that this meeting was devoted to understanding
mechanisms. In this respect we have heard many fine contributions
at this meeting. But I also have a general impression that we in
too many cases are only able to give graphical or pictorial description of the results. We are often at a loss to explain the mechanisms involved. This is an unhappy state of affairs. One also
begins to wonder when one hears of different laboratories reacting
the same type of specimens under seemingly identical conditions
and receiving "opposite" results.
This prompts me to ask the following questions: 1) Are we doing
too many complicated and costly engineering-type experiments in this
field with so many experimental parameters that it virtually
becomes impossible to unravel the mechanisms? Do we have sufficient
control of the experiment parameters in our experiments? Do we
recognize what are the critical parameters that we should have
under control and do we monitor these well enough?
My four minutes are at the end, but my main point is that I think
we ought to devote just a small part of the often very expensive
and large outlays we have on more engineering-type measurements to
more basic, relevant and critical studies in order to secure
a platform from which we can interpret the many data that are being
obtained. Thank you.

B. Ilschner

Mr. Chairman, ladies and gentlemen, I also try to put down my impressions from this conference and I would like to make three points.

1. The effect of high temperature corrosion on creep rates is indirect or secondary in nature: It is due to changes in chemical composition and subsequent micro-structure changes, in a reaction zone adjacent to the corrosion surface, and to reduction of load-bearing cross-section in some cases. If the depth of the reaction zone is small compared to specimens or component thickness, there will be no significant effect on secondary creep rates. Due to the difference in diffusion coëfficients, there is a different effect of oxidizing compared with carborizing atmospheres.

2. The major engineering hazard is mechanical damage due to cracking. It is documented by reduced rupture times. Attention of future research should therefore be focussed on corrosion induced crack initiation and corrosion enhanced crack propagation. A third important point on mechanism of crack healing. To separate these application of the experimental methods and theoretical methods of fracture mechanics are strongly recommended.

3. It appears to be of great practical importance to investigate creep-corrosion interaction not only at constant stress and constant temperature, but also under loading/unloading as well as heating/cooling cycles. This implies a stronger concentration on problems of scale adherence and of LCF of layer structures; both topics are practically not mentioned at this conference.

V. Guttmann

The mechanical properties of structural materials are usually determined under closely specified laboratory testing conditions. In
many cases these conditions do not correspond very well to those
encountered in technical applications. In order to permit the use
of laboratory results in design data, corrections have to be made
and safety factors to be introduced. Most of them however are based
only on more or less qualitative experience.
The subject of this conference (the conjoint action of mechanical
stressing of materials and aggressive environment exposure) provides a typical example where significant difference of the material behaviour can be expected under laboratory and service conditions. Such a situation automatically provides the question as to
how far it becomes necessary or at least preferable to conduct component testing in addition to laboratory testing in order to determine the material properties in a more realistic way and in how
far investigations should be carried out under conditions which
are more similar to those encountered in service.
This type of improved testing of course has the general disadvan-
tage of being a very expensive business. Thus, in a justification
it is essential to judge if the output of such type of tests can
be from such a benefit that the high investigation costs can be
balanced, at least after many years of material services. Thus
besides purely technical aspects the economical view point has
to be considered as well.
The following survey presents a list of problems arising when
usual laboratory testing results must be transfered into design
data. Although preferentially considered under the aspect of simultaneously mechanical and chemical exposure, in some cases the
problems are related also to a "non-aggressive atmosphere". The
evaluated consequences should enable a discussion about the importance of component testing and should probably give an idea how far
a revision of laboratory testing methods and specifications should
be recommended.

Technical application Design aspects	Laboratory testing
Component size	<u>Situation</u>: Specimen size rarely related to component size. <u>Consequence</u>: Under- or overestimation of environmental influence, especially in the case of pure surface effects.
Component shape	<u>Situation</u>: Stress distribution in service differs from that in laboratory experiments. <u>Consequence</u>: Stresses not always well simulated (see also below: thermal stresses).
Composition and material structure	<u>Situation</u>: Differences possible between laboratory specimens and technical components particularly in cast materials. <u>Consequence</u>: Under- or overestimation of strength and environmental effects.
Surface condition of components	<u>Situation</u>: Quality of specimen surface rarely corresponding to that of components. <u>Consequence</u>: Corrosive attack can differ from that in industrial application.
Environment flow rate	<u>Situation</u>: Frequently not exactly simulated in laboratory tests. <u>Consequence</u>: Sometimes differences existing in chemical reaction rate. Possibility of high cycle fatigue due to vibrations not taken into account.
Applied stress variation	<u>Situation</u>: Creep-fatigue interaction tests are conducted but very often no realistic simulation of the stress regime. <u>Consequence</u>: Stresses not exactly simulated and under- or overestimation of environmental influence possible.

Technical application Design aspects	Laboratory testing
Thermal stress	<u>Situation</u>: Usually simulated in standardized tests but sample shape rarely corresponding to component shape. <u>Consequence</u>: Shape dependence of thermal stresses may lead to inaccurate stress simulation.
Test duration	<u>Situation</u>: In most cases relatively low test durations. <u>Consequence</u>: Determined creep mechanisms may differ from those governing the long term component behaviour. Structural changes during long term service not quantitatively described. Corrosive attack for long term service eventually over- or under-estimated. <u>In all cases</u>: Uncertainty of creep data extrapolation.

F. Schubert

As the last speaker of this panel, I would like to summarise briefly the results obtained and to pinpoint areas which have not been mentioned. Further, I would like to make some proposals for future research and development.

The outcome of this meeting has been:

- ideas for understanding of mechanisms,
- basic data provision,
- indication of scatterband for mechanical properties in corrosive environments,
- some results of component testing.

Less emphasis has been put on:

- data transfer from experiments to design of components,
- implication of the data for design codes,
- safety recommendations,
- effects of multi-axial-stresses,
- fracture problems,
- creep-fatigue interaction,
- extensive component testing.

I think it is worthwhile to draw now some conclusions and to consider and analyse possible future trends for studies to be carried out on the interaction of environmental and mechanical exposure to high temperature materials.

Most of the contributions to this symposium fall into columns II and III (see attached table) bridging between fundamental research (column I) and applied research (column IV). Future activities should enter the fields described in columns I and IV, i.e. the better understanding of mechanisms leading to a deeper knowledge of the materials behaviour and the testing of component in technical plants.

The described objectives make task splitting between involved
institutions necessary. Universities and other organisations with
a main interest in fundamental research should concentrate on the
tasks described in column I. Industry should enter work laid down
in column IV and receive support from all other fields (columns I,
II, III).

table

	I	II	III	IV
Investigations	Model environments (parametric studies) Standard specimens (uniaxial) from simplified (model) alloys	Simulated service environments Standard specimens (uniaxial) from commercial alloys	Service environments Specimens from components (multiaxial, weldments, etc.)	Service environments, exposure history full scale components, prototype facilities — components in service
Objectives	Understanding of mechanisms	Data generation scatter bands	Life prediction	plant design, feasibility, relevance of predictions — component behaviour and failure
Essential future requirements		Understanding of basic material behaviour ←	Transfer of data to design and safety requirements, design code development →	

DISCUSSION

<u>T. Boniszewski</u>:   (Comment on Ilschners' recommendation on application of fracture mechanics)
1)   Fracture mechanics deals with a pre-existing crack therefore it cannot deal with crack initiation, but only with the initiation of failure from a pre-existing crack, therefore it deals with crack propogation.

2)   Cracks induced in fracture mechanics specimens are generally transgranular, but in creep cracks are intergranular. There can be a problem where at the crack tip the mechanism or crack path must change.

<u>B. Ilschner</u>: Your point number one: I did not say that fracture mechanics is suitable for studying crack initiation but I did say that it is suitable for separating effects of crack initiation and crack propagation. If you make any experiments aiming at rupture times, you have the combined effect of crack initiation and crack propagation. To separate these two also in non-corrosive environments, it is to be recommended to apply fracture mechanics because this would give information on crack propagation only. This comes to your point two: Of course fracture mechanics does give this information only if you have a starting crack. Crack initiation by fatigue very often will result in a transgranular crack. However, there are other methods to introduce cracks into non-metallic material: by thermal stresses for instance. There is quite some amount of literature on this topic, and there is no reason to doubt that it is too difficult to measure sub-critical crack growth in oxide scales and oxide layers in general. A lot of work of this type is going on in ceramic industry, particularly in materials like silicon nitride at high temperatures. So I do not think it is impossible. Your warning against misleading interpretation has to be taken serious, as serious as any warning against misinterpretation of any experimental data. I think it is common feature that one has to be very cautious with interpretation, but I am sure that fracture mechanics is a useful tool amongst others, if combined with micro-structural investigations. Otherwise not! If you take it only as a continium mechanical approach without micro-structural evidence, the risk of mis-interpretation would indeed be high.

<u>D.B. Meadowcroft</u>:   I support strongly the comments by the panel dealing with the two topics of the conference seperately. I will deal with the effect of mechanical stress on corrosion first. Here I am reasonably happy with the directions in which we are moving and certainly the CERL work has been able to cover the range from basic measurements to their use in estimating plant lives. I agree

with Kofstad that more basic properties of the oxide need to be
measured, but I would stress that the properties must be measured
on oxide films and not on bulk ceramics.

On the effects of corrosion on mechanical properties, I feel there
is a need for workers to think very carefully about what are the
best experiments to carry out. The experiments to date have
indicated that crack initiation or propagation must be
studied , and experiments should be designed to study these in
detail. For instance, we are trying to use thermal fatigue
Glenny disk   to study the effect of hot corrosion on initiation,
and for initiation, I agree with Ilschner that fracture mechanics
techniques should be used. However, thought must be given to the
experimental design. Simply starting a crack by fatigue and
then following its propagation in a hot corrosion atmosphere is
not enough, because hot corrosion has an incubation time before
rapid corrosion occurs, and there could not be the relevant
sulphides in the crack produced by the above technique. In
summary, detailed thought is needed to choose the best experiments.

R. Rolls: The conference theme has underlined the need to
identify the important sources and mechanisms of instability, both
mechanical and chemical, in designing experiments which will
provide the crucial data (like creep of oxides and crack initiation
and growth) for the application of our knowledge to solve
practical problems.

C. Coddet: Thermal oxide layers are essentially inhomogeneous,
their mechanical properties are moreover, strongly dependent upon
their composition. Therefore, I think that it is not worthwhile
to study oxides alone. On the contrary, we have to manage with
the metal oxide system which is much more complicated but also
more realistic. Consequently, creep data on oxides are not easy
to obtain and also to use.

P. Kofstad: C. Coddet and D.B. Meadocraft have stressed the need
for creep studies on oxide scales or films rather than on bulk
ceramics. My own opnion is that further knowledge is needed for
both scales and bulk ceramics. Growing scales will generally
exhibit large gradients in defect concentrations from the gas/scale
to the metal/scale interface. Accordingly one may expect large
differences in creep behaviour of the scale at the two interfaces and throughout the scale. Thus the creep behaviour of
growing scales reflects a complex process. In order to make
the proper analysis it would be highly desirable to know the creep
behaviour of oxides as a function defect concentration. Such
data are obtained from specimens of equilibrated bulk ceramics.

<u>E. Erdös</u>:   I am very much in favour of the opinion brought forward
by P. Kofstad that we need simple experiments where the results are
understandable.   An example is the influence of chlorides which
has been studied with sophisticated apparatus with the aim to
simulate service conditions.   However, the results have been
contradictory and are not easily understood.   Only if we
distinguish between the role of sulphur and chlorine and design
specific experiments then we can hope to solve this problem, which
might be the problem of high temperature stress corrosion by
hologenides.   It is our duty to serve the engineer with data but
this data should be understood.

ERICSSON T.   Linköping University, S

FRANCK F.   JRC, Petten Establishment, NL

FRANSEN T.   TH Twente, Enschede, NL

FOUCAULT M.   Electricité de France, Chatou, F

GALSWORTHY J.C. (Author)   AMTE(HH) Holton Heath, Poole,
                                      Dorset, UK

GAROT W.   Rolled Alloys BV, Zwyndrecht, NL

GEENEN P.V.   KSLA, Amsterdam, NL

GIBS V.   Kernforschungszentrum Karlsruhe, D

GOODISON D.   HM Nuclear Installations Inspectorate,
                                      Health & Safety Exec., London, UK

GÖTZMANN O. (Author)   Kernforschungszentrum Karlsruhe, D

GRABKE H.J.   Max-Planck-Institut für Eisenfor-
                                      schung, Düsseldorf, D

GRÜNLING H.W.   Brown, Boveri & Cie AG, Mannheim, D

GUTTMANN V. (Author)   JRC, Petten Establishment, NL
          (Panel Member)

HALPIN K.   Inst. for Industrial Research &
                                      Standards, Dublin, Ireland

HARTNAGEL W. (Author)   Brown, Boveri & Cie AG, Mannheim, D

HEMPTENMACHER J. (Author)   Max-Planck-Institut für Eisenfor-
                                      schung, Düsseldorf, D

HOFFELNER W. (Author)   BBC-Brown Boveri Research Centre,
                                      Baden, CH

HOFMANN P. (Author)   Kernforschungszentrum Karlsruhe, D

HUMPHRIES M.J. (Author)   Exxon Research & Engineering Co.,
                                      New Jersey, USA

ILSCHNER B. (Panel Member)   Universität Erlangen-Nürnberg, Inst.
                                      für Werkstoffwissensch. 1, D

JELGERSMA J.H.N.	N.V. KEMA, Arnhem, NL
KEMENY G.	JRC, Petten Establishment, NL
KOFSTAD P. (Panel Member)	University of Oslo, Dept. of Chemistry, N
KRIKKE R.H.	Shell Internationale, The Hague, NL
KRÖCKEL H.	JRC, Petten Establishment, NL
KRUTZ S.	M.A.N., Maschinenfabrik Augsburg-Nürnberg, München D
LANG E.	JRC, Petten Establishment, NL
LEISTIKOW S. (Author)	Kernforschungszentrum Karlsruhe, D
LUNDBERG L.R.	Swedish Institute for Metals Research, Stockholm, S
MARRIOTT J.B.	JRC, Petten Establishment, NL
MCLAUCHLIN I.R. (Author)	CEGB, Berkeley Nuclear Labs., Glos, UK
MANNING M.I. (Author)	CERL, Leatherhead, Surrey, UK
MARSCH H.D.	UHDE GmbH, Dortmund, D
MAYER C.	Dow Chemical (Ned) B.V., Rotterdam, NL
MEADOWCROFT D.B. (Chairman) (Author)	CERL, Leatherhead, Surrey, UK
MONSAU T.	Ministry of Economics, Small Business & Transport, Düsseldorf, D
MÜHLRATZER A.	M.A.N., Maschinenfabrik Augsburg-Nürnberg, Neue Technologie, München, D
NORTON J.R.	JRC, Petten Establishment, NL
ÖNEL K.	Max-Planck-Institut für Eisenforschung, Düsseldorf, D
OTTOSSON C.	Institute for Radiation Protection, Helsinki, Finland
PALAZZI A. (Chairman)	CSM, Rome, I

PAMDETH   N.M.   Burmeister & Wain Damp (Power Group)
                                        Virum, DK

PENKALLA   H. (Author)   Inst. für Reaktorwerkstoffe, Kern-
                                        forschungsanlage Jülich GmbH, D

PHEASANT   N.J.   JRC, Petten Establishment, NL

PIERAGGI   B. (Author)   Lab. de Met. Physique, ENSCT,
                                        Toulouse, F

PILOUS   V.   SKODA, Zentralforschungsinstitut,
                                        Plzen, Czechoslovakia

RAHMEL   A. (Chairman)   Dechema Institut, Frankfurt/Main, D

ROLLS   R. (Author)   University of Manchester, Inst. of
                                        Science & Technology, UK

RONCHETTI   C.   CISE, Milano, I

SANNIER   J.   CEA Centre Etudes Nucleaires,
                                        Fontenay-aux-Roses, F

SCHEEPENS   C.P.   Metaal Instituut TNO, Apeldoorn, NL

SCHENDLER   W.   Mannesmann-Forschungsinstitut,
                                        Duisburg, D

SCHINKEL   J.W.   Metaal Instituut TNO, Apeldoorn, NL

SCHNAAS   A.   Sandvik, Düsseldorf, D

SCHMIDT   M. (Author)   Dechema Institut, Frankfurt/Main, D

SCHMITZ   F. (Author)   Kraftwerk Union, Mülheim/Rühr, D

SCHUBERT   F. (Panel Member)   Inst. für Reaktorwerkstoffe, Kern-
                                        forschungsanlage GmbH, D

SCHULZ   H.   Gesellschaft für Reaktorsicherheit
                                        (GRS) GmbH, Köln, D

SCHÜTZE   M. (Author)   Dechema Institut, Frankfurt, D

SIMONTHOMAS   J.P.   Technische Hogeschool, Gebouw Warmte
                                        & Stroming, Eindhoven, NL

THON   H.   Central Institute for Industrial
                                        Research, Blindern, Oslo, Norway

TIMM   J.   JRC, Petten Establishment, NL

TOESCA S. (Author)   Université de Dijon, F

VAN DER BIEST O.   JRC, Petten Establishment, NL

VAN DEN BRUCK U.   Schmidt + Clemens, Lindlar, D

VAN DER HOEVEN A.   Dow Chemical (Ned) BV, Rotterdam, NL

VAN DER LINDE A.   Netherlands Energy Research Foundation
                                         ECN, Petten NL

VAN DER SCHAAF B.   Netherlands Energy Research Foundation
                                         ECN, Petten, NL

VAN DE VOORDE M.   JRC, Petten Establishment, NL

VAN DER ZWAN C.   Hoogovens-IJmuiden, NL

VICTORIA M. (Author)   University of Gent, Belgium

WAGNER D.   UHDE GmbH, Dortmund, D

WARD D.M. (Author)   INCO Europe Ltd, Birmingham, UK

<u>JRC CONFERENCE STAFF</u>

ARMSTRONG K.A.

MERZ M.

SOMMER S.

VAN DER SMAAL F.